COOLANT FLOW INSTABILITIES IN POWER EQUIPMENT

T0225439

COOLANT FLOW INSTABILITIES IN POWER EQUIPMENT

Vladimir B. Khabensky
Vladimir A. Gerliga

CRC Press
Taylor & Francis Group
Boca Raton London New York

CRC Press is an imprint of the
Taylor & Francis Group, an **informa** business

CRC Press
Taylor & Francis Group
6000 Broken Sound Parkway NW, Suite 300
Boca Raton, FL 33487-2742

First issued in paperback 2017

© 2013 by Taylor & Francis Group, LLC
CRC Press is an imprint of Taylor & Francis Group, an Informa business

No claim to original U.S. Government works

ISBN-13: 978-1-4665-6704-7 (hbk)
ISBN-13: 978-1-138-07361-6 (pbk)

This book contains information obtained from authentic and highly regarded sources. Reasonable efforts have been made to publish reliable data and information, but the author and publisher cannot assume responsibility for the validity of all materials or the consequences of their use. The authors and publishers have attempted to trace the copyright holders of all material reproduced in this publication and apologize to copyright holders if permission to publish in this form has not been obtained. If any copyright material has not been acknowledged please write and let us know so we may rectify in any future reprint.

Except as permitted under U.S. Copyright Law, no part of this book may be reprinted, reproduced, transmitted, or utilized in any form by any electronic, mechanical, or other means, now known or hereafter invented, including photocopying, microfilming, and recording, or in any information storage or retrieval system, without written permission from the publishers.

For permission to photocopy or use material electronically from this work, please access www.copyright.com (http://www.copyright.com/) or contact the Copyright Clearance Center, Inc. (CCC), 222 Rosewood Drive, Danvers, MA 01923, 978-750-8400. CCC is a not-for-profit organization that provides licenses and registration for a variety of users. For organizations that have been granted a photocopy license by the CCC, a separate system of payment has been arranged.

Trademark Notice: Product or corporate names may be trademarks or registered trademarks, and are used only for identification and explanation without intent to infringe.

Visit the Taylor & Francis Web site at
http://www.taylorandfrancis.com

and the CRC Press Web site at
http://www.crcpress.com

Contents

Foreword

The problems of thermal-hydraulic stability are found to occur in almost all areas of engineering where transfer of heat by moving gas, steam–liquid, and liquid–metal fluids takes place. This is of particular importance for thermal and nuclear power plants, space technology, chemical and petroleum industries, radio and electronic and computer cooling systems, to mention but a few.

This book represents an attempt to generalize experimental and predictive results of investigations that dealt with various types of thermal-hydraulic flow instabilities in components of power equipment of thermal and nuclear installations reported elsewhere in the world. It also seeks to make the reader familiar with the state of the art in the field.

The book includes material that, until recently, has not been reflected in monographs and education literature. Since 1960, over a thousand papers have been published on flow stability in equipment components in various branches of engineering. The authors gave up on the idea of a systematic overview of all of them; rather, they used individual examples illustrating sufficiently common regularities in flow instability investigations.

Though the range of problems and types of thermal-hydraulic instabilities is rather wide, a general approach has been applied. It largely concerns qualitative consideration of a phenomenon and determination of the influence of various factors based on the analysis of the physical mechanism of a process. As far as possible, an attempt was made to combine an illustrative manner of material presentation with a sufficient degree of physical correctness. We hope that such an approach will be instrumental in

- refining the notion of mechanisms and major regularities of the thermal-hydraulic flow instability, and understanding the unity and simplicity of the reasons determining the loss of stability
- systematizing the available data on thermal-hydraulic instability and master predictive techniques for concrete types of instabilities in real systems
- selecting the appropriate method of experimental or predictive investigation and planning the respective investigation
- interpreting the results and verifying their extrapolation to full-scale systems
- selecting the optimal ways and means for increasing flow stability of the newly designed systems

In our opinion, the book will be useful for students, practical engineers, designers, and researchers involved in designing and operating power equipment and investigating problems of thermal-hydraulic instability.

Introduction

It is customary to regard flow instability as a periodic or single change of thermal-hydraulic parameters in channels or circulating loops when a plant operates under steady-state conditions.

Thermal-hydraulic instability may impair thermal reliability of reactor core or other power equipment components. It has been shown in numerous publications that a sudden decrease in flow rate can lead to superheating and burnout in a heated channel. Also, periodic oscillations may cause failure of the channel wall or fuel element cladding because of the periodic magnetite layer deterioration. Flow rate or pressure fluctuations may lead to low-frequency vibrations.

Various types of instabilities of two- and single-phase nonadiabatic coolant flows have been intensively investigated in recent decades. First, these investigations dealt with hydraulic schemes of steam boilers and steam generators with forced and natural coolant circulation. With the advent of nuclear power plants, especially those using boiling water reactors and coolant natural circulation, emphasis has been placed on flow stability in circulation loops of these plants. Cases of coolant flow instability at supercritical pressure and at the two-phase flow condensation have been discovered and made the subject of research.

As a result of numerous experiments and predictions, different types of instabilities with regard to their mechanisms have been revealed. For some of them, analytical and numerical methods and correlations have been developed to determine the stability boundary and to establish the time-dependent behavior of coolant parameters in the unstable region, as well as the effect of flow and design parameters' variation on the stability boundary. However, a variety of types of instabilities, sophistication of circulation loops and designs of reactor core process channels, higher heat fluxes and possible superposition, and interaction between different instabilities frequently complicate not only quantitative but also qualitative analysis. It should be noted that the phenomenon in question has not been studied in detail yet. Higher heat fluxes and new circulation schemes may result in unknown types of coolant flow instabilities or in a combination of conventional types of instabilities.

Modern theoretical investigations of coolant flow instabilities employ various methods [1–9] based on the study of unsteady-differential or integro-differential partial equations, together with respective boundary and initial conditions. With reference to differential equation studies, two directions are followed.

The first one, aimed at the determination of flow stability boundary and the effect of various parameters, involves a study of behavior of a complete set of problem solutions with no search for exact or approximate solution. To this end, the direct Lyapunov's method and frequency response methods are used. With reference to real plants and vivid presentation of information of the effect of parameters on flow stability, a distributed parameters model requires substantial simplification and a multiplicity of assumptions. Though quantitatively satisfactory with regard to determination of flow stability regions, the methods require experimental verification. That is why they are usually used for qualitative determination of the effect of separate parameters on the stability boundary.

The second direction of investigations, which permits quantifying self-oscillating regimes more precisely, represents a study of system parameters' behavior in the instability region by direct numerical solution of the nonlinear dynamic model. Numerous investigations [8–11] showed that, for a concrete system, this method yields a sufficiently reliable quantification of the coolant flow instability region for low-frequency thermal processes, if there are adequate constitutive quasistatic correlations for heat transfer coefficients, hydraulic resistance, steam quality, etc. The approach is also advantageous for investigating the interaction of processes and parameters in the instability region. This makes it possible to conduct a numerical experiment and "watch" time variation of different distributed parameters, including those difficult to determine experimentally. The numerical solution of the distributed parameter model helps in establishing specific features of physical mechanisms and some regularities in the influence of design and flow parameters on the stability boundary. In some cases, a nomogram [12] or an algebraic dependence [13] can be developed for an approximate determination of the stability boundary.

The orientation toward numerical methods only is not optimal for the flow stability studies. A more efficient approach supposes the use of analytical techniques or a nomogram, given in, say, Lokshin, Peterson, and Shvarts [12], for an approximate evaluation of the instability region and finding ways of optimal improvement of the system stability. Additionally, experimental investigations and numerical solution of a distributed parameter model are required for the quantitative refinement of the flow stability boundary.

These methods are described in sufficient detail in monographs and papers aimed at mathematically well-equipped researchers and designers.

However, a practicing engineer needs a simple method requiring no special mathematical knowledge and permitting qualitative evaluation of the effect of design and flow parameters, taken separately or as a whole, on the flow stability boundary. This would be instrumental in selecting optimal diagram and design solutions promoting flow stability in the system and components. It is exactly the way that was used to apply the minimum of mathematics to present the material on flow instability in heat exchange systems in nuclear and thermal power plants. Qualitative analysis of the effect of parameters on the stability boundary is based on detailed consideration of the physical mechanism of various types of instabilities, taking into account complex interaction of parameters of a concrete heat exchanger. For example, throttling of the reactor core channels with natural circulation promotes flow stability and at the same time decreases the loop flow rate, which in turn substantially undermines throttling efficiency, etc.

Much attention is given to the analysis of conditions that are the cause of instability in nuclear reactors and power equipment.

The presented qualitative analysis of flow instability, especially the integrated, comprehensive consideration of the effect of interrelated parameters in a circulation loop, is used for specifying requirements to simulation of thermal-hydraulic flow instability, interpretation of experimental data, and validation of the appropriateness of experimental data extrapolation to real power plants.

Such approach may be useful for designers for the early identification of the type of instability and selection of the required degree of mathematical model specification.

As for the flow instability under diabatic conditions, mostly thermal-hydraulic processes in a channel or loop are taken into account. The effect of such feedback as neutron kinetics, operation of automatic control systems, various circulation promoters, disturbances due to natural oscillations of loop components, etc. is not considered here as being dependent on specific features of concrete power plants, while the possible effect of such processes on flow stability is shown in detail in references 1, 3, 5, and 6. Also, no analysis has been made of thermal-hydraulic instability of loop circulation for complex geometries because it is sufficiently treated in Mitenkov, Motorov, and Motorova [5].

It should also be mentioned that the treatment of thermal-hydraulic instability covers macroinstability only, as it determines changes of coolant integral parameters in the entire channel or loop. The local thermal-hydraulic instability governed by the local thermal free convection, by Taylor and Helmholtz instability initiated at the interphase boundary, etc., is not analyzed individually, and is mentioned only when its effect on system macroinstability is considered.

To increase the efficiency of qualitative analysis, a well-justified classification of known types of thermal-hydraulic instability and typical boundary conditions is of importance. In this case, classification should be reasonably general and boundary conditions sufficiently universal. Excessive detalization some subtypes of instability is, as a rule, a result of the influence of individual constructional peculiarities of a hydraulic system, or of the complex interaction of interrelated parameters of a loop, which lead to a change in boundary conditions. These factors can be easily analyzed within the framework of a more general classification of instability types and boundary conditions, while the increased number of instability subtypes considerably complicates qualitative analysis.

Particular attention has been paid to oscillatory (low-frequency and thermal-acoustic) and static thermal-hydraulic coolant flow instability.

To this end, the physical mechanism of instability has been considered in detail, especially its peculiarities at low and high exit qualities. Also, the effect of change of design and flow parameters on the flow stability boundary has been shown.

Wide application of two-phase flows in nuclear and thermal power plant heat exchangers leads to the appearance of numerous types of instability caused by different mechanisms. Here, deaerators, separator-superheaters, bubblers, etc. may be mentioned.

When such types of two-phase flow instabilities are analyzed, their physical nature and possibility of occurrence in modern designs of nuclear and thermal power plants are considered briefly.

Nomenclature

A: Cross-sectional area, m^2
A: Amplitude, %
A: Work performed by a bubble during the oscillation cycle (Equation 6.1)
A: $= \dfrac{\partial \Delta P_p}{\partial G}\big|_0 - \dfrac{\partial P_c}{\partial G}\big|_0$, Equation (5.5)
A: $= \rho_G/\rho_{sh}$
a: Acoustic speed, m/s
a$_p$, a$_v$:   Complex amplitude of the bubble pressure and of the bubble volume oscillations, respectively
c$_p$:   Specific heat under constant pressure, J/(kg·K)
C: Coefficient in Equation (3.5), derived from Figure 3.2
c: Constant in Equation (5.35)
D$_h$, d$_h$:   Hydraulic diameter, m
D: Stability criterion in Equation (3.8)
D: Heated medium flow rate, kg/s (Section 5.3.4)
D, d: Max and min screw diameters, m (Section 8.1)
E: Acoustic energy flux per bubble unit surface
F$_G$:   Bubble surface, m^2
G: Flow rate, heating medium flow rate (Section 5.3.4), kg/s
G$_{by}$:   Bypass flow rate, kg/s
g: Gravitational acceleration, m/s
H: Length, height, m
h: Step in a screw pump, m
h$_m$:   Mass transfer coefficient in Equation (6.4)
ΔH$_{ec}$, ΔG$_{in}$:   Amplitude of oscillations at the economizer section boundary and of the inlet flow rate, respectively
i: Enthalpy, J/kg
i$_{LG}$:   Latent evaporation heat, J/kg
i$_c$:   Enthalpy at the onset of pseudoboiling under supercritical pressure (Equation 3.16)
I, T: Heating medium enthalpy and temperature in case of convective heating, J/kg, °C (Section 5.3.4)
i, t: Heated medium enthalpy and temperature in case of convective heating, J/kg, °C (Section 5.3.4)
Im [ ]:   Complex number imaginary part
j: $= \sqrt{-1}$
K, ∏:   Stability criteria, formulas (1.18) and (1.29), (3.14) and (3.15), respectively

K:	$= \Delta P_{in.t}/0.5\rho W_L^2$, inlet throttling
K:	$= G_{by}/G_C$, bypass factor
K:	$= |\Delta G_{c1}/\Delta G_{c2}|$, ratio of flow rate deviations in stable and unstable channels, Equation (5.13)
K_p:	Correction factor for pressure (Figure 3.1)
KK:	Number of the bubble oscillation mode (harmonic) (Section 6.1.1)
M, m:	Steam, liquid mass, kg
N:	Power, MW
N_p:	Number of economizer section decomposition parts
$\bar{n}$:	Relative neutron flux ($= n/n_0$)
n:	Number of channels, of pump revolutions
P:	Pressure, MPA
ΔP:	Pressure drop, MPA
$\Delta P_1, \Delta P_2$:	Friction pressure drop in the liquid and in the two-phase section, respectively, MPA
P_0, P_{oo}:	Pressure at the channel inlet and exit, respectively, MPA
$\Delta P_{L,t}, \Delta P_{G,i}$:	Local resistance pressure drops across the economizer and evaporating/superheating sections, respectively (Equation 1.30), MPA
$\Delta P_p(G)$:	Pump head characteristic, MPA
ΔP_f:	Friction pressure drop, Equation (7.1), MPA
P_{eav}:	Pressure at the beginning of cavitation, MPA
P_{cr}:	Pressure at which the pump head starts decreasing during cavitation, MPA
P_{br}:	Pressure at which a sharp head drop occurs, MPa
P_1:	Condensate pressure in the line to deaerator, MPA
P_{2r}:	Condensate pressure at the deaerator inlet, MPA
P_2:	Steam pressure at the deaerator condensate inlet, MPA
q:	Heat flux density, MW/m^2
Q:	Heat transferred in convective heating (Section 5.3.4)
Q_1, Q_2:	Fluid volumetric flow rate in the feed and discharge pipelines, respectively, m^3/s (Section 8.1)
Re [ ]:	Complex number real component
s:	Characteristic equation root
$S_G\Delta_P$:	Reciprocal spectral density of noise powers
S_{GG}:	Self spectral power density
T:	Temperature, °C
$\Delta T_{in}, \Delta T_{sub}$:	Inlet subcooling, °C ($= T_s - T_{in}$)
ΔT, Δt:	Minimal temperature difference between the heating and heated media in the superheating and economizer/evaporating sections' ballast zone, respectively, °C (Section 5.3.4)
$\bar{T_c}$:	Coolant average temperature in the channel, °C
T_{up}:	Coolant temperature in the upper chamber, °C

t:	Time, s
u:	Screw-centrifugal pump leading edges peripheral velocity
V:	Volume, m^3
W:	Velocity, m/s
$(\rho W)_{o, p}$:	Boundary mass velocity in a horizontal channel, $kg/m^2 \cdot s$ $[= (\rho W)_0 \cdot K_P,$ Equation (3.2)]
$(\rho W)_0$:	Boundary mass velocity in a horizontal channel at fixed pressure, $kg/m^2 \cdot s$, Figure 3.1
$(\rho W)_G{}^{bou}$:	Steam boundary mass velocity during bubbling, $kg/m^2 \cdot s$
X:	Mass steam quality
$\bar{\tilde{o}}$:	Parameter relative value $(= X/_0)$
z:	Distance, m
∏:	Perimeter, m
∏(s):	Characteristic equation, transfer function

Greek Symbols

α:	Steam void fraction
β:	Screw leading edges angle of inclination
β:	Volumetric expansion coefficient
γ:	Angle of flow incidence in the screw-centrifugal pump
δ:	Parameter deviation
$\tilde{\delta}\pi$:	Laplace transform of parameter π
Δ:	Difference in parameter values
ξ:	Hydraulic resistance coefficient
θ:	Particle residence time, s
λ:	Thermal conductivity, $W/(m \cdot k)$
v:	Kinematic viscosity, m^2/s
v:	Oscillations' frequency
ρ:	Density, kg/m^3
ρ', ρ":	Densities of liquid and steam on the saturation line, respectively, kg/m^3
τ, τ_{cyc}:	Oscillation cycle, s
τ_w:	Friction shear stress
τ:	Flow coefficient for a pump, Section 8.1
ϕ, φ:	Oscillation phase shift
$\bar{\psi}$:	Two-phase flow inhomogeneity correction factor
ω:	Circular frequency, rad/s

Indexes

L, G:	Liquid and steam (gas), respectively
in, e:	Inlet and exit, respectively
C:	Channel
Σ:	Accumulative value
st, 0:	Stationary value
l:	Loop
ec:	Economizer section
TP:	Two-phase section
ev:	Evaporating section
sh:	Superheating section
d:	Riser section
non:	Nonheated section
hd:	Header
up:	Upper chamber
bn:	Boundary value
s:	Saturation
nom:	Parameter value in the nominal operation mode
gr:	Gravity
t:	Plate, throttle
h:	Horizontal
v:	Vertical
el, con:	Joule and convective heating, respectively
con:	Condensate
cz:	Instability boundary determined by the ballast zone, relationship (5.31)
IGS:	A point in stream where steam generation starts
Beg:	Initial value
***f*:**	Forced movement
w:	Wall
p:	Pump
m:	Maximum heat capacity zone at supercritical parameters
Bc:	Steam that has passed through the bubbler
ev:	Fresh steam
Gc:	Steam condensing in jets
cr:	Critical
M:	Model installation
fs:	Full-scale installation

Dimensionless Quantities

Subcooling parameter:

$$N_{SUB} = \frac{(\rho_L - \rho_G)\Delta i_{in}}{\rho_G i_{LG}}$$

Equilibrium frequency parameter:

$$N_{pch.eq} = \frac{10^3 \overline{q}\Pi H (\rho_L - \rho_G)}{A i_{LG}(\rho W)\rho_G}$$

Reynolds number:

$$Re = \frac{W d_h}{\nu}$$

Grashof number:

$$Gr_q = \frac{g\beta q d_h^4}{\lambda \nu^2}$$

Grashof number:

$$Gr_{\Delta t} = \frac{g\beta(\overline{T}_c - T_{up})d_h^3}{\nu^2}$$

Prandtl number:

$$Pr = \frac{C_p \rho_L \nu}{\lambda}$$

Rayleigh number:

$$Ra = Gr_q \cdot Pr$$

1

Two-Phase Flow Oscillatory Thermal-Hydraulic Instability

1.1 Classification of Types of Thermal-Hydraulic Instability and Typical Thermal and Hydrodynamic Boundary Conditions

Numerous works dedicated to flow instability [8,9,14,15] offer classifications of instability types that are generally consistent with regard to a sufficiently general subdivision into static and dynamic instabilities, but differ when a more detailed classification is made. In this case, thermal-hydraulic instability of the same physical nature is related by different authors to different types. Often, flow instability in coolant natural circulation systems is considered as an independent type, though, as one can see later, natural circulation does not determine the physical nature of instability, but rather affects the system of interconnected parameters (i.e., quantitatively influences the stability boundary due to changes of boundary conditions).

Some authors make attempts to subdivide complex types of instabilities into independent ones. Such complex types occur during the interaction between several thermal-hydraulic instabilities of different physical natures. This classification is of small practical importance because it is impossible to reveal inherent features of these complex types and the analysis becomes more complicated.

Our classification of coolant flow instabilities caused by thermal-hydraulics coincides in general with most accepted types of instability of different physical natures. Thus, the present book recognizes the following thermal-hydraulic instabilities of the coolant flow:

1. Oscillatory (dynamic) instability in a heated channel or circulation loop under steady-state conditions (This category includes the low-frequency two-phase thermal-hydraulic instability at low exit qualities and the instability at high exit qualities, which in foreign sources is often called the density-wave instability.)

2. Static instability in a heated channel or circulation loop, which occurs in the region of ambiguous hydraulic characteristics

   3. Thermal-acoustic instability
   4. Flow instability during condensation of steam and steam–liquid coolant
   5. Two-phase coolant flow instability caused by flow structural changes (sometimes called the relaxation instability), incorporating periodic change of flow patterns, unstable steam generation, explosive boiling, and "geysering," resulting in periodic flow rate disturbances in a system
   6. Adiabatic flashing flows instability

In analyzing these types of instabilities, special attention has been paid to differences and specific features of their physical mechanisms, which determine the different nature of the effect of design and flow parameters on the stability boundary and flow oscillation amplitude.

The adopted classification is applicable to both natural and forced coolant circulation systems. The first three types of instabilities are considered in more detail as they most frequently occur in practice.

To simplify the analysis of coolant flow thermal-hydraulic instability, three types of boundary conditions, typical of nuclear and thermal power plants and equally relating to both forced and natural coolant flow circulation systems, are generally considered:

- Thermal-hydraulic flow instability in a single channel at constant pressure drop (instability of hydrodynamically isolated channel) [ΔP_C = const., T_{in} = const.]
- Parallel-channel thermal-hydraulic instability (interchannel instability) [G_Σ = const., T_{in} = const., $\Delta P_C \ne$ const.]
- Circulation loop thermal-hydraulic instability (loop instability)

The first type of boundary condition assumes consideration of instability of a single channel whose parameters are the worst from a stability viewpoint and that is a part of a system with a large number of stably operating parallel channels. In this case, the fixed inlet temperature and pressure drop are preset for this channel due to the system of stable parallel channels. The total flow rate through the system shall be significantly higher than that through the channel in question to obviate the influence of flow rate change in the unstable channel on the total flow rate. This case corresponds to real conditions where, for example, the number of unstable channels in the core is considerably smaller than the total number of channels.

The second type of boundary condition means instability of a group of channels constituting the major part of the total number of parallel channels, or instability of the operating in parallel nonidentical steam generators. The overall parallel channel flow rate and coolant inlet temperature are assumed constant, with occurrence of hydrodynamic interaction between stable and unstable channels leading to periodic redistribution of flow rates and fluctuating pressure drop in unstable channels with flow fluctuations.

Under real conditions, this type of boundary condition takes place when pressure drop disturbances in the parallel-channel system are considerably lower than those in the circulation loop and have practically no effect on the loop flow rate.

If the unit is made up of parallel thermally and hydraulically identical channels, or of a system of pairwise identical channels, then the flow rate oscillations with interchannel thermal-hydraulic instability start in all identical channels or in pairwise identical channels. In such a case, the phase shift of flow rate oscillations in unstable channels ensures constant overall flow rate through the unit and equal pressure drop there (i.e., the first- and second-type boundary conditions coincide).

Since in a multichannel steam generator system each channel has generally at least one channel with similar thermal and hydraulic characteristics, the first boundary condition is the governing one when analyzing the system stability and ensuring thermal technology reliability.

In the case of circulation loop thermal-hydraulic instability corresponding to the third type of boundary condition, investigation is concentrated on the interaction between thermal-hydraulic parameters of all closed-loop components of the system with coolant natural circulation, as well as on the interaction between thermal-hydraulic parameters of loop components and the head dynamic characteristics of the pump in forced circulation systems. The loop and parallel-channel instabilities may exist simultaneously with, in a general case, different oscillation cycles and amplitudes. Because the loop and interchannel instabilities are found to interact, either augmenting or suppressing each other, in the case of loop instability, the units with parallel nonidentical channels should be considered as a general case taking into account hydrodynamic interaction of parallel channels, but not as a single equivalent channel, as is a common case. Under real conditions, this type of boundary condition appears when a pressure drop perturbation in the system of parallel channels in the case of interchannel instability is commensurable with the pressure drop across the circulation loop and affects the loop flow rate oscillations the loop flow instability precedes the interchannel instability of the unit made up of a system of parallel heated channels.

These boundary conditions are listed in increasing complexity with regard to investigation of thermal-hydraulic instability.

Depending on particular aims of investigation and specific features of the considered circulation loops, system parameters, and type of thermal-hydraulic instability, one or several of the preceding boundary conditions should be treated.

It should be noted that the first two boundary conditions are applied as approximate in practice and represent ideal conditions close to the real ones. Therefore, sometimes it is useful to investigate to what extent the stability boundary is sensitive to variation of boundary conditions.

This chapter deals with a detailed consideration of basic mechanisms of low-frequency (density-wave) oscillatory thermal-hydraulic two-phase flow instability.

For different ranges of design and flow parameters, individual components of pressure drop may depend in different manners on the determining parameters, and their contribution to the total pressure drop in the channel or the unit under consideration may also be different. As a result, the conditions of oscillatory thermal-hydraulic instability may be a function of both static and dynamic relationships of parameters.

The oscillatory (dynamic) flow instability generally takes place when a channel or circulation loop operates under conditions of the stable section of the static hydraulic characteristic. Coolant flow oscillations are known to be induced by dynamic interaction between flow parameters (velocity, density, pressure, enthalpy, and their distribution) due to delays and feedbacks.

Depending on the range and combination of thermal-hydraulic parameters, different pressure drop components may play a governing role in self-sustaining flow rate oscillations (including the case of constant channel pressure drop). This leads to a different influence of design and flow parameters on the flow stability boundary, depending on the pressure drop component that governs flow oscillations. The physical mechanism of instability, though common in nature, also has some specific features.

Numerous experiments and theoretical studies permitted identification of two most typical cases of low dynamic thermal-hydraulic instability. The first case is observed in the range of zero or low exit qualities, when oscillations of the gravity component of pressure drop play a decisive role. The second one occurs at high and moderate exit qualities, when the role of oscillations of the friction component of pressure drop is predominant. This is usually called density-wave hydrodynamic instability. Though flow oscillations both at low and high qualities are the function of density waves, we shall traditionally apply the term "thermal-hydraulic instability initiated by density waves" only to the instability at high steam qualities.

These two options of dynamic two-phase flow instability may occur in both natural and forced coolant circulation. Figure 1.1 illustrates the typical location of regions of dynamic instabilities of Fucuda and Kabori [14,16]. The experimental loop test section represents two vertical parallel electrically heated channels with nonheated outlet vertical sections.

It could operate with both forced and natural circulation of coolant. For the forced circulation, the loop flow rate was kept constant with varying input heat power, while with natural circulation it varied in accordance with variations of the moving head. With the increased test section power when the channel outlet coolant enthalpy was $x_e \approx 0$, flow fluctuations were observed. With the further increase in power, these oscillations first developed with an increasing amplitude and then became smaller; at $x_e \approx 1\%-3\%$, the flow stabilized. The further power increase in a wide range proceeded

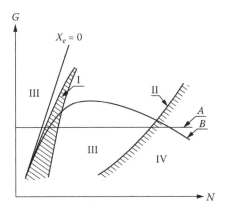

FIGURE 1.1
Region of a steam-generating channel unstable operation [14,16]. A: Forced circulation; B: natural circulation. I: instability region at low exit qualities; II: density waves' instability boundary; III: stable operation region; IV: density waves' instability region.

at a stable coolant flow rate. However, at some boundary value of exit quality ($x_e \geq X_{Bn}$), flow rate fluctuations started to occur in the system with an amplitude increasing along with the increasing power. Figure 1.1 shows that, first, at low qualities the instability region becomes narrower in terms of the x_e variation interval at an increase in flow velocity, when specific contribution of the friction pressure drop increases, and second, a wide stability region exists between two instability regions, corresponding to low and high exit qualities. Therefore, in the instability region corresponding to low exit qualities and with no possibility of increasing the system flow rate, the flow can be stabilized by increasing the exit quality, say, by increasing power, or decreasing flow rate or inlet subcooling, and also by decreasing pressure.

Now, let us consider the two cases of two-phase flow dynamic instability in more detail.

1.2 Two-Phase Flow Instability at Low Exit Qualities

For the previously stated reasons, flow oscillations at low exit qualities are known to occur mostly in vertical heated channels with relatively low coolant velocities, where pressure drop perturbations and flow rate disturbances are predominantly caused by gravity. A channel with the nonheated outlet section, the so-called riser, may serve as an example. In this case, flow oscillations are characterized by the following factors: equiphase changes in the channel inlet and outlet flow rates, sometimes lack of harmonic oscillations though periodicity is sufficiently clear, and the oscillation cycle equals the overall time during which the liquid particle passes both the channel heated section and the riser.

The experimentally obtained results concerning flow rate fluctuations at low exit qualities [14,16,17] are illustrated in Figure 1.2, where Figure 1.2(a) and (b) show the results of Fucuda and Kabori [14,16] for two parallel vertical identical joule-heated channels and individual upflow sections with natural and forced circulation, respectively. Figure 1.2(c) presents the data of Hayama [17] for the natural-circulation loop and one heated channel. It is obvious from Figure 1.2(b) that oscillations are periodic, but not harmonic. The tests with natural circulation and parallel channels (Figure 1.2a) have demonstrated two types of oscillation with different cycles: namely, interchannel oscillations with antiphase flow rate oscillations at parallel-channel system inlet (G_1 and G_2) and loop flow rate oscillations with a large cycle (G_1).

The mechanism of thermal-hydraulic oscillatory instability at low exit qualities, as in references 7, 14, 15, and 16, can be explained as follows. Small flow rate perturbations cause the channel outlet section quality (α) to change. In the range of low bulk qualities (x), especially at low pressures, ($\partial \alpha / \partial X$), or $\partial \rho / \partial i$ (which is the same), strongly depends on enthalpy. Therefore, even a small change in the exit enthalpy due to the change of flow rate results in an appreciable change of coolant density at the channel outlet (ρ_e). In the case of the nonheated individual upflow section of a considerable height, the coolant exit density changes over the entire height of the upflow section, causing a substantial change of the gravity pressure drop. At a preset constant pressure drop, the decreased coolant flow rate owing to any perturbation and respective considerable decrease in the gravity component of pressure drop because of the increased exit quality leads, with a certain delay, to an increase in flow rate above the steady-state level. The increased flow causes, with a certain delay, a decrease in exit quality (or complete disappearance of the boiling section) and the respective increase of the gravity component of pressure drop with the resultant decrease in coolant flow rate below the steady-state level, considering thermal delay. Thus, the process is repeated periodically.

Let us consider the initiation of thermal-hydraulic oscillatory instability in a steam-generating channel with a riser in more detail. For the sake of clarity and simplicity, let us assume that the residence times of coolant particle in the heated section (θ_1) and in the riser (θ_2) are equal ($\theta_1 = \theta_2 = \theta$); the length of the two-phase section of the channel heated portion is considerably smaller than that of the liquid section with practically the same residence time of the coolant particle in the heated and single-phase sections; and that α_d steam-void fraction under steady-state conditions is constant over the riser height.

Figure 1.3 shows a simplified diagram of oscillations development in the considered channel with a riser. Let, say, the coolant flow rate suddenly change at the initial period of time ($t = 0$) during a short period of time (Δt). This will lead to an increased enthalpy over the entire channel heated section due to the stepwise flow rate decrease during Δt and, hence, to an increased

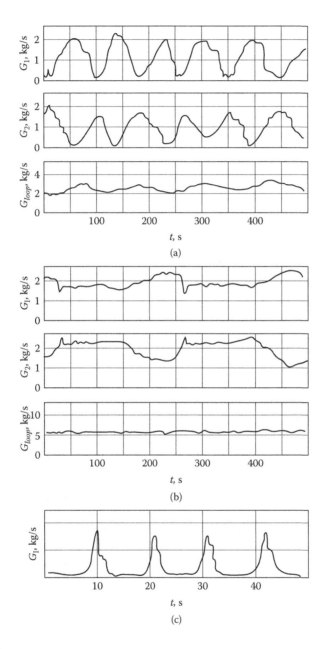

FIGURE 1.2
Time variations of loop and channel inlet flow rates at flow steady-state fluctuations. G_1: First channel flow rate; G_2: second channel flow rate; G_{loop}: circulation loop flow rate. (a) Natural circulation [16]; (b) forced circulation [14]; (c) loop natural circulation [17].

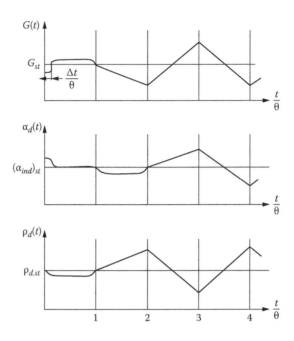

FIGURE 1.3
Simplified qualitative process of self-oscillations at low exit qualities.

riser inlet quality and respective decrease in the two-phase mixture density at the riser inlet. The reduced average two-phase mixture density in the riser causes an increased and then stabilized channel coolant flow rate at a level corresponding to the exit quality reduction (or exit enthalpy) down to the previous steady-state value. In this case, the average two-phase coolant density in the riser is stabilized at the level of density change with the disturbed exit enthalpy. The mass of the two-phase mixture head decreases during $t \in [0, \theta]$ by the value of $g\,(\partial \rho / \partial i)_d\,\delta i_d$. In this case, $(\partial \rho / \partial i)_d$ is a mixture density derivative of the riser enthalpy, and δi_d is the enthalpy perturbation at the end of the channel heated section. With the constant channel pressure drop, the increase in flow rate by a constant value during $t \in [0, \theta]$ causes, in turn, a decreased riser inlet quality during the next period $t \in [\theta, 2\theta]$ (i.e., after the time required for the transportation of enthalpy negative disturbance over the heated section). During this period of time, positive disturbance of coolant density will be accumulating in the riser, which during the time interval $[\theta, 2\theta]$ leads to an increased mass of the two-phase flow and decreased coolant flow rate.

At $t > 2\theta$, the flow rate starts increasing, since the previous flow rate reduction that had begun at $t > \theta$ resulted in the increased enthalpy, which reaches the riser inlet at $t > 2\theta$, causing an increased quality and decreased mass of the riser two-phase flow. The process will continue until $t = 3\theta$.

At t > 3θ, the growth of coolant flow rate during the interval [2θ, 3θ] starts manifesting. This phenomenon, with θ delay, initiates a reduction in steam quality at the riser inlet, thus leading to a reduction of the mass of the two-phase flow mixture head and, hence, coolant flow rate, etc. The simplified qualitative analysis and Figure 1.3 show that the process under consideration transforms into the oscillatory process with oscillation cycle of 2θ. This being the case, each rise/reduction in coolant flow rate during the period θ, with θ delay, corresponding to the transport time of enthalpy disturbance over the heated section, causes an opposite change of flow rate (i.e., reduction/rise). As a result, the time 2θ is in fact the cycle of flow parameter oscillations in the steam-generating channel at the constant pressure drop. It is evident that the process will proceed with oscillation amplitude increment if the same coolant flow rate disturbance causes a larger modular change of the liquid head as compared to the modular change of hydraulic friction.

For the analysis of the mechanism of oscillations occurrence in the steam-generating channel, simplified mathematical models are of certain practical importance. Let us obtain a characteristic equation for the steam-generating channel with a riser and low exit quality. The single-phase section energy equation may be written as

$$A\rho_L \frac{\partial i}{\partial t} + G \frac{\partial i}{\partial z} = \Pi q, \tag{1.1}$$

where G(t) is the coolant flow rate constant over the channel length, q is the specific heat flux assumed constant over the length and in time (channel wall thermal lag is disregarded), ρ_L is the liquid density constant over the channel length and in time, and Π is the channel heated perimeter.

The momentum equation for $\Delta P_{ec} \gg \Delta P_{TP}$, neglecting losses due to acceleration with respect to length and time, is

$$g \int_0^{H_1} \rho \, dz + g \int_{H_1}^{H} \rho_d \, dz + \Delta P_{ec} = \Delta P_c, \tag{1.2}$$

where ΔP_C is the channel pressure drop, ΔP_{ec} is the economizer section loss due to friction, ΔP_{TP} is the two-phase section loss due to friction, A_d is the riser flow pass area, H is the channel total length, and H_1 is the channel heated section length.

Upon linearization of Equations (1.1) and (1.2), and assuming constancy of the liquid head mass in the channel heated sections, we obtain

$$A\rho_L \frac{\partial \delta i(z,t)}{\partial t} + \delta G(t) \frac{\partial i(z)}{\partial z} + G \frac{\partial \delta i(z,t)}{\partial z} = 0 \tag{1.3}$$

for the liquid section,

$$A_d \rho_d \frac{\partial \delta i_d(z,t)}{\partial t} + G \frac{\partial \delta i_d(z,t)}{\partial z} = 0 \tag{1.4}$$

for the riser section, and

$$g \frac{\partial \rho_d}{\partial i} \int_{H_1}^{H} \delta i_d(z,t)\, dz + \frac{d\Delta P_{ec}}{dG} \delta\, G(t) = 0 \tag{1.5}$$

for the entire channel.

In Equations (1.3) to (1.5), the quantities without disturbance sign (δ) correspond to steady-state conditions.

The boundary conditions for Equations (1.3) and (1.4) are as follows:

$$\delta i(t, 0) = 0 \quad \text{with } z = 0 \tag{1.6}$$

and

$$\delta i(t, H_1), = \delta i_d\,(H_1, t) \quad \text{with } z = H_1. \tag{1.7}$$

Let us further consider the time behavior of a partial solution of (1.3) to (1.7) written as

$$\delta i = \tilde{\delta i}\,(z)\cdot e^{st},$$

$$\delta i_d = \tilde{\delta i}\,_d(z)\, e^{st}, \tag{1.8}$$

$$\delta G = \tilde{\delta G} e^{st}.$$

If all S, which are the roots of the characteristic equation of the system (1.3) to (1.7), meet the condition Res < 0, the channel will be stable "in the small."

After substituting (1.8) into the system of (1.3) to (1.7), we obtain

$$\frac{d\tilde{\delta i}}{dz} + \frac{As}{G}\tilde{\delta i} = -\frac{di}{dz} \times \frac{\tilde{\delta G}}{G}, \tag{1.9}$$

$$\frac{d\tilde{\delta i}_d}{dz} + \frac{A_d s}{G}\tilde{\delta i}_d = 0, \tag{1.10}$$

$$g\left(\frac{\partial \rho}{\partial i}\right)_d \int_{H_1}^{H} \tilde{\delta i}_d\, dz + \frac{d\Delta P_{ec}}{dG}\tilde{\delta G} = 0, \tag{1.11}$$

$$\delta \tilde{i}(0) = 0, \tag{1.12}$$

and

$$\delta \tilde{i}(H_1) = \delta \tilde{i}_d(H_1). \tag{1.13}$$

With $z = H_1$, the solution of the system of (1.9) to (1.12) is

$$\tilde{\delta i}(H_1) = -\frac{di}{dz} \cdot \frac{\delta G}{G} \cdot \frac{W_L}{s} \left(1 - e^{-s\theta_1}\right), \tag{1.14}$$

where W_L is the liquid velocity and θ_1 is the coolant particle residence time in the single-phase section.

Similarly, the solution of (1.10) is

$$\tilde{\delta i}_d(z) = \tilde{\delta i}_d(H_1) \exp[-s(z - H_1)/W_{TP}], \tag{1.15}$$

where W_{TP} is velocity in the two-phase section.

Taking (1.13)–(1.15) into consideration, it follows from (1.11) that the derived characteristic equation of the system (1.3) to (1.7) is

$$-g\left(\frac{\partial \rho}{\partial i}\right)_d \frac{di}{dz} \frac{W_L}{Gs}\left(1 - e^{-s\theta_1}\right) \int_{H_1}^{H} e^{-s(z-H_1)/W_{TP}} \, dz + \frac{d\Delta P_{ec}}{dG} = 0, \tag{1.16}$$

which, after rearrangements, becomes

$$1 - e^{-s\theta_1} - e^{-s\theta_2} + e^{-s(\theta_1 + \theta_2)} + s^2 \theta_1 \theta_2 K = 0, \tag{1.17}$$

where

$$K = \frac{d\Delta P_{ec}}{dG} \bigg/ \frac{d|\Delta P_{TP}|}{dG}; \qquad \frac{d|\Delta P_{TP}|}{dG} = g\left|\frac{\partial \rho}{\partial i}\right|\frac{di}{dz}\frac{H_1 H_2}{G};$$

$$\theta_1 = \frac{H_1}{W_L}; \qquad \theta_2 = \frac{H_2}{W_{TP}}; \qquad \Delta P_{TP} = g\rho_d(i)H_2; \qquad H_2 = H - H_1.$$

The flow stability boundary corresponds to condition $s = j\omega$, and it follows from (1.17) that

$$K = \left[1 + \cos\left(\pi \frac{\theta_2/\theta_1 - 1}{\theta_2/\theta_1 + 1}\right)\right] \bigg/ 2\left(\frac{\pi}{1 + \theta_2/\theta_1}\right)^2 \frac{\theta_2}{\theta_1}. \tag{1.18}$$

See Figure 1.4.

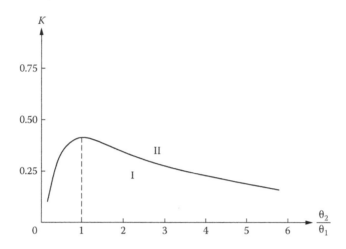

FIGURE 1.4
Flow stability boundary in coordinates $[K - \theta_2/\theta_1]$ from Equation (1.18). I: Stable region; II: unstable region.

Under real conditions, the flow oscillation cycle and amplitude also depend on mechanical and thermal channel coolant inertia and on the nature of the density wave movement in the channel nonheated upflow section.

In a general case, as is shown in references 7 and 16, thermal-hydraulic instability at low exit qualities is observed on the positive branch of the channel hydraulic characteristic. However, the previously cited works also show that in a narrow range of geometries and flow parameters, especially at small heights of individual risers, the hydraulic characteristic may have an ambiguous portion. In such a case, flow rate oscillations due to static instability may be observed, or static and dynamic instabilities may be superimposed. The mentioned particular cases of instabilities had led some authors to the erroneous conclusion about mandatory existence of an ambiguous hydraulic channel characteristic for the initiation of oscillatory flow instability at low exit qualities.

The previously described instability mechanism makes it possible to perform a sufficiently simple analysis of the qualitative effect of different parameters on the flow stability boundary at low exit qualities.

1.2.1 Effect of the Individual Upflow Section Height

The change of height of the nonheated individual upflow section produces an essential effect on the channel flow stability boundary, as it leads to a change of the oscillation amplitude of the gravity pressure drop when flow oscillation occurs. So, with the decreased/increased height of the section, the flow stabilizes/destabilizes, respectively. The experimental investigations on a thermophysical facility with natural coolant circulation described in Mitenkov et al. [7] showed that an increase of height of the individual riser sections from 1 to 1.35 m at the constant total size of the upflow section of the circulation loop extended the range of interchannel instability and increased the amplitude of flow oscillations, while the decreased height from 1 down 0.6 m resulted in a 1.5 to 2 times lower amplitude. In accordance with the physical mechanism of instability, it has been experimentally obtained in Kotani et al. [21] that the increased length of the nonheated horizontal channel outlet section with no height rise had, under otherwise equal conditions, no effect on the stability boundary at low exit qualities because horizontal sections are of no influence on the change of the gravity pressure drop. Similar results can be found in Fucuda and Kabori [14,16].

The described influence of the unheated upflow section height on flow instability is observed in both natural and forced circulation systems, and under different boundary conditions, such as (a) channel flow instability with the constant pressure drop, (b) interchannel instability with the constant loop flow rate, and (c) loop flow instability.

Shortening of an individual upflow section may be used to stabilize channel flows. However, according to Mitenkov et al. [7], certain constraints should be taken into account for natural circulation systems, as the former may affect other thermal-hydraulic reliability parameters of the system:

1. The channel flow rate profile becomes impaired and the circulation loop flow rate reduced.
2. The ambiguity of channel hydraulic characteristics may reveal itself in the case of considerable height reduction of individual sections leading to static instability or flow rate oscillations induced by static instability.

It should be noted that the change of one parameter in the coolant natural circulation loop may lead to changes in a number of interrelated parameters, which produce the reverse effect on flow stability. Therefore, the net effect on the stability boundary in some real systems may be inconsistent with the local effect produced by the initially changed parameter on flow instability. For example, it has been experimentally found in Bailey [9] and Kotani et al. [21] that when the total length of the heated and nonheated upflow sections does not change, while the length ratios are different, the natural circulation loop flow instability changes insignificantly. This may be a consequence of the integrated influence of parameters in a natural circulation loop. Indeed, when the power input is constant, the length of the individual riser section, increased due to the heated section shortening, as well as the associated increased heat flux surface density (q) in the heated section, should destabilize the system, which will be shown later.

On the other hand, the increased head of natural circulation due to an increased height of the individual riser section results in the increased loop flow rate and, hence, a decreased contribution of the gravity pressure drop to the system and a decreased exit bulk channel quality, which cause flow stabilization if the exit quality is near the lower instability boundary. The effect of these mutually opposed factors may become balanced in some systems and changes of the section length ratios may have a weak effect on flow instability. No mutual compensation of such factors may be observed in other systems with a different range of parameters.

The coolant flow may be stabilized by installing connecting pipes or "breathing" headers between the nonheated upflow sections. The method is applicable to channels with the constant pressure drop and to interchannel flows, but is of no effect in the case of loop instability.

Physically, the mechanism of the method is as follows. As a result of pressure equalization in parallel channels in the place of a connecting pipe or "breathing" header attachment, transportation of the single-phase coolant through the breathing header or the connecting pipe starts from other channels into the unstable ones at the appearance of the steam phase in the latter, thereby decreasing the oscillation amplitude of the gravity pressure drop in the case of interchannel flow rate oscillations. The effect of complete pressure equalization on the stability boundary by breathing headers is equivalent to a reduction of the height of individual upflow sections down to the point where the header is connected; the resistance between channels and the header yields a positive effect that is growing smaller with the decreasing pressure equalization. In Fucuda and Kabori [16], opening of a connecting pipe between the individual upflow sections of two identical channels in the natural circulation system caused a reduced amplitude and cycle of the interchannel flow oscillations (Figure 1.5).

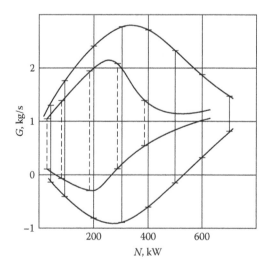

FIGURE 1.5

Effect of junction between upflow sections on the amplitude of flow rate fluctuations at an instability corresponding to low exit qualities [16]. Solid vertical lines: With "closed" connecting pipe; Dashed vertical lines: with "open" connecting pipe.

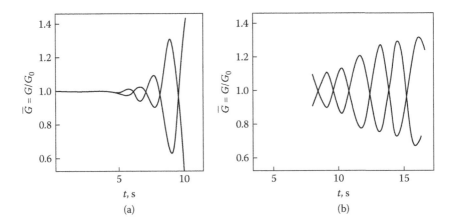

FIGURE 1.6
Effect of finite resistance of the breathing header on the interchannel instability [7]. (a) Without header; (b) with header.

The further upflow sections shortening at the expense of the connecting pipe in forced circulation experiments caused the disappearance of flow oscillations at low exit qualities.

Figure 1.6 offers similar results from Mitenkov et al. [7] as an illustration of a numerical solution using the distributed mathematical model assessing the effect of the breathing header simulator with the finite resistance (5% of the friction resistance in the channel heated section) on the initiation of oscillations in parallel channels.

To achieve utmost efficiency, the breathing header should be located at the beginning of the individual upflow section.

1.2.1.1 Throttling Effect

Inlet throttling at the isolated channel inlet improves flow stability at the constant pressure drop and stabilizes interchannel instability in a system of parallel channels because of the increased contribution of friction pressure drop to the total channel pressure drop. In a rather wide range of parameter variation, when flow rate oscillations are the same and other conditions are identical, the amplitude of gravity pressure drop fluctuations substantially exceeds that of friction pressure drop fluctuations; the influence of inlet throttling on channel stability improvement is not so appreciable.

Figure 1.7 [14] shows that in the case of coolant natural circulation under experimental conditions of Fucuda and Kabori [14], the increase of inlet throttling from $K = 60$ to $K = 400$ (where $K = \Delta P_{in.t}/0.5\rho W_L^2$, $\Delta P_{in.t}$ is the orifice plate pressure drop) had almost no effect on the location of the stability boundary at low exit qualities, but substantially improved stability (the so-called density-wave instability) at high exit qualities.

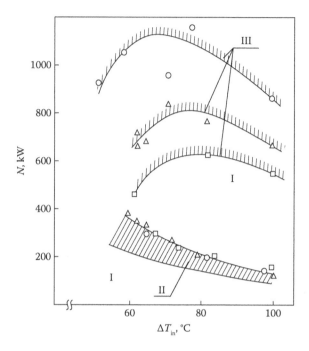

FIGURE 1.7

Effect of inlet throttling on flow instability boundary [14]. Squares, triangles, circles: experimental data; squares: K = 60; triangles: K = 120; circles: K = 400. I: Stability region; II: instability region with low exit qualities; III: instability boundaries of density waves.

The effect of inlet throttling on better flow stability is expected to be enhanced at a higher coolant flow rate, when the amplitude of the friction pressure drop oscillations becomes commensurate with that of the gravity pressure drop fluctuations. The stabilizing effect of inlet throttling was observed in the experiments described in Mitenkov et al. [7] and Zavalsky et al. [19].

The overall downflow section throttling in natural-circulation systems was found to improve loop flow rate stability, but impair interchannel flow stability in parallel channels. This is due to the fact that, with no direct influence on the friction pressure drop in parallel-channel systems, throttling causes reduction of this pressure drop by decreasing circulation loop flow rate with unchanged exit quality. It has been experimentally obtained [7] that, with the constant total resistance of the entire downflow section and channels (G = const., x_e = const.), the increasing resistance ratio between the total downflow section and channels from 0.7 to 1.5 was causing a two- to three fold increase of the interchannel flow instability at low exit qualities.

Thus, referring to references 7–9, the equivalent throttling of the total downflow section and channels, with the resultant change of the loop flow rate by the same amount, may cause an opposite effect on the interchannel instability at low exit qualities, destabilizing or stabilizing channel flow, respectively.

It was shown in Bailey [9] that similar phenomena may be observed in closely spaced rod bundle channels, in the cells of which interconnected oscillations are possible. Here, throttling through the entire channel of the natural-circulation system may impair cell flow rate stability because of the reduced channel flow.

If a natural-circulation loop is considered as a single channel with the upstream nonheated downflow section of high resistance, the region of interchannel instability, as is noted in references 7, 9, and 14, is substantially wider as compared to the circulation loop instability (if height reduction of individual riser sections does not eliminate the interchannel instability).

Channel outlet or riser section throttling at the constant loop flow rate shall, in principle, like the inlet throttling, enhance flow stability due to the equiphase change of channel inlet and outlet flow rates in presence of oscillations. However, in the case of natural-circulation systems, outlet throttling has a stronger effect on the circulation loop flow rate, as compared to the inlet throttling, and causes other interrelated parameters to change with the resultant impaired interchannel and loop stability [7].

The use of displacers in individual riser sections, apart from the increasing outlet resistance, also decreases the time constant of the coolant transit through the riser section. This may stabilize the flow, if the riser section time constant becomes lower (or higher) than that of the coolant in the channel heated section, thereby reducing the amplitude of oscillations of the pressure drop and flow rate. Otherwise, it may destabilize the flow, if the riser section time constant, when decreasing/increasing, approaches the time constant of the coolant in the channel heated section, thus involving the increasingly larger height of the riser section influenced by the gravity pressure during the oscillation cycle (Figure 1.4). In practice, most often the parameters ratio is manipulated, so flow stabilization is achieved through a decrease in the riser section time constant.

In natural-circulation systems, throttling is not an optimal method of increasing flow stability because it impairs other loop characteristics (e.g., reduces coolant flow rate).

1.2.1.2 Effect of Exit Quality Increase

The exit quality increase above a certain value (x > 1%–3%), due to a higher power, lower flow rate, inlet subcooling, or lower loop pressure, stabilizes the flow because the region of sharp dependence of coolant density on enthalpy shifts down from the channel outlet section. Therefore, flow rate oscillations produce no abrupt change of coolant exit density and, hence, of coolant density in the individual riser section. This considerably damps the gravity pressure drop oscillations' amplitude and, in accordance with the instability physical mechanism, suppresses flow oscillations at low exit qualities. The exit quality may be increased by reducing the inlet coolant subcooling in nuclear reactors with integral layout of equipment and steam pressurizer. Here, the level is somewhat decreased, with a result that part of the surface of secondary heat exchangers is in the reactor steam space and starts operating in the condensing mode.

When instability is suppressed by increasing the exit quality, the constraint mentioned in Mitenkov et al. [7] should be taken into account. This is the increased steam reactivity effect that may lead to the neutron-thermal-hydraulic instability in the region where thermal-hydraulic and neutron stability in the system are ensured separately.

1.2.1.3 Effect of Coolant Flow Rate and Pressure in the System

The increased coolant flow rate and pressure in the system at the same relative qualities stabilize the flow, since specific contribution of resistance increases and the $\partial p/\partial X$ value decreases at the heated channel exit correspondingly. For example, the results of Mitenkov et al. [7] showed that the pressure increase from 1 to 2.5 MPa results in a threefold reduction of the flow rate oscillations' amplitude.

1.2.1.4 Effect of Heat Flux Surface Density

Urusov, Treshchev, and Sukhov [20] experimentally obtained that with the decreased heat flux density, the region of the exit bulk quality decreased, too. This is the region with thermal-hydraulic instability and minimum q value, at which instability does not occur at any subcooling. This is clearly shown in Figure 1.8 (from Urusov et al. [20]). Similar results have

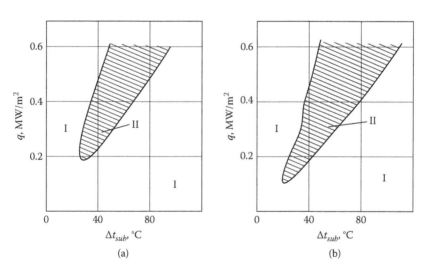

FIGURE 1.8
Instability boundary of the natural-circulation loop [20]. (a) P = 8.82 MPa; (b) P = 4.9 MPa. I: Stable region; II: unstable region.

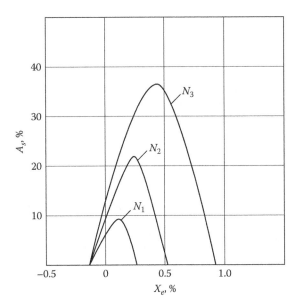

FIGURE 1.9
Dependence of the flow rate fluctuations' amplitude on the exit quality at different heat inputs [19] ($N_1 < N_2 < N_3$).

been obtained in Zavalsky et al. [19] on a natural circulation system model. Figure 1.9 from the above-cited work shows that with the increasing input power, which is equivalent to the increase of the heat flux surface density, the region of exit bulk qualities (with thermal-hydraulic flow instability) increases, as well as the oscillations' amplitude, in spite of an increase in the loop flow rate.

With the increasing q at low exit qualities, the gradient of coolant density changes in the channel heated exit section because of the increased steam nonequilibrium and a faster increment of bulk quality per unit length of the channel. As a result, the flow rate disturbance is the same, while variation of the channel exit coolant density increases along with the increasing q, and in the presence of the individual upflow section leads to the increase of the gravity pressure drop oscillations' amplitude.

This effect is characteristic of both interchannel and loop instabilities.

1.2.1.5 Effect of Power Distribution along the Height

It is shown in Mitenkov et al. [7] that when the maximum of power distribution is shifted to the heated channel inlet (characteristic of the early stage of a reactor's lifetime), the upper portion of the channel starts playing the role of an additional nonheated riser section, while the heat flux surface density in the region of intensive energy release increases essentially. Both factors may enhance the interchannel instability or may cause its appearance even with no individual riser sections.

The shift of the energy release maximum to the heated channel exit in the case of the individual riser section may also destabilize the flow as compared with the uniform energy distribution along the height in the region of low exit qualities because of the heat flux density increase. However, the destabilizing effect in this case shall be weaker against the shift of energy release maximum to the channel inlet, if the destabilizing effect of the additional individual riser section is not compensated for partially by the increased flow rate in natural-circulation loops.

1.2.1.6 Effect of Coolant Flashing in Individual Riser Section

Coolant flashing occurs at low pressures (of up to 3 MPA), when a small increase of bulk quality with the decreasing hydrostatic pressure along the height of the individual riser section results in a noticeable increase of the steam void fraction.

The effect of coolant flashing is ambiguous. For instance, it is reported in Kramerov and Shevelev [1] that coolant flashing in individual riser sections may cause thermal-hydraulic instability, while in other work [7] it is stated that the numerical solution of a distributed mathematical model (Figure 1.10) shows that the account for flashing reduces the flow rate oscillation amplitude (i.e., stabilizes the flow).

Based on the physical mechanism of instability at low exit qualities, it may be concluded that flashing stabilizes the flow, when taking it into account reduces the fluctuation amplitude of the average steam void fraction in the riser section in presence of flow rate oscillations, and to destabilize the flow when the amplitude of quality fluctuations increases. The first case may occur at high exit bulk qualities close to the upper instability boundary, with exit quality being about the same as maximum limiting quality,

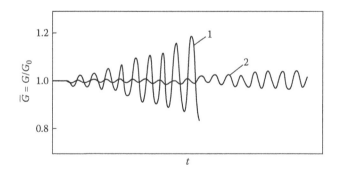

FIGURE 1.10
Effect of superheated coolant flashing on flow fluctuations [7]. 1: Flashing neglected; 2: flashing taken into account.

when the actual exit void fraction is sufficiently large and its increase due to flashing tends to decrease the amplitude of its fluctuation in the riser section, where flow rate oscillations occur because of the decreased $(\partial\alpha/\partial x)_e$ gradient. The second case may take place at exit bulk qualities close to the lower instability boundary (with x being about the same as x_e), when the exit quality is sufficiently small and its increase in the riser section as a result of flashing tends to increase the amplitude of its fluctuations at flow rate oscillations due to increasing gradient $(\partial\alpha/\partial x)_e$.

1.2.1.7 Effect of Nonidentical Parallel Channels

The results of an experimental investigation of the effect of nonidentical parallel channels on flow instability are reported in Mitenkov et al. [7]. The natural-circulation loop incorporated a heated section made up by four channels with pairwise identical channels. The increase in thermal difference between two pairs at constant total power was decreasing the amplitude of flow rate fluctuations up to their complete disappearance (Figure 1.11).

This effect is observed in cases of both interchannel and loop instabilities. (The interchannel instability in the latter case was eliminated by decreasing the length of the individual riser sections.)

A reason of said influence of nonidenticality may be that an increase of thermal difference at the constant input power shifts the channel exit parameters out from the instability region. For high-power channels, the exit bulk

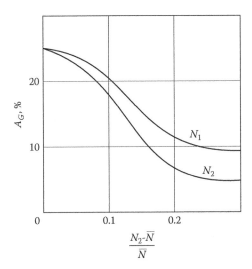

FIGURE 1.11
Dependence of the amplitude of interchannel flow rate fluctuations on the parallel channel thermal nonuniformity [7], $(N_1 < N_2)$. $\overline{N} = 0.5\,(N_1 + N_2)$.

quality surpasses the upper instability boundary $\left(x_e^h > x_{bn}^{\max}\right)$, while for lower power channels it goes below the lower boundary $\left(x_e^h < x_{bn}^{\min}\right)$.

For the interchannel instability (at least for a small number of channels in the case of a test facility), forced fluctuations of pressure drop or loop flow rate, which are in antiphase with the channel flow rate fluctuations and damp the flow fluctuations amplitude in the channel, appear due to nonharmonicity of flow rate fluctuations (even at small amplitudes) and serve as an additional stabilizing factor. For example, in the case of a system of two thermally nonidentical parallel channels, an increase of power (or decrease of the inlet subcooling) is accompanied by flow instability first in the hottest channel, causing forced antiphase flow rate fluctuations in the less heat-loaded channel. These forced flow rate fluctuations give birth to fluctuations of the pressure drop between parallel channels, which are in antiphase with flow rate fluctuations in the unstable heat-loaded channel and lead to the decreased amplitude of flow rate fluctuations as compared to a dynamically isolated channel (i.e., stabilize the system). If thermal difference between parallel channels is not big, the further increase in power initiates instability in the less heat-loaded channel, while the heat-loaded channel becomes stable, starts operating with forced and antiphase (with respect to the less heat-loaded channel) flow rate fluctuations, and decreases the flow rate fluctuations' amplitude in the unstable channel. Such a sequence of instability development may produce a false notion that nonidentical parallel channels extend the region of flow instability as compared to identical channels (i.e., they seem to destabilize the system, which is not always really the case).

The presented qualitative analysis of the effect of parameters on the instability boundary at low exit qualities is based on the well-known published experimental and prediction results. Some conclusions, however, require numerical and experimental validation. The qualitative analysis allows one to determine sufficiently and simply the conditions that enhance stable operation of thermal-hydraulic schemes at early design stages and to plan both experimental instability studies on test facilities and investigations of numerical mathematical models.

1.3 Two-Phase Flow Oscillatory Instability at High Exit Qualities (Density-Wave Instability)

This type of instability is characterized by regularity of mode, cycle, and amplitude of flow fluctuations. The fluctuation mode is close to sinusoidal near the instability boundary at low amplitudes. With the increase of the fluctuations' amplitude due to nonlinear effects, the mode of flow rate fluctuations deviates from the sinusoidal mode and stays regular in cycle. Specific features of the oscillatory density-wave instability are the antiphase change of channel inlet and exit flow rates (Figure 1.12), a comparatively high exit quality, and the oscillation cycle $\tau \approx (1 - 2)\theta_1$ (where θ_1 is the time required by a liquid particle to pass through the heated channel).

When analyzing the density-wave instability in natural- and forced-circulation systems, three types of the previously mentioned boundary conditions are usually considered.

Most simple for analysis is the first type of boundary condition, where instability is considered in an isolated channel with the constant pressure drop within the system of stable parallel channels. At the same time, this case attracts great attention. First, the instability under these boundary conditions is frequently observed before instabilities with the two other types of boundary conditions occur. Second, experimental detection of a single unstable channel among the multiplicity of other stable parallel channels prior to its failure is practically impossible. That is why such a channel is very dangerous from the point of view of thermal reliability. At the same time, investigations of oscillations in an isolated channel are easy to organize on a test facility.

For the other two most typical cases—that is, with periodic oscillations of boundary conditions (channel pressure drop, absolute pressure drop in a system, inlet enthalpy, etc.), it is more difficult to organize and analyze experimental and numerical investigations.

Nevertheless, the physical mechanism of the two-phase flow density-wave instability and qualitative effect of different parameters on the stability boundary are identical for the three most characteristic types of boundary conditions. Therefore, let us first consider flow instability in an isolated

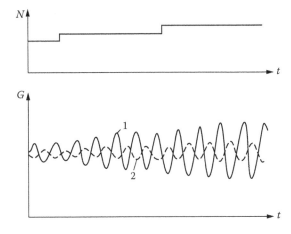

FIGURE 1.12
Development of flow instability in the channel with increasing power. 1: G_{in}—channel inlet flow rate; 2: G_e—channel outlet flow rate.

channel with the constant pressure drop as the most simple and illustrative case. Then, peculiarities of instability in hydraulically interconnected parallel channels and circulation loop will be revealed.

1.3.1 Instability Mechanism

Physically, such a mechanism has been studied in sufficient detail using approximate analytical models [2–9,14–16,18,22–24], numerical analysis of distributed parameter mathematical models [10,25–30], and numerous experimental results. Let a steam-generating channel operate in the region of parameters close to the stability boundary. The flow always features small-amplitude fluctuations of parameters. Any change of the inlet flow rate in magnitude causes the equiphase change of flow rate and friction pressure drop in the economizer section. However, the two-phase region flow rate changes with a delay, as compared to the channel inlet flow rate, due to the time-dependent change of accumulated coolant mass in the economizer section and partly at the entrance to the evaporating section, with the resultant gradual shift of the incipient boiling boundary. Evaporation of this accumulated coolant mass at the decrease of the inlet flow rate yields a two-phase mixture exit flow rate that stays unchanged during the initial period of time and reduces subsequently.

This may be illustrated by a simple model of the steam-generating channel that, apart from numerous assumptions used for deriving (1.3), is characterized by channel low pressure and large exit quality. This permits us to consider the coolant mass in the liquid section many times as high as that in the two-phase section and the particle channel residence time to be approximately equal to the particle residence time in the liquid (economizer) section.

According to (1.3), the dependence of enthalpy perturbation on the flow rate perturbation at $\delta i\,(0, t) = 0$ on the steady-state incipient boiling boundary (H_1) has the following form:

$$\delta i(H_1,t) = \frac{W_L}{G}\frac{di}{dz}\int_{t-H_1/W_L}^{t} \delta G(\tau)d\tau. \qquad (1.19)$$

With adopted assumptions, it follows from the equation of the channel liquid mass balance that perturbations of the two-phase and liquid flow rates, as well as the liquid section lengths, are linked by

$$\delta G_e(t) = \delta G_L(t) - A\rho_L\frac{d\delta H_1}{dt}, \qquad (1.20)$$

where δG_e is the disturbance of the channel exit two-phase mixture flow rate.

The shift of the incipient boiling boundary and enthalpy perturbation on the nondisturbed incipient boiling boundary are linked by

$$\delta i(H_1, t) = -\delta H_1 \frac{di}{dz}. \tag{1.21}$$

From Equations (1.19) to (1.21), it follows that

$$\delta G_e(t) = \delta G_L(t - \theta), \tag{1.22}$$

where $\theta = H_1/W_L$ is the time it takes a liquid particle to pass through the single-phase section.

Irrespective of rather rough assumptions used for deriving (1.22), the latter is qualitatively valid with regard to the relation between the boiling channel inlet and exit flow rates.

Taking into account the importance of the delayed change of the two-phase mixture exit flow rate with respect to the coolant flow rate perturbation at the channel inlet to ascertain the mechanism of the density-wave oscillatory instability initiation, let us find out a relation between the preceding flow rates by using other, more illustrative considerations.

The liquid temperature distribution along the economizer section length $T(z)$ in the nondisturbed state at the constant specific heat flux (Figure 1.13, curve 1) is defined by

$$T(z) = T_0 + \frac{\Pi q}{GC_{P_L}} z, \quad 0 \le z \le H_1. \tag{1.23}$$

Let the channel inlet liquid flow rate change stepwise and in the time interval Δt return to the initial state. Let us determine the time during which the channel exit two-phase mixture flow rate will increase. With the increased

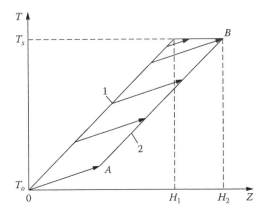

FIGURE 1.13
Coolant temperature distribution along the channel length with the Π-shaped flow rate disturbance. 1: Coolant temperature distribution in the undisturbed state; 2: coolant temperature distribution after the disturbance.

liquid flow rate, the slope of the curve according to (1.23) will be reduced, and during the time interval Δt, liquid particles located at $t = t_0$ on line 1 (Figure 1.13) will be moving along the lines designated by arrows to reach curve 2 at $t = t_0 + \Delta t$. At this time, the coolant temperature distribution in the channel will be represented by the dashed line OAB. Figure 1.13 shows that a stepwise flow rise during the time interval Δt causes an increase of the economizer section length from H_1 up to H_2 without changing the two-phase mixture flow rate.

At subsequent time points, the distribution curves T(z) will be distorted as is shown in Figure 1.14. The inclined step 3 is transported at the nondisturbed velocity to the channel section with H_2 coordinate, where the coolant temperature is T_s, and reaches the channel section in the time interval $\theta(\theta = H_1/W_L)$ after the initiation. During this time interval, the flow rate at the economizer section exit and through the entire evaporating section, in accordance with (1.23), will be equal to the inlet nondisturbed flow rate, since the economizer section length H_2 stays constant. The temperature gradient of temperature distribution 2 (Figure 1.13) is equal to the nondisturbed gradient 1 (i.e., steady-state conditions are kept at the end of the economizer section).

From the moment the inclined area 3 intersects with the T_s line (Figure 1.14), the incipient boiling boundary shifts from H_2 to H_1 (i.e., the liquid will tend to boil off more intensively and the two-phase mixture flow rate will increase).

Let us determine the velocity at which the point of intersection of area 3 with line T_s (Figure 1.14) moves. To this end, a fragment of Figure 1.14 presented in Figure 1.15 should be considered.

Within the time interval Δt, a particle moving from point C at the velocity W_{1L} (W_{1L} is the coolant inlet velocity during a stepwise disturbance) passes the distance Δz_2, or distance Δz_1, if it moves at the velocity W_L. (W_L is the inlet coolant nondisturbed velocity.) Obviously, the incipient boiling point will

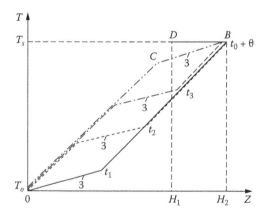

FIGURE 1.14
Transport of the coolant temperature disturbance in the channel over time ($t_0 < t_1 < t_2 < t_0 + \theta$).

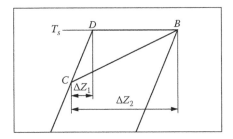

FIGURE 1.15
Diagram for determining the velocity of the incipient boiling boundary movement.

cover the distance DB when moving from point B to point D within the same time period Δt at the velocity

$$\frac{dB}{dt} = \frac{\Delta z_2 - \Delta z_1}{\Delta t} = W_{1L} - W_L .$$

Thus, the velocity of the incipient boiling boundary displacement (the boil-off rate) will be

$$\frac{dH_{ec}}{dt} = -\left(W_{1L} - W_L\right),$$

where H_{ec} is the economizer section length.

In this case, the two-phase mixture flow rate during the time interval when the incipient boiling boundary moves leftward, at $\rho_L \gg \rho_e$, equals

$$G_e = G + \rho_L A \left(W_{1L} - W_L\right) = \rho_L A W_{1L} = G_1.$$

Hence, the disturbance of the two-phase flow at $t = t_0 + \theta$ is $\delta G_e(t_0 + \theta) = A\rho_L(W_{1L} - W_L)$, which is equal to the disturbance of the liquid inlet flow rate at $t = t_0$. Thus, as before, we have

$$\delta G_e(t) = \delta G_{in}(t - \theta). \tag{1.24}$$

This simplified analysis not only yields a qualitative relation between the channel inlet and exit flow rates, but also shows the responsible mechanism.

The performed analysis disregards thermal inertia of the channel heated wall, finite heat transfer coefficient for the economizer section, distribution of the density, and flow rate along the evaporating section length, and therefore represents a qualitative analysis. The real flow rate distribution along the channel length within different time intervals at stabilized flow oscillations, obtained by numerical integration of the distributed parameter mathematical model from Khabensky, Baldina, and Kalinin [26],

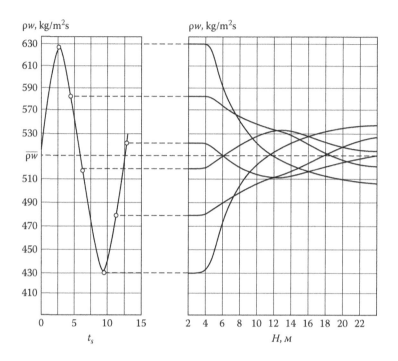

FIGURE 1.16
Flow rate variation along the channel length at flow steady-state fluctuations at different time moments [26].

is presented in Figure 1.16. The figure shows that the flow rate distribution along the length and delay in flow rate disturbance propagation from the start of the evaporating section up to the channel exit are of a more complex nature than it follows from the simplified analysis. However, qualitative consistency that yields close quantitative results at low pressures is evident. The most intensive flow rate change along the channel length occurs at fixed moments in time in the evaporating section area adjacent to the incipient boiling boundary (i.e., in the region of the strongest dependence of density on enthalpy). As one can see in Figure 1.16, the change of flow rate along the channel length in the process of oscillations at fixed moments in sections with a large bulk quality, as well as in the superheating and economizer sections, is not very significant because of the comparatively weak dependence of density on enthalpy in these regions.

A delay in the flow rate disturbance in the evaporating section caused by the delayed movement of the incipient boiling boundary in case of the inlet flow rate perturbation leads to a delay in the friction pressure drop change in the evaporating section (which is the governing factor for the given type of thermal-hydraulic instability) and the gravity pressure drop as compared to the change of the friction pressure drop in the economizer section.

For the sake of obviousness, let us consider a more simple case of flow rate fluctuations in a horizontal steam-generating channel where there is no gravity pressure drop and (1.20) is valid. In this case, the steam-generating channel may be presented as an oscillatory system in the following way.

Let the inlet liquid flow rate in the steam-generating channel at time t_0 [$G_L(t_0)$] increase randomly in the Π-like manner. In time θ, according to (1.20), the two-phase mixture flow rate will increase in the channel two-phase section G_e ($t_0 + \theta$). With constant pressures downstream (P_{00}) and upstream (P_0) in the channel, it will cause an increased pressure drop along the two-phase section (i.e., a higher pressure in the incipient boiling region P_{Bo} ($t_0 + \theta$). The rise of pressure P_{Bo} ($t_0 + \theta$) at the constant pressure before the channel (P_0) will induce a drop in the inlet liquid flow rate G_L ($t_0 + \theta$), which, in turn, in a time θ will cause the two-phase mixture flow rate drop [G_e ($t_0 + 2\theta$)]. This change, with due account for P_{00} = const., will result in the decreased channel pressure at a time $t_0 + 2\theta$ and increased liquid flow rate at the same time $G_L(t_0 + 2\theta)$, and so on. Thus, the initial liquid flow rate change will repeat in time 2θ. Similar reasoning is valid for the initial decrease of inlet flow rate, as well as for the initial perturbation of any other flow parameter of a steam-generating channel. So, this is a case of an oscillatory system because any disturbance in the system causes oscillations with a 2θ cycle. In this case, the friction pressure drop in the two-phase section has a phase shift π as compared to the pressure drop in the economizer section.

If the amplitude of oscillations of such pressure drops is identical, then the steady-state flow rate oscillations of certain amplitude will be maintained if the pressure drop stays constant in the channel. If the antiphase oscillations' amplitude of the friction pressure drop in the evaporating section is smaller than that in the economizer section, the flow rate oscillations will damp in time and the system become stable. If the preceding amplitude is larger, then the flow rate oscillations' amplitude increases and stabilizes at a higher level.

Using this instability mechanism (first published in 1969 in Morozov and Gerliga [2]) and a simple mathematical model, one can arrive at a simple condition for determining the stability boundary.

A simplified mathematical model of the steam-generating channel may be described by (1.22) and the following correlations for friction pressure drops ΔP_1 and ΔP_2 in the liquid and two-phase sections, respectively:

$$P_0 - P(t) = \Delta P_2(G_L(t)); \tag{1.25}$$

$$P(t) - P_{00} = \Delta P_2(G_e(t)). \tag{1.26}$$

After rewriting and linearizing Equations (1.25) and (1.26), from (1.22), (1.25), and (1.26) we obtain

$$\delta G_e(t) = \delta G_L(t - \theta)$$

$$-\delta P(t) = \frac{d\Delta P_1}{dG} \delta G_L(t) \qquad (1.27)$$

$$\delta P(t) = \frac{d\Delta P_2}{dG} \delta G_e(t).$$

From the system of (1.27), there follows a homogeneous equation

$$\delta G_L(t) + \frac{d\Delta P_2/dG}{d\Delta P_1/dG} \cdot \delta G_L(t - \theta) = 0,$$

with the characteristic equation based thereon:

$$1 + \frac{d\Delta P_2/dG}{d\Delta P_1/dG} \cdot e^{-s\theta} = 0, \qquad (1.28)$$

where s is the equation root.

The steam-generating channel will be stable with Re s < 0, which is fulfilled with

$$K = \frac{d\Delta P_2/dG}{d\Delta P_1/dG} < 1. \qquad (1.29)$$

Though (1.29) is based on a rather rough model, it may be useful in some cases for determining the qualitative effect of parameters on the steam-generating channel stability.

In the case of the two-phase flow instability in vertical channels, the stability boundary is under a noticeable influence of the gravity pressure drop. The changes of pressure drop components and other parameters in the case of established flow oscillations in the vertical channel (obtained in Khabensky, Baldina, and Kalinin [28] by numerical integration of the distributed mathematical model) are shown in Figure 1.17. One can see that oscillations of the incipient boiling boundary and of the gravity pressure drop are close in terms of phase with the channel exit flow rate oscillations. Therefore, with steady-state oscillations of the two-phase vertical channel flow, the amplitude of friction pressure drop oscillations in the evaporating section may be smaller (to maintain constancy of the pressure drop) than that in the economizer section by the magnitude of oscillations' amplitude of the gravity pressure drop. In other words, the steady-state flow rate oscillations in the vertical channel may exist at a smaller length ratio between the evaporating and economizer sections (or at a larger coolant mass flow), as compared to the identical horizontal channel. As a result, flow in the vertical channel is less stable.

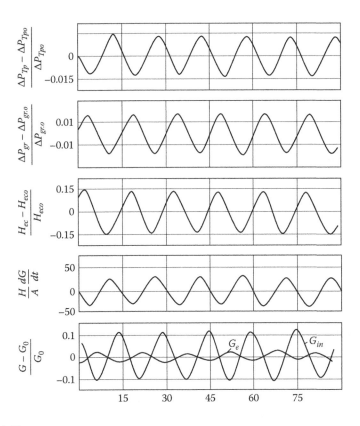

FIGURE 1.17
Variation of the channel input and output flow rates, economizer section length, and pressure drop components in the vertical channel in the region of flow steady-state fluctuations [28].

Thus, initiation of the two-phase flow dynamic instability due to density waves depends on two factors. First, it requires a delay in the flow rate change in the evaporating section relative to the flow rate change in the economizer section with the π phase shift defined by the intensity of the incipient boiling boundary displacement. Second, an appreciable pressure drop must exist in the evaporating-superheating section of the channel, which, in presence of antiphase inlet and exit flow rate oscillations, should yield antiphase and equal-amplitude oscillations of pressure drop components in the evaporating and economizer sections (taking oscillations of the gravity pressure drop component into account). It is evident that the larger the maximum enthalpy density gradient at the beginning of the two-phase section is, the larger is the amplitude of antiphase two-phase mixture flow rate oscillations, and the smaller may be the share of friction pressure drop in the evaporating section in the total pressure drop to maintain stable oscillations with a preset amplitude.

On the basis of the mechanism of the density-wave instability development, let us consider the qualitative effect of different design and flow parameters on the two-phase flow stability boundary.

1.3.2 Effect of Pressure

Under otherwise equal conditions (the maintained preset inlet subcooling included), the system pressure rise stabilizes and improves flow stability. This result has been proven by numerous experimental and predictive investigations. If the pressure is rising, several factors contribute to stability. First, it is the reduced relative share of friction pressure drop in the evaporating section in the total pressure drop due to higher average density of the fluid in the evaporating section. Second, it is the decreased maximum density gradient in the two-phase section because of the reduced gradient of change of the two-phase coolant density because of enthalpy and, hence, the reduced amplitude of the outlet flow rate oscillations. Third, it is the reduced amplitude of the gravity pressure drop in the vertical channel.

These factors lead to a reduced amplitude of pressure drop oscillations in the evaporating section, as a result of which pressure drop oscillations in the single-phase region start dominating (i.e., the system becomes stable). To approach the stability boundary again, the share of the pressure drop in the evaporating section in the total pressure drop shall be increased, and it may be achieved, say, by increasing power or reducing the flow rate. This will increase the evaporating section length, exit bulk quality, and amplitude of the channel exit flow rate oscillations.

To take the effect of pressure on the stability boundary into account (under otherwise equal conditions), the ρ_G/ρ_L parameter can be used, as is recommended in Boure [8]. Since the gradient of ρ_G/ρ_L change with respect to pressure is big at low and medium pressures, the influence of pressure on stability improvement is especially pronounced at $P < 10$ MPa. It should be noted that the influence of pressure on the stability boundary is most effective in forced-circulation systems, because in natural-circulation systems a pressure rise leads to a somewhat decreased moving head and, hence, to a lower mass velocity in the channel with a potentially lower stabilizing effect. The positive effect of pressure rise in the natural-circulation loop of a boiling nuclear reactor may also manifest itself in a higher neutron and thermal-hydraulic stability in core channels due to decreased steam reactivity. A pressure rise exerts a similar qualitative influence on the interchannel and loop circulation stability.

1.3.3 Effect of Local and Distributed Hydraulic Resistance

An increase in flow stability due to higher local and distributed hydraulic resistances in the economizer section was established long ago. The previously described instability mechanism helps in proving it. With the increased resistance across the economizer section, the share of its pressure drop in the total channel pressure drop will increase. Therefore, under otherwise equal conditions, the amplitude of the pressure drop antiphase oscillations in the evaporating section fails to compensate for the large-amplitude pressure drop oscillations in the economizer section. This leads to damping of flow rate oscillations. Position of local and distributed resistances in the economizer section has no effect on channel flow stability because of the equiphase change of flow rate in the single-phase system. Experimental and predictive results [13] show that the inlet throttling has the strongest effect at low and medium subcooling, which somewhat weakens at high subcooling. The reason is that, at low subcooling, because of small economizer section length, throttling sharply increases the contribution of the economizer section resistance to the total pressure drop. This also explains the reduced relative throttling effect on flow stability with increasing extent of throttling.

A reverse effect is attained when hydraulic resistance is increasing in the evaporating and superheating sections. The cause of flow destabilizing lies in the increased share of pressure drop in the evaporating and superheating sections, with the resultant increased amplitudes of their oscillations. Position of the local (or distributed) resistance in the evaporating section substantially influences flow stability because the closer to the end of the evaporating section it is located, the larger the pressure drop effect is. Location of resistance in the superheating section is of almost no effect on flow stability due to the constant density in the superheating section. On the basis of the previously mentioned regularity, it is recommended [12] to use smaller diameter step channels for the economizer section and larger diameter channels for evaporating and superheating sections as a flow stabilizing method, as it ensures an increased share of the pressure drop in the economizer section.

To take the effect of local and distributed resistances on the flow stability boundary into account, the Petrov's parameter,

$$\Pi = \frac{\Delta P_{ec} + \Delta P_{Lt}}{\Delta P_{TP} + \Delta P_{Gt}}, \tag{1.30}$$

may be used, where ΔP_{ec}, ΔP_{TP} are friction drops in the economizer and evaporating/superheating sections, respectively; and ΔP_{Lt}, ΔP_{Gt} are local resistance pressure drops across the economizer and evaporating/superheating sections, respectively.

It has been shown [2,10,13,22,26,31, etc.] that with the constant pressure and inlet subcooling at the flow stability boundary in the investigated channel, the Π parameter is practically constant when the exit quality is constant. For example, at the stability boundary with the increased local resistance in the evaporating section and with other parameters unchanged, the channel starts operating in the domain of developed flow oscillations. To restore the stability boundary for the channel, an increased inlet throttling is required

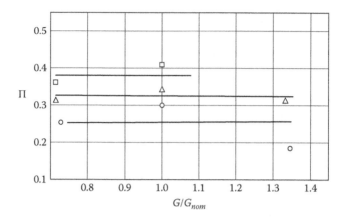

FIGURE 1.18

Effect of relative flow rate (G/G_{nom}) on the P parameter boundary value at different exit qualities [31]. Circle: $x_e = 0.8$; triangle: $x_e = 0.9$; square: $x_e = 1.0$.

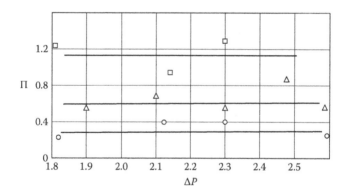

FIGURE 1.19

Effect of the steam section resistance on the Π parameter boundary value at different exit qualities [31]. Circle: $x_e = 0.8$; triangle: $x_e = 0.9$; square: $x_e = 1.0$.

so that the equivalent increased resistance in the economizer section restores the previous Π parameter value (see Figure 1.18).

The effect of local and distributed resistances on loop instability is similar. For example, experimental investigations of instability in the steam-generating loop [31] provided evidence that a change of the loop flow rate or an increase of local resistances at the stability boundary leaves the parameter Π practically constant (Figures 1.18 and 1.19) if the exit quality remains practically unchanged in presence of the constant pressure and inlet subcooling. A decreased Π value at the stability boundary means a lower flow stability in a channel or loop.

Inlet throttling aimed at increasing flow stability is generally applied in forced-circulation systems. In natural-circulation systems, the inlet throttling is less efficient due to simultaneous reduction of the channel coolant flow rate with the resulting reduced stabilizing effect.

1.3.4 Effect of Nonheated Inlet and Exit Sections

An upstream nonheated section at the channel inlet stabilizes the flow. Apart from serving as the inlet resistance, an upstream section of sufficiently large capacity increases liquid inertia.

A contribution of the increased pressure drop due to friction losses along the upstream section is easily taken into account by using parameter (1.30) by introducing an equivalent pressure drop into the numerator. To evaluate the effect of coolant inertia, the results of Takitani and Fakemura [22,24] and Khabensky et al. [26] can be used by applying a dimensionless parameter $\bar{\tau} = \tau^*/(\tau_{ec}/2 + \tau_{ev})$, where

$\tau^* = H_1 G/A\Delta P_{ec}$ is the inertia time constant

$\tau_{ec} = A\rho_L H_{ec}/G$ and $\tau_{ev} = A\rho_L(1 - \alpha) H_{ev}/G$ are the time intervals a liquid particle requires to pass through the economizer and evaporating sections, respectively

H_{ec} and H_{ev} are the lengths of the economizer and evaporating section, respectively

H_1 is the total length of the upstream and economizer sections

ΔP_{ec} is the friction pressure drop in the economizer and upstream sections, respectively

It has been shown [22–24] that the effect of liquid inertia on the stability boundary is insignificant with $\bar{\tau} < 1$. With $\bar{\tau} > 1$, it reveals itself in a sharp flow stabilization, which is well illustrated in Takitani and Fakemura [23] by a diagram (Figure 1.20).

An upstream nonheated section is also found to stabilize loop circulation. If a natural-circulation loop is considered as a channel where the downflow section is regarded as an upstream section with high resistance, its stabilizing effect will be generally manifested in a loop instability that would be delayed in comparison with the thermal-hydraulic interchannel instability. As in the case with thermal-hydraulic instability at low exit qualities, an increased resistance of the downflow section of the loop increases loop circulation stability, but impairs interchannel flow instability because of the reduced loop flow rate. According to Bailey [9], similar effects may be observed in core channels with closely packed rod bundles where interrelated oscillations may occur in cells. This being the case, channel throttling will improve channel flow stability, but may impair intercell flow stability.

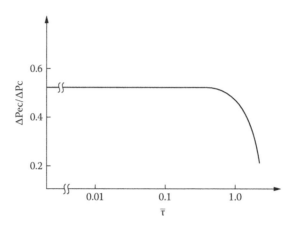

FIGURE 1.20
Effect of coolant inertia on the stability boundary [23].

The presence of the channel exit nonheated section destabilizes the flow because oscillations of the nonheated section flow rate occur at high qualities in phase with oscillations of the heated channel exit flow rate. Therefore, the pressure drop in the nonheated section may be considered as an equivalent pressure drop at the channel exit. The effect of the exit nonheated section may be assessed, using criterion (1.30), by introducing an equivalent pressure drop across the nonheated section into the denominator.

Flow destabilization in a channel is most sharply manifested in the presence of the exit nonheated section in the forced-circulation systems where flow rate is kept constant. In natural-circulation systems, the nonheated riser section results in a higher loop flow rate and therefore it somewhat compensates for the destabilizing effect of the exit pressure drop increase. However, if thermal-hydraulic instability is a limiting factor with regard to thermal reliability, a decreased height of individual riser sections may be positive irrespective of the reduced flow rates through the loop and the hottest channels. Each case requires a detailed investigation.

1.3.5 Effect of Inlet Coolant Subcooling

The effect of subcooling on the stability boundary is ambiguous [8–10,13–16,22,26]. Under otherwise equal conditions, the increasing subcooling has a destabilizing effect at low subcooling and a stabilizing effect at high subcooling. Figure 1.21 shows an example from Khabensky et al. [26] that offers experimental data and a numerical prediction made using the distributed parameter model for a system in a stable state, when a decreased inlet coolant enthalpy resulted in channel flow instability, while the further reduction of enthalpy stabilized the flow. This can be explained as follows. With the increased inlet subcooling, two opposed factors influence flow stability.

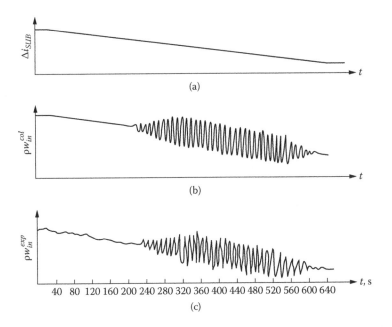

FIGURE 1.21
Effect of inlet subcooling on the channel flow stability [26]. (a) Inlet enthalpy variation; (b) inlet flow rate calculated variation; (c) inlet flow rate experimental variation.

First, the economizer section increases and the evaporating section reduces with the resultant increased stability of the system because of the higher share of pressure drop in the economizer section. Second, the amplitude of oscillations of the incipient boiling boundary increases and leads to flow destabilization due to the increased amplitude of the antiphase flow oscillations in the evaporating section. At low subcooling values [8–10], the destabilizing factor prevails when subcooling increases. At a certain subcooling value, the effects of stabilizing and destabilizing factors are equipotent, and here the minimum system stability is observed. With the further increase in subcooling, the stabilizing effect starts to dominate and flow stability improves.

Some results [8–10,26,28,32] show that for vertical channels, under otherwise equal conditions, the minimum flow stability is most often observed at ΔT_{SUB} of about 20–40 K, and for horizontal channels at ΔT_{SUB} of about 15–25 K (where ΔT_{SUB} is the inlet coolant subcooling below the saturation temperature at the channel inlet). This is shown in Kramerov and Shevelev [1], and it follows from the instability mechanism that with zero subcooling, thermal-hydraulic flow oscillations do not take place.

The effect of the extent of inlet coolant subcooling on flow stability, under otherwise equal conditions, may be evaluated using the dimensionless subcooling parameter $\Delta i_{SUB}/i_{LG}$, where $\Delta i_{SUB} = i_s - i_{in}$, and i_{LG} is the latent heat of evaporation.

1.3.6 Effect of Specific Heat Flux, Mass Velocity, Channel Length, and Equivalent Diameter

Under otherwise equal conditions, an increased heat flux or decreased mass velocity results in flow destabilization. This fact can be explained by the instability mechanism. Indeed, under otherwise equal conditions, an increased heat flux or decreased mass velocity leads to a shorter economizer section and a longer evaporating section. It decreases the economizer section pressure drop contribution to the channel total drop. If the increase of heat flux is accompanied by a proportional rise in mass velocity, the parallel change is of almost no effect on flow stability because the ratio between pressure drops across the economizer and evaporating sections remains almost constant.

Figure 1.22 illustrates experimental data from Takitani and Fakemura [22] on the flow stability boundary in coordinates of mass velocity (ρW) and dimensionless enthalpy (N/Gi_{LG}), where N is the channel power input, G is the flow rate, and i_{LG} is the latent heat of evaporation. It is obvious from the figure that with the change of ρW, N/Gi_{LG} remains practically constant at the flow stability boundary (i.e., when the channel geometry, pressure, and inlet enthalpy stay unchanged, the heat flux changes almost proportionally to mass velocity). Some deviation from the linear dependence between the heat flux and mass velocity may take place at the stability boundary because, with the increasing mass velocity (especially in the low-value region), the ratio between friction factors in the real and two-phase homogeneous flows [12] decreases and, with identical exit bulk qualities, causes a decrease of the relative share of pressure drop in the evaporating section. However, for rough estimations, the N/Gi_{LG} constancy can be used for determining the effect of mass velocity and heat flux on the stability boundary.

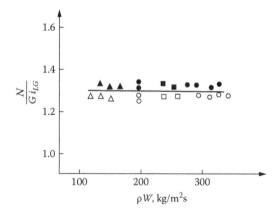

FIGURE 1.22

Flow stability boundary depending on mass flow rate (ρW) at the constant N/Gi_{LG} value [22]. Light triangles, circles, squares: experimental data on stable regimes; dark symbols: same, for unstable regimes.

It is evident from Khabensky et al. [26] that the effect of the heated length and equivalent diameter of the channel on the stability boundary is also determined by N/Gi_{LG}.

An increase in channel length, under otherwise equal conditions including the constant specific heat flux, leads to an increased length of the evaporating or superheating section (i.e., to an increased relative share of pressure drop along the evaporating-superheating section). For instance, a flow may be stabilized by decreasing the specific heat flux or increasing mass velocity. If the system is at the stability boundary, then to stay there with the increased channel length and to keep the relative pressure drops ratio in the channel approximately constant, either the specific heat flux should be proportionally reduced or mass velocity increased.

This expresses constancy of N/Gi_{LG}, which may be written as

$$\frac{N}{Gi_{LG}} = \frac{4qH}{(\rho W)D_h i_{LG}}, \tag{1.31}$$

where H is the channel heated length, D_h is the equivalent diameter, and q is specific heat flux.

Thus, $\left(\dfrac{N}{Gi_{LG}}\right)_{Bn}$ at the stability boundary is found to be constant. This means that respective factors in (1.31) are changed, providing the constant ratio between pressure drops in economizer and evaporating sections.

The increase of flow stability by installing intermediate and breathing headers, where pressure between parallel channels is equalized completely or partially, has the same effect as the decreasing of the channel heated length. In this case, intermediate headers cause direct reduction of the heated length, while the effect of breathing headers is a function of the junction resistance between the channels and the breathing header. The smaller the extent of pressure equalizing by the breathing header in the parallel-channel system is, the lower is its influence on flow stabilization. Intermediate and breathing headers should be placed at the beginning of the evaporating section [12].

Similar reasoning may be applied to prove that the influence of equivalent diameter on the stability boundary may be also approximately evaluated by N/Gi_{LG}. Indeed, an increase of, say, channel diameter, under otherwise equal conditions (constant ρW, q, H, P, Δi_{SUB}) increases flow stability because (1.31) shows that the enthalpy increment in the channel reduces and, at the preset inlet enthalpy, corresponds to a reduction of exit enthalpy, when the economizer section length is increased and that of the evaporating section is decreased. Generally speaking, this mechanism is sufficiently correct for simple geometry channels. For channels of complex geometry, an increasing equivalent diameter may produce certain effects on the distribution of quality and resistance in the evaporating section, which may distort the influence of N/Gi_{LG}.

The preceding regularities in the influence of N/Gi_{LG} were fully proven by experimental and predictive results [2,8–10,13,22–24,26,27].

It should be noted that, despite the fact that the design and flow parameters ρW, q, H, and d_h, with the dimensionless criteria parameter N/Gi_{LG} maintained constant have little influence on the location of the flow stability boundary, the oscillation cycle is found to vary substantially when such parameters are changed. Thus, the preceding instability mechanism shows that with N/Gi_{LG} = const., the oscillation cycle reduces at the increase of heat flux and mass velocity and increases when the heated length grows.

1.3.7 Effect of Kind and Height Distribution of Heating

Let us consider a case of a heat flux that is nonuniform along the channel height but constant over time. The numerical and experimental investigations described in Morozov and Gerliga [2], Khabensky [11], Khabensky et al. [26], etc. showed that in comparison with the uniform heat distribution along the channel height (with identical channel input power), the increasing intensity of heating from the beginning to the end of the channel promotes flow stability, and vice versa. This is explained by the increased (decreased) length of the economizer section and, hence, its contribution to the channel total pressure drop. The extent of the effect on stability depends on the relative heat flux height nonuniformity, nature of heat flux distribution, and inlet subcooling. For example, with the cosine height distribution and different fixed inlet subcooling, flow stability may increase or decrease as compared to the uniform heat flux height distribution. This depends on the economizer section boundary location at the uniform heat flux distribution and the direction of its shift, as well as on the change of the pressure drop distribution along the evaporating section in case of the cosine heat flux distribution.

Now, let us consider the effect of the kind of heating on flow stability and its thermal and inertia characteristics that defines the nature of variation of the heat flux from the heat transfer surface over time. This is mostly manifested in the change of the heat flux over time in the economizer section. In this case, it follows from the thermal and hydrodynamic mechanism of density wave instability that if the unsteady heat flux from the channel heat transfer surface of the economizer section contributes to damping of the amplitude of the incipient boiling boundary oscillations in the economizer section, it will increase the two-phase channel flow stability. If the unsteady heat flux leads to the increased amplitude of the incipient boiling boundary oscillations, the flow destabilization occurs.

It has been shown [22,23] that the use of linear approximation makes it possible to represent this effect by coefficient b relating the change of the heat flux in the economizer section and that of coolant velocity (q ~ bG). In the case when b < 1/2, instability in the channel under investigation was not observed, while with b > 1/2, an increase of the coefficient up to 1 resulted in the growing flow destabilization. It should be noted that the greater the relative length of the economizer section is—that is, the larger θ_{ec}/θ_{ev} (where θ_{ec}/θ_{ev} are the time intervals a particle requires to pass through the economizer and evaporating sections, respectively) is, the stronger the stabilizing and destabilizing effects are.

The nature of the unsteady heat flux dependence on the coolant velocity in the economizer section is determined by the flow regime (which in turn influences the effect of velocity on the heat transfer coefficient), by thermal and inertia characteristics of a fuel element or channel wall, kind of heating (electrical, convective, nuclear), and cycle of channel flow oscillations.

It has been demonstrated [2] that in the case of the laminar regime, the channel coolant flow is more stable as compared to the turbulent regime. This is explained by independence of the heat transfer coefficient from coolant velocity in the case of the laminar regime and almost directly proportional dependence of such parameters in the case of the turbulent regime.

It has been shown [24] that if the half-cycle of flow oscillations, which is proportional to the time it takes a particle to pass through the heated channel, is much less than the time of heat transport from the heating element to the coolant, thermal and inertial characteristics of the heating element and kind of heating produce an insignificant effect on the stability boundary as compared to the case of the heat flux constant over time. If these time intervals are commensurate, the effect of heating element characteristics and of the kind of heating on flow stability may be either positive or negative.

For instance, it follows [9,22–24, etc.] that the convective heating (e.g., in liquid metal steam generators) renders a system more thermohydrodynamically stable as compared to similar electrically heated channels. This is explained by different dynamic dependences of heat input to the economizer section on flow velocity when liquid metal and electric current are used for heating. In the latter case, the channel wall heat flux stays constant. However, with convective heating, the heat flux depends not only on thermal inertia of the dividing wall but also, to a great extent, on transport delay of heat transfer of the heating and heated coolant.

The work of Takitani and Fakemura [24] presents experimental data on the behavior of basic thermal-hydraulic parameters on the flow stability boundary in identical channels, differing only by the kind of heating (liquid sodium convective heating or electrical heating). The following correlations have been obtained:

$$(\Delta P_{ec} / \Delta P_c)^{el}_{bn} / (\Delta P_{ec} / P_c)^{con}_{bn} \approx 7,$$

$$(\Delta H_{ec} / H_{ec})^{el}_{bn} / (\Delta H_{ec} / H_{ec})^{con}_{bn} \approx 3,$$

and

$$(\Delta G_{in} / G_{in})^{el}_{bn} / (\Delta G_{in} / G_{in})^{con}_{bn} \approx 3,$$

where ΔH_{ec}, ΔG_{in} are the amplitude of oscillations of the economizer section boundary and the channel inlet coolant flow rate, respectively; ΔP_{ec}, ΔP_C are the pressure drop across the economizer section and the total pressure drop in the channel, respectively; and $(\ )_{bn}^{el}$ $(\ )_{bn}^{con}$ are the stability boundary parameters with electrical and liquid sodium convective heating, respectively.

The correlations show a substantially higher stability of the investigated convectively heated channel than of the identical electrically heated one. In a general case, there are no reasons to insist that, with all possible combinations of geometrical and flow parameters in identical channels, the system is more stable with convective than with electrical heating. This circumstance is of importance and needs further investigation because, quite often, convectively heated channels are simulated using electrically heated channels in stability studies.

The effect of the kind of heating and thermal and inertial characteristics of a heater on flow stability may be assessed with the help of simplified models used in investigations of heat transfer in economizer sections with preset forced harmonic flow rate oscillations within the range of frequencies typical of the given channel in the case of density wave instability.

1.3.8 Effect of Channel Orientation

Vertical channel flows are less stable as compared to horizontal channels due to the influence of the gravity pressure drop (see Figure 1.17). For illustration, Figure 1.23 shows the power ratio in identical vertical and horizontal channels on the stability boundary at different subcooling and pressure values with low inlet throttling.

The results, presented in Khabensky et al. [26], were obtained by numerical solution of the distributed mathematical model and proved by experimental data. One can see from the figure that the largest flow destabilization in the vertical channel in comparison with the horizontal channel is observed at low pressures and inlet subcooling of about $\Delta T_{SUB} \approx 30\text{--}50$ K. With the increasing pressure and subcooling departure from said range, the destabilizing effect weakens. At subcooling of about $\Delta T_{SUB} \approx 5$ K and the system pressure $P > 5$ MPa, flow stability is almost equal in both types of channels. The results are consistent with the experimental data of Koshelev, Surnov, and Nikitina [32]. The destabilizing effect of the gravity pressure drop in the vertical channel weakens with the increasing inlet throttling because of the reduced share of the gravity pressure drop in the channel total pressure drop.

For channels with the upflow-downflow coolant motion, flow stability may approach that of the identical horizontal or vertical channel, depending on the location of the incipient boiling boundary.

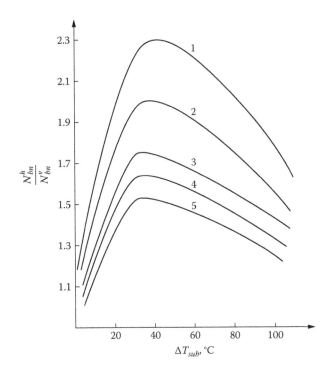

FIGURE 1.23
Horizontal channel power to vertical channel power ratio at the stability boundary versus inlet subcooling [26]. 1: P = 4 MPa; 2: P = 6 MPa; 3: P = 8 MPa; 4: P = 12 MPa; 5: P = 18 MPa.

1.4 Simplifying Assumptions Underlying Mathematical Model and Their Effect on Accuracy of Thermal-Hydraulic Stability Boundary Prediction

The effect of some simplifying assumptions underlying a mathematical model on the accuracy of the stability boundary prediction may be assessed using the physical mechanism of thermal-hydraulic instability. Among the most important assumptions are (1) the effect of parameters' distribution along the economizer and evaporating sections, (2) the effect of surface boiling, (3) the two-phase slip, (4) the two-phase incipient boiling and condensation in individual riser sections at low exit qualities, and (5) the height distribution of static thermal-hydraulic characteristics in the element under investigation. Such assumptions are considered in references 7, 13, 15, 20, 22–24, 38, etc.

It has been shown [24] that taking the distribution of unsteady enthalpy in the economizer section into account in the energy equation has

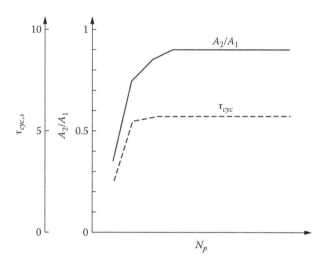

FIGURE 1.24
Effect of the number of economizer section cells on the cycle and amplitude of channel flow fluctuations in a numerical investigation of instability [24]. N_p: Number of economizer section cells; A_1: inlet flow rate fluctuations' amplitude at time t; A_2: inlet flow rate fluctuations' amplitude at time $(t + \tau_{CYC})$.

a destabilizing effect on flow stability, since it increases the flow oscillations' amplitude growth rate about three times with the increasing economizer section cells from 1 to 4 (Figure 1.24). This is also accompanied by the increased cycle of oscillations. The further increase in the control volumes' number for conditions of Takitani and Fakemura [24] exerts no influence on flow stability.

An explanation of this effect is that the description of coolant enthalpy in the economizer section with the distribution taken into account leads to a larger delay with regard to the displacement of the incipient boiling boundary and to the growth of the amplitude of its oscillations in the onset of instability.

The destabilizing effect also reveals itself when the description of enthalpy in the boiling region takes distribution into account, especially in the section adjacent to the incipient boiling area. However, the effect is not so strong (under conditions of Takitani and Fakemura [24], the rate of change of flow oscillations increases about 1.3 times) and exerts no influence on the oscillation cycle (Figure 1.25).

The stabilizing effect may take place when the distribution is taken into account in heat transfer over the thickness of the heating element in the economizer section. In this case, the stabilizing effect is due to the increased amplitude of oscillations of the unsteady heat flux from the heating element surface in the event of flow rate oscillations. In the case of a small phase shift between the flow rate and the heat flux, it leads to damping of the incipient boiling boundary oscillations.

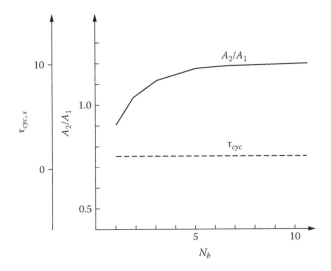

FIGURE 1.25

Effect of evaporation section cells on the cycle and amplitude of channel flow rate fluctuations in a numerical investigation of instability [24]. N_b: Number of evaporation section cells; A_1: inlet flow rate fluctuations' amplitude at time t; A_2: inlet flow rate fluctuations amplitude at time $(t + \tau_{CYC})$.

An opposite effect of these two factors on the stability boundary may lead to a situation when, under certain conditions, a calculation employing the lumped parameter model will provide a satisfactory description of the experimental stability boundary. For example, a comparison of experimental data with the results of the distributed and lumped parameter models' application is offered in Takitani and Fakemura [24]. It showed the two models to be consistent with the experiment concerning the density-wave boundary determination; for the oscillation cycle, the experimental data agreed with the distributed model predictions and differed greatly from those of the lumped parameter model. This may be explained by the fact that the neglect in the lumped parameter model of the coolant enthalpy distribution across the economizer section, which stabilizes the system, and the neglect in the distributed parameter model of the wall heat flux distribution, with its destabilizing effect for a particular system, result in their cancellation.

Such a qualitative effect of taking parameters' distribution into account on the flow stability boundary is valid for the constant volumetric heat release of the heating wall, typically for an electrically heated system. With thermal-hydraulic instability in the nuclear reactor core channel, volumetric heat release in fuel elements will vary and depend on the fuel element and coolant thermal inertia, in case of negative density coolant reactivity and fuel temperature reactivity coefficients.

Under certain combinations of the channel inertial characteristics and density reactivity coefficients, the accounting for parameter distribution in the mathematical model on the predicted stability boundary may have an opposite effect as compared to the previously described electrically heated systems. A significant distortion of the qualitative effect of the considered accounting for parameters' distribution may also occur in convectively heated systems.

The performed qualitative analysis shows that care should be exercised when justifying assumptions made in developing a mathematical model for investigating, say, flow instability in reactor core channels using the experimental results obtained on the test facility with the electrically heated channels.

When surface boiling is taken into account, an increased evaporating and decreased economizer section are obtained. This, in line with Section 1.3, should result in flow stabilization at low inlet subcooling and flow destabilization at large subcooling. The results of Saha, Ishii, and Zuber [13] fully confirm this nature of surface boiling effect.

Also, the flow stability boundary is affected by dynamics of incipient boiling boundary change. Therefore, dynamic equations describing surface boiling shall adequately reflect the real quality distribution along the height and over time under both steady and unsteady conditions.

An investigation performed by R. D. Filin [41] analyzed the effect that different techniques of describing surface boiling have on the flow stability boundary. It has been shown that the use of techniques from Plyutinsky and Fishgoit [39] and Molochnikov and Batashova [40] for describing surface boiling in a specific system with low inlet coolant subcooling, under otherwise equal conditions, has yielded substantially different results concerning the flow stability boundary. The static height quality distribution across the areas of surface and developed boiling, determined by techniques from these authors [39,40], has been selected almost identically with the help of a small adjustment of the equation constant factors. All other equations and boundary conditions in the comparative study of Filin [41] were identical. The difference in results is due to different dynamic characteristics of the change in the surface boiling boundary and in quality in the sections of surface and developed boiling. For the conditions under investigation, the technique from Plyutinsky and Fishgoit [39] yielded a lower inertial change in quality and a wider area of stability, which better agreed with the experimental data (Figure 1.26).

Since the predictive determination of the stability boundary is very sensitive to surface boiling description at low subcooling, the experimental results on flow stability may be applied for validating dynamic mathematical models of the subcooled liquid boiling.

The nonequilibrium quality produces a big effect on the boundary and propagation of instability at low exit qualities. Taking surface boiling into account in a mathematical model shifts the lower stability boundary toward the lower exit bulk quality. On the other hand, it leads to a smoother quality

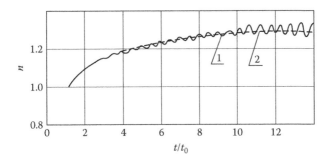

FIGURE 1.26
Effect of different techniques used for describing surface boiling on flow instability [41]; $n =$ relative neutron flux. 1: Calculated according to Filin [41]; 2: calculated according to Plyutinsky and Fishgoit [39].

distribution along the channel height as compared to the equilibrium model. The dependence of real quality on enthalpy enters a flatter section of the saturation line at lower bulk qualities. Since the instability at low exit qualities depends on the value of the $(\partial \rho / \partial i)_e$ derivative [or $(\partial \alpha / \partial i)_e$], the preceding effect of surface boiling on the $(\partial \alpha / \partial i)_e$ value may result in a decreased amplitude of flow rate oscillations in the instability region and a shift of the lower stability boundary into the domain of lower (or negative) bulk qualities.

It has been demonstrated [7,13,24, etc.] that if the phase slip in the two-phase flow is properly taken into account, it strongly influences the predicted flow stability boundary. When density waves are instable at high exit qualities, taking the steam slip into account results in the decreased real quality in the evaporating section, thereby decreasing the friction pressure drop in the evaporating section and the gravity pressure drop oscillations' amplitude. The mentioned effects promote flow stability. It has been noted [24] that taking the slip into account increases the oscillation cycle due to the increased time it takes a particle to pass through the evaporating section because of the reduced real quality.

According to Mitenkov et al. [7], taking the slip into account in a mathematical model is of stabilizing effect even under instability at low exit qualities. In this case, the stabilizing effect of the slip is due to the decreased value of the $(\partial \alpha / \partial i)_e$ derivative. The experimental and numerical investigations described [7] dealt with interchannel instability in a system of identical channels and showed that the model that ignores the slip overestimates the flow rate oscillations' amplitude eight times as compared with the experimental data, while the introduction of a slip factor of 3 decreases the predicted flow rate oscillations' amplitude four times.

The effect of coolant self-boiling on the stability boundary at low exit qualities has already been considered in Section 1.2, so here we only remind that the noticeable effect is observed at pressures below (P < 3 MPa) and that when self-boiling is taken into account in a mathematical model, the lower and upper stability boundaries are shifted toward the domain of low exit bulk qualities. According to Urusov et al. [20], the instability region may decrease due to the nonuniform shifting of the lower and upper stability boundaries.

Condensation of the nonequilibrium steam in the nonheated individual riser sections may increase oscillations of the gravity pressure drop because with the decreased exit quality during flow rate oscillations (due to the increased steam nonequilibrium) steam condensation increases, thereby contributing to a more sudden decrease in quality in the riser section. When the exit quality during flow rate oscillations increases because of a lower degree of nonequilibrium, steam condensation decreases. It follows from the instability mechanism at low exit qualities that the increasing oscillation amplitude of the gravity pressure drop destabilizes the flow. Its regularity is also mentioned in Urusov et al. [20].

It should be noted that self-boiling and condensation of the nonequilibrium steam exert opposed effects on the stability boundary and may partially compensate each other. At present, descriptions of the nonequilibrium steam condensation in nonheated upflow sections are not sufficiently reliable.

As has been stated before, the accuracy of predictive determination of the stability boundary depends to a great extent on the accuracy of specifying static thermal-hydraulic dependencies and characteristics in a mathematical model. However, for large-scale plants or even for thermal-physical models of modern plants, steady-state thermal-hydraulic characteristics are known approximately due to complex geometry of circulation loops and process channels, as well as due to location and technological tolerances. The accuracy of specifying such characteristics in a predictive model is usually checked by comparing them with such experimental parameters as coolant flow rates in a loop and in separate experimental channels.

Therefore, in case of a discrepancy between the experimental static thermal-hydraulic parameters of a real plant or its physical model and static parameters obtained using dynamic models thereof, the dynamic model is generally adjusted by fitting predicted flow rates to experimental ones. This can be made (if the real cause of discrepancy is not clear) in different ways: by changing either coefficients of local and distributed real resistances across the cold and hot loop sections, or quality dependencies, or power at the heat input section, or heat distribution along the height, etc. However, each technique may have a different effect on the system thermal-hydraulic instability. As a result of the incorrect fitting, large-scale plant steady-state integral parameters and those of a mathematical model may coincide, with the latter being inadequate with regard to the thermal-hydraulic flow instability region of the system under investigation.

To determine the real cause of a discrepancy, specially planned experimental investigations are conducted. Quite often, they employ full-scale plants or model reactor loops constructed for the purpose.

If the cause of discrepancy cannot be detected due to some constraints, then the parameters deteriorating the system stability should be changed when adjusting static parameters of a mathematical model. This permits determination of the stability boundary with a safe margin.

2

Oscillatory Stability Boundary in Hydrodynamic Interaction of Parallel Channels and Requirements to Simulate Unstable Processes on Test Facilities

2.1 Qualitative Effect of Hydrodynamic Interaction of Parallel Channels on Oscillatory Stability Boundary

When the number of unstable channels in the system of parallel steam-generating channels becomes commensurate with their total number, or the amplitude of oscillations in the unstable channels reaches some substantial value, the stability boundary starts to be affected by the interchannel hydrodynamic interaction.

For the second type of boundary conditions, the physical mechanism on density-wave interchannel instability and qualitative effect of parameters on the stability boundary are as those for instability of the isolated channel with constant preset pressure drop. However, hydrodynamic interchannel interaction may quantitatively influence the stability boundary and the amplitude of parameters' oscillation in the unstable region.

An analysis of this effect is of practical and methodological importance, as it permits determination of requirements to instability flow simulation in multichannel systems on test facilities and assessment of the effect of flow interchannel instability on loop instability and vice versa.

Usually, the options of parallel channels employed for simulating interchannel instability in natural and forced circulation loops are as follows:

1. A heated channel with the inlet and outlet compressible "volumes"
2. A heated channel with the nonheated bypass with different flow rate ratios between the channel and bypass
3. Two parallel heated channels

4. Two parallel hydraulically identical heated channels with the nonheated bypass

5. Several (n above 2) parallel heated channels

The results of experiments and predictions of these cases of interchannel instability can be found in references 7–9, 22–24, and 31–37.

2.1.1 Use of Compressible Volumes

In some experiments, the channel constant pressure drop on a test facility is ensured by means of pressurizers or dampers for smoothing pressure (or flow rate) fluctuations caused by feed pumps, pressure release into the atmosphere, etc. This may have an appreciable effect on the channel flow stability, and, if the following relationships are not observed, the results of stability boundary determination may disagree with the real conditions where such components are not used.

It should be remembered that this simulation technique is used only for determining the stability boundary in the presence of boundary conditions of the first type (i.e., for the dynamically isolated channel), where P_{in} = const., P_e = const., and T_{in} = const.

A test facility employing compressible volumes for the channel dynamic isolation is illustrated in Figure 2.1.

Inlet tank 3 has gas space V_G and is partly filled with liquid. A tank filled with the steam–liquid mixture is located at the steam-generating channel outlet.

The volume V_G defines the level of pressure fluctuations p_{in} when self-oscillations develop in the channel. Pressure fluctuations may greatly influence the stability boundary. Let us approximately determine V_G to achieve either the absence or a low level of inlet pressure fluctuations.

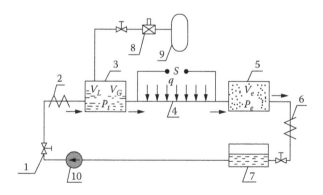

FIGURE 2.1

Test facility schematic: 1: control valves; 2: heater; 3: upstream tank; 4: steam generator; 5: downstream tank; 6: cooler; 7: drain tank; 8: pressure-reducing valve; 9: gas cylinder; 10: pump.

The inlet tank process can be described as follows:

$$\rho_L \frac{dV_L}{dt} = G_1 - G_{in},$$

(2.1)

$$V_L + V_G = \text{const.},$$

(2.2)

and

$$P_{in} V_G = \text{const.},$$

(2.3)

where G_1 and G_{in} are coolant flow rates past the pump and at the heated channel inlet, respectively.

When sufficient throttling of the line between pump 10 and tank 3 is ensured, which is the usual case, one may assume that $G = \text{const.}$, and, after linearization, it follows from (2.1)–(2.3) that

$$\frac{\Delta P_c}{P_{in}} \frac{\rho_L V_G}{G} \frac{d}{dt}\left(\frac{\delta P_{in}}{\Delta P_c}\right) = -\frac{\delta G_{in}}{G},$$

(2.4)

where ΔP_C is the steady-state channel pressure drop and G is the steady-state coolant flow rate.

Let the system have harmonic oscillations of coolant flow rate disturbances at the channel inlet

$$\delta G_{in}/G = A_G \sin \omega t,$$

which cause pressure fluctuations in the inlet tank

$$\frac{\delta P_{in}}{\Delta P_c} = A_P \sin (\omega t + \phi),$$

where A_G, A_P are amplitudes of relative oscillations of the coolant inlet flow rate and inlet tank pressure, respectively; ω is the angular oscillation frequency; and ϕ is the phase shift of flow rate and pressure fluctuations.

In this case, (2.4) yields the following relation for A_G and A_P:

$$A_P = A_G \frac{G \cdot P_{in}}{V_G \Delta P_c \rho_L \omega}.$$

(2.5)

For a limiting case, where $A_G = 1$ and pressure fluctuations in the inlet tank do not exceed 5% of the steady-state channel pressure drop, (2.5) yields the following inequality to evaluate the gas space volume:

$$V_G \geq \frac{G \cdot \Delta P_{in}}{0.05 \Delta P_c \rho_L \omega}. \qquad (2.6)$$

The volume of the outlet tank is considered for the limiting case of $i_e > i_G$ at the steam-generating channel outlet. The process can be approximately described as follows:

$$\frac{dM}{dt} = G_e - G_2 \qquad (2.7)$$

and

$$\frac{P_e \cdot V_e}{M} = const., \qquad (2.8)$$

where M is the steam mass in the outlet tank with the constant volume V_e and G_e and G_2 are coolant flow rates from the heated channel to the outlet tank and from the outlet tank to the steam line, respectively.

Assuming the steam flow rate from the outlet tank to be constant, it follows from (2.7) and (2.8) that

$$\frac{M}{P_e} \frac{d\delta P_e}{dt} = \delta G_e$$

or, similarly to the preceding,

$$A_P = A_e \frac{G \cdot P_e}{M \cdot \omega \cdot \Delta Pc}. \qquad (2.9)$$

Taking into account $M = V_e \rho_G$ and the conditions adopted for obtaining (2.6), from (2.9) we get

$$V_e \geq \frac{G \cdot P_e}{0.05 \cdot \Delta P_c \cdot \rho_G \cdot \omega}. \qquad (2.10)$$

It is obvious from (2.6) and (2.10) that when the frequency of oscillations decreases, volumes of the damping tanks increase and their use becomes practically impossible. In the limiting case of static instability, when $\omega \to 0$, the use of damping tanks is unacceptable. Moreover, the presence of compressible volumes when hydraulic characteristics are ambiguous may cause low-frequency oscillatory instability, the so-called pressure drop-induced oscillation instability [15] that is not observed in the absence of compressible volumes.

From (2.6) and (2.10) it also follows that

$$\frac{V_e}{V_{in}} = \frac{\rho_L}{\rho_G}$$

and permissible value for V_e turns out to be rather high even at large oscillation frequencies. This also limits the area of application for this channel dynamic isolation method. However, with the increased steam moisture at the channel outlet and the decreased $P_e/\Delta P_c$ ratio, the permissible value of V_e reduces. With the exit quality $x < 1$, (2.10) may be written as

$$V_e \geq \frac{x \cdot G \cdot P_e}{0.05 \cdot \rho_G \cdot \Delta P_c \cdot \omega}. \tag{2.11}$$

It should be recognized, however, that the use of damping volumes is undesirable. They may be used only for small-scale channels at relatively high pressures to perfect methodology or for investigating loop processes where the appearance of compressible volumes is possible in real conditions.

The real designs employed in heat power and nuclear power engineering generally have no compressible volumes at the core inlet and outlet. However, this situation in principle may occur under emergency conditions, when steam and gas volumes may appear in some reactor areas, or in the case of core increased power density, when the presence of a gas pressurizer at the reactor inlet yields a two-phase flow at the core exit.

2.1.2 Use of Bypass

Investigations of flow instability in heated channels with the nonheated bypass have established that the flow stability boundary in the heated channel substantially depends on the ratio between flow rates in the bypass (G_{By}) and channel (G_c). A high G_{By}/G_c ratio (>10–15) signifies the limiting case for the isolated channel operating at the constant pressure drop. When the G_{By}/G_c ratio decreases and the channel/bypass hydrodynamic interaction reveals itself, stability of the heated channel increases. Figure 2.2 offers experimental results [22] that illustrate the effects of stability rise in the channel with the reduced G_{By}/G_c.

The cause of a higher stability at the decrease of G_{By}/G_c is as follows. A reduction of the heated channel flow rate near the stability boundary simultaneously increases (with the second type of boundary conditions, when the total flow through the system is constant) the bypass flow rate more, the lower the G_{By}/G_c ratio becomes. The increased bypass flow

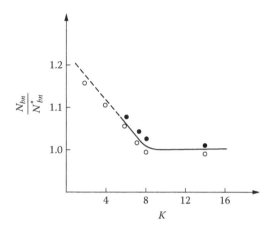

FIGURE 2.2
Dependence of the relative channel power at the instability boundary on the bypass factor [22]. $K = G_{By}/G_c$: the bypass factor; N^*_{bn}: the instability boundary channel power with K tending to infinity.

rate causes a pressure drop increase in the heated channel-bypass system. Thus, when the flow rate is decreasing, an in-phase increase in the channel pressure drop occurs, thereby preventing the inlet flow rate reduction and facilitating flow stabilization. It has been shown [9] that the bypass effect may be conventionally represented as additional inlet throttling in the heated channel with the constant pressure drop, which, in the case of the inlet flow rate change by the δG_c value, results in the inlet pressure drop—the change of which is equivalent to the pressure drop change in the bypass $\delta \Delta P_{By}$, which corresponds to the flow rate change δG_c.

Investigations on this test facility approximately simulate real conditions with a specified constant total coolant flow rate through the system of parallel channels, when it is necessary to determine stability of an isolated channel or of a single channel in a system with a small number of a priori stable channels. This is also true for a system of several parallel operating steam generators, one of which is close to the stability boundary.

The "heated channel-nonheated bypass" system is mainly used for investigating flow instability in isolated channels with the constant pressure drop in forced-circulation loops with high coolant velocity (>1 m/s). It is also used to investigate the interchannel instability (one unstable channel and a small total number of channels) in forced-circulation and natural-circulation loops, if the coolant upflow in the bypass is ensured and fluctuations of parameters in the component with the unstable channel are of little effect on the change in loop parameters. As a rule, such test facilities are unsuitable for investigations of the isolated channel instability at the constant pressure drop in natural-circulation loops and with forced circulation at low rates. This is because, at such rates, which are especially typical of the natural circulation, it is frequently impossible to ensure high G_{By}/G_c ratios and eliminate the bypass hydrodynamic effect that leads to channel pressure fluctuations. This is due to the fact that the operating area of the nonheated bypass with upflow is determined by

$$\Delta P_c - g\rho_L H_c \geq 0. \qquad (2.12)$$

It has been shown [35] that a limitation of the upflow velocity in a bypass is observed in a wide area of natural-circulation loops operation. In the case of (2.12) violation, the nonheated bypass starts operating as the downflow branch, and the channel-bypass system switches to the natural circulation mode. The latter mode completely eliminates the possibility of modeling flow oscillations in the isolated channel and with interchannel instability.

Another disadvantage is observed when a hydraulic bypass is used for investigating flow stability in the heated channel. When the steam–water mixture (or steam) is mixed at the steam generator channel exit with cold water fed from the bypass, intensive steam void collapse occurs and leads to strong hydraulic shocks, noise, and vibration.

2.1.3 Use of a System of Two Parallel Heated Channels

In experimental investigations of the two-phase flow instability on the facility with two parallel channels, two limiting situations are possible: (1) one of the channels is a priori stable because of the high inlet throttling or due to power that is lower than in the other channel, so stability of the other parallel channel is investigated; or (2) two channels are thermally and hydraulically identical.

The first situation is similar to the case of the heated channel and parallel bypass, when their ratio is of the order of unity. It has been shown earlier that, with such a rigid hydrodynamic connection, stability of the tested channel is many times higher than under conditions with the constant pressure drop in the channel.

The second situation is identical to the case of the isolated channel with the constant pressure drop. Indeed, two completely identical channels have identical characteristics with respect to the stability boundary and the change of parameters in the unstable region. Therefore, when such channels happen to be in the unstable parameters region, inlet flow rate fluctuations in them are in antiphase. Exit flow rate oscillations also occur in antiphase and with identical amplitude, causing no fluctuations of the exit total quality and pressure in the system. Consequently, two identical channels feature antiphase self-sustaining identical flow rate oscillations and oscillations of other parameters under the constant pressure drop. There is no hydrodynamic connection between the channels. Identical conditions occur in the case of several parallel identical channels.

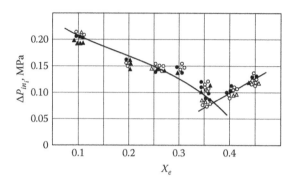

FIGURE 2.3

Interchannel stability boundary [33]. Triangle: three-channel steam generator; dark circle: two-channel steam generator; light circle: single dynamically isolated channel.

As an example, Figure 2.3 illustrates the results of an experimental investigation [33] concerning stability boundary of an isolated steam-generating channel and a system of two and three parallel identical channels. All the channels were identical and had the following geometries and flow parameters: a tube with an internal diameter of 6 mm, the total length of 1400 mm and the heated section of 1100 mm, an average flow rate in each tube at the stability boundary (G_j) of 15 g/s, and fluid pressure (P_{in}) of 1.5 MPa.

When determining the stability boundary of a single channel, compressible volumes located at the inlet and outlet were used for its dynamic isolation and elimination of pressure drop fluctuations. The stability boundary is constructed in Figure 2.3 in the coordinates ΔP_{inj}; x_e, where ΔP_{inj} are pressure drops across the inlet throttles in the single dynamically isolated channel ($j = 1$), the arithmetic mean inlet drop for the two- ($j = 2$) and three- ($j = 3$) channel systems. In the cases of $j = 2$ and $j = 3$, the inlet pressure drops at the moment of attaining the stability boundary differed by 3%–5%, on the average. The statistical processing of experimental data showed that stability boundaries of a single dynamically isolated channel and interchannel stability of two- and three-channel systems composed of the same channels are statistically indistinguishable. The measurements indicated that both in the stability boundary region and with the development of small-amplitude self-oscillations, the total coolant flow rates before and after the steam-generating channel and header pressures were constant.

Thus, the use of identical parallel channels makes it possible experimentally to determine in a sufficiently simple manner the stability boundary of an isolated channel with the constant pressure drop in the parallel channel system in forced and natural circulation loops (e.g., channels in reactor core with natural coolant circulation). In this case, the thermophysical test facility should represent a working section with two parallel identical channels of maximum similarity to the full-scale design, and a loop with forced circulation enabling flow rate control through the test section to match the flow rate through the real natural circulation loop channels.

It should be noted that when density-wave thermal-hydraulic instability is simulated on a two parallel-channel test facility, the results may be close to real conditions at the constant pressure drop in the channel only for the stability boundary. However, concerning the development of the amplitude of fluctuations in the instability region, the results may be close to real conditions only for small amplitudes of flow fluctuations. This is due to the fact that with the increasing amplitude of fluctuations, their shape increasingly deviates from the sinusoidal [25] because of nonlinear effects, and there appear fluctuations of the pressure drop or of the total flow rate, which influence the amplitude of fluctuations as a stabilizing factor. These fluctuations are typical of a thermophysical test facility but not of the real conditions because loop characteristics of a thermophysical test facility differ from those of a full-scale loop.

The previous constraints are most pronounced with thermal-hydraulic instability corresponding to low exit qualities when flow fluctuations deviate from sinusoidal even at a very small amplitude [7]. Figure 2.4 illustrates the influence of nonlinear effects on the loop flow rate at the development of low-exit quality instability in two identical channels [7]. The figure shows that the interchannel flow rate fluctuation in two identical channels caused by nonlinear effects leads to the appearance of double frequency fluctuations in the loop.

Taking into account that the stability boundary is experimentally found at a certain amplitude of fluctuations, the influence of nonlinear effects for this type of instability may yield errors even during experimental determination of the stability boundary on the test facility with two identical channels.

The investigations conducted by the authors show that the nonidentity of thermal and hydraulic characteristics of parallel channels in a multichannel system of 5% max is of insignificant effect on the interchannel stability boundaries.

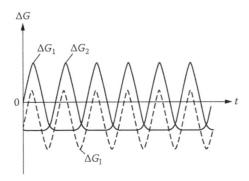

FIGURE 2.4
Effect of nonlinear interchannel fluctuations in identical channels on the loop flow rate [7]. ΔG_1, ΔG_2: flow rate deviations in unstable parallel channels; ΔG_l: loop flow rate deviation.

2.1.4 Use of Two Hydraulically Identical Parallel Channels with Nonheated Bypass

In order to determine more exactly the stability boundary of an isolated channel and the nature of flow fluctuations' development in the unstable region in the natural-circulation system and at low flow rates in forced circulation systems, sometimes a nonheated bypass is connected in parallel to two identical channels on a thermophysical test facility. Although the ratio between bypass and channel flow rates may be small, it suffices to eliminate hydrodynamic effects of channels if they are incompletely identical due to process inadequacies. It also reduces pressure drop fluctuations due to nonlinear distortions of the flow fluctuation sinusoidal form. Figure 2.5 shows flow rate fluctuations in two parallel, almost identical channels (with small hydrodynamic deviations) with closed and open bypasses [22]. All other characteristics of parallel channels were identical. One can see that the use of a bypass somewhat impairs stability by increasing the flow rate fluctuations' amplitude due to a weakened influence of channels' hydrodynamic coupling.

It is typical that the bypass ratio ($K = G_{By}/G_c$), as has been experimentally shown [22,23], influences the phase shift between the parallel channel inlet velocities. With no bypass, the phase difference amounts to 180°, while with $K = 10$, the phase difference is 50° (Figure 2.5). It changes from 45 to 90°

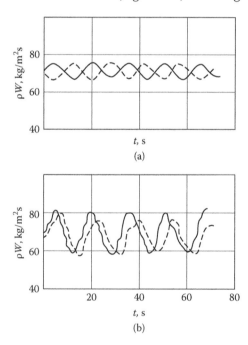

FIGURE 2.5

Phase difference between inlet velocities in two parallel identical channels [22]. (a) Zero bypass flow rate ($G_{By}/G_c = 0$); (b) $G_{By}/G_c = 10$; solid line: channel no. 1; dashed line: channel no. 2.

for K > 6 and depends on inequality of boundary conditions between two channels, such as the inlet resistance and power. Evidently, the phase shift of inlet flow rate fluctuations gets established in the channels to ensure minimum disturbance of the system pressure drop (with G_1 = const.) and in the limits at the infinite bypass ratio, the phase shift of inlet flow rate fluctuations in unstable channels will be arbitrary.

2.1.5 Use of Test Facilities with Multichannel Systems

In full-scale plants, the units with parallel channels (e.g., the reactor core) are in fact complex multichannel systems incorporating groups of channels that differ hydrodynamically and thermally and have practically identical channels in each of these groups. To reveal the nature of interaction of identical and nonidentical parallel channels of the same system, as well as to determine the manner of parameters' fluctuation in unstable identical channels, thermal-physical test facilities with several (n > 2) parallel channels are used.

The results of experimental and numerical investigations of flow instability under conditions of a thermophysical test facility with several parallel channels are presented in references 7, 15, 18, 22–24, 31, 33, and others. According to them, mutual hydrodynamic effect of parallel channels on stability is qualitatively the same as the one described before. However, investigations on such test facilities provide additional information on the behavior of integral thermal-hydrodynamic parameters of the parallel-channel system in the region of interchannel instability and on the effect of such fluctuations on the change in the loop parameters (e.g., on the neutron-physical processes in the reactor core).

If all channels are identical in a multichannel system, the results of the preceding work show that at small amplitudes there is no hydrodynamic influence between them in the unstable region, and the flow stability boundary corresponds to the stability boundary of an isolated channel at the constant pressure drop. In this case, the phase shift of the parallel channel inlet flow rate fluctuations and, respectively, fluctuation parameters may have several equiprobable regimes. Their number becomes larger, the larger the number of identical channels becomes. For example, Gerliga, Dulevsky, and Mokhrachev [33] offer theoretical and experimental results that confirm the existence of two fluctuation regimes on the three-channel test facility with identical channels:

1. In three channels, inlet flow rate fluctuations had identical amplitudes and a phase shift of 120° ($e^{-2/3\pi j}$, 1, $e^{2/3\pi j}$).
2. The flow rates in two channels fluctuated in antiphase, while in the third channel it was stable (1, –1, 0).

The latter mode is practically impossible, since the third channel is in the unstable region and in real conditions there always exist fluctuating parameters inducing the self-oscillations mode in the channel.

For the three-channel test facility investigated in Fukuda and Hasegawa [18], three theoretically equiprobable fluctuation modes were obtained: $(1, -1, 0)$, $(1, 1, -2)$, and $(e^{-2/3\pi j}, 1, e^{2/3\pi j})$. The same work also presents the experimental results of test facilities with four and five identical channels. For the four-channel case, two channel inlet flow rates were found to fluctuate in phase, whereas in two other channels, flow rates fluctuated in antiphase. During the experiments with five identical channels, the inlet flow rate fluctuation phase shift was 72°. In all the experiments, parameters' behavior in the upper and lower plenums was stable and the total flow rate of coolant through the parallel-channel system was constant.

The provided examples make it clear that, in full-scale conditions, it is difficult to determine which of the equiprobable flow fluctuation modes will be activated in identical parallel channels. Evidently, it depends on minor differences in channels caused by process deviations, small power differences, channel location in the core, nonuniform pressure distribution in common headers, etc. In a real system, the most energetically optimal mode is selected automatically, ensuring minimum fluctuations of the pressure drop and other parameters in the instability region of the system. It may be supposed that during different equiprobable modes of flow fluctuations in almost identical parallel channels, fluctuations of integral parameters in the system do not differ much. Thus, in order to take account of, say, neutron-physical processes on stability in the first approximation, the channel inlet velocity fluctuation phase shift may be assumed to be 360°/n (where n is the number of channels). However, validation of this supposition requires additional research.

2.2 Simulation of Thermal-Hydraulic Instability in Complex Systems

If a multichannel system (like the reactor core parallel channels) contains several channel groups that differ thermally and hydrodynamically, while in each group channels are identical, the boundary of interchannel stability of such a system may be close to instability of an isolated channel with the constant pressure drop. Indeed, when the core parameters approach the stability boundary, the identical channels of the group closest to the stability boundary are found to fluctuate first. According to the previous discussion, flow fluctuations in such a group at small amplitudes occur at practically constant pressure drops, total flow rate, and other integral parameters. When departing from the stability boundary, the amplitude in the first group of unstable channels increases and fluctuations start in the next group of identical channels, and so on.

The described succession of flow fluctuations' development in different groups of identical channels takes place at the core constant pressure drop until the rising flow fluctuations' amplitude causes—due to nonlinear effects—the fluctuations to form to deviate from the sinusoidal with the resultant pressure drop fluctuations in the channels and fluctuations of integral characteristics (e.g., coolant density) in the core. This may, first, limit the fluctuations' amplitude and, second, cause forced fluctuations of the loop pressure and loop circulation with the resultant interaction between thermal-hydraulic and neutron-physical processes.

Based on the suggested mechanism of the reactor core interchannel fluctuations' development, it may be recommended that the boundary of the hottest isolated channel under the constant pressure drop should be taken for determining the stability boundary, while several groups (5–10) of equivalent nonidentical channels, each of which consists of two identical channels, may be used for simulating core processes in the instability region.

These results are valid for the conditions ensuring thermal-hydraulic stability of loop circulation, when possible pressure drop disturbances in the core with the interchannel instability have no appreciable effect on the loop flow rate.

Now, let us consider the conditions of loop instability and interaction between the interchannel instability and fluctuations of the loop flow rate.

In the case of the loop circulation instability, fluctuations of the loop flow rate are accompanied by the in-phase flow rate fluctuations in identical channels of the core. The amplitude of flow rate fluctuations in the channel is n times lower (where n is the number of channels) if compared to that of the loop flow rate fluctuations [33]. Such a consideration of the loop instability is valid for the case of no interchannel instability in the core.

It has been shown [7,9,15,31,33, etc.] that under the preceding conditions, for the analysis of thermal-hydraulic instability in the natural-circulation loop, say, of boiling water reactor with natural coolant circulation, it may be considered as an equivalent isolated core channel with a large nonheated section (represented by the downflow section of the loop) and a nonheated riser (a common riser section). Therefore, the qualitative effect of parameters on the thermal-hydraulic flow stability boundary in the circulation loop is as that for the isolated channel considered before.

It has been shown previously that flow stability in the circulation loop equivalent channel is generally higher due to high friction and large inertia of the downflow section capacity, as compared to the interchannel instability in the core channels. Therefore, the interchannel instability usually precedes that of the loop and the analysis of the core isolated channel instability under the constant pressure drop yields the lower stability boundary of the system. Quite frequently, when providing stability of the isolated channel, the interchannel and circulation loop stabilities are ensured as well.

In principle, it may happen that boundaries of the loop and interchannel instabilities in the natural-circulation loops will coincide, or the loop instability occurs before the interchannel one. It has been shown [7] that this may take place (e.g., when the height of individual riser tubes is reduced and the total riser section is increased appropriately), with the resultant increased interchannel and impaired loop instabilities. However, such an approach impairs natural-circulation loop characteristics and its application is not optimal.

In real situations, when the interchannel instability precedes that of the loop, a study of loop processes should be accompanied by simulation of the interchannel instability due to their strong mutual effect. If the regions of the interchannel and loop instabilities are sufficiently close to each other, then the interchannel instability is immediately followed by the loop instability. If the interchannel instability starts long before that in the loop, then with the increasing amplitude of interchannel fluctuations, forced fluctuations of the loop flow rate start, which have mutual effect, too.

Namestnikov [34] has stated that the interaction of nonlinear fluctuations of different types can lead to the suppression of one type of fluctuation by another; mutual amplification of fluctuations, beat regimes when the differences in cycles yield time-dependent periodic suppression, and self-synchronizing of different nonlinear fluctuations (frequently occurring in practice fluctuation "packages" appear).

The same work uses a simplified mathematical model to show the possibility in principle of suppressing loop fluctuations by interchannel fluctuations in reactors with natural coolant circulation (Figure 2.6).

As noted previously, a joint investigation of both interchannel and loop instabilities may provide satisfactory accuracy if the core is modeled by several groups of equivalent pairwise identical channels.

Simultaneous experimental investigation of the loop and interchannel instabilities should be conducted on thermophysical test facilities that are, in fact, loops with natural coolant circulation and parallel heated channels, geometrically close to those of the core. These facilities represent semicommercial plants similar in height to the full-scale ones and ensure obtaining nominal parameters of coolant temperature, pressure, and velocity.

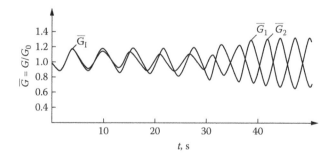

FIGURE 2.6

Loop and interchannel flow instabilities interaction [34]. G_l: relative loop flow rate; G_1, G_2: relative parallel channel flow rates.

Besides stability investigations, the tests are also used for verifying distribution of static thermal-hydraulic loop parameters at different power levels, which is also necessary for numerical studies of stability.

In flow instability studies on thermophysical test facilities, simulation of a full-scale plant requires not only identity of static thermal-hydraulic characteristics to those of a full-scale plant and presentation of the core by several groups of pairwise identical channels, but also identity of thermal inertia characteristics of all loop elements to those of the full-scale plant. The latter condition is rather strict and is not always satisfied. If the region of the loop instability is to be determined in the absence of interchannel instability, the core may be simulated by a single equivalent channel.

It should be noted that the study of interchannel and loop instabilities interaction is at the initial phase and requires further numerical and experimental investigations.

Let us consider some examples of unstable processes' simulation on sufficiently simple test facilities; results may be extrapolated to the full-scale plants.

Generally, once-through steam generators are multichannel, modular systems. For the steam generator in Figure 2.7, three types of instabilities are possible: interchannel in modules, intermodule, and the system-wide ones. Steam generator tubes and modules are assumed to be identical.

If there exists only the intermodule instability (the interchannel instability in the module is absent due to, say, inlet tube throttling), parameters in all steam-generating tubes of the modules fluctuate in phase. In this case, each module behaves as a single channel.

When studying intermodule instability, each module may be replaced by a single steam-generating tube with additional inlet and outlet resistances that simulate liquid and steam connecting pipelines. Such a "pipelines-channel" system is called "reduced." The additional inlet and outlet resistances are selected to make pressure drops across them respectively equal to those in full-scale pipelines connecting the modules. Generally speaking, in addition to equality of the pressure drop in the full-scale pipeline (ΔP_e (G_t), $G_1 = NG_t$) and the reduced resistance ($\Delta P_{e\,red}(G_1)$), the exit section requires equality of

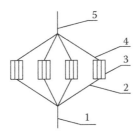

FIGURE 2.7
Multimodule steam generator; 1, 5: Headers; 2: upstream line; 3: steam-generating module channel; 4: downstream line.

derivatives $\partial\Delta P_e/\partial x$ and $\partial\Delta P_{e\,red}/\partial x$. The experience in selecting reduced resistances (throttling orifices) shows that if the first condition is fulfilled, then the second condition is also satisfied in the vicinity of x_e.

Figures 2.8 and 2.9 [33] show the intermodule fluctuation boundaries for two and three-channel modules and stability boundaries of respective reduced channels. The experimental results are plotted in coordinates $[(\Delta P_{in} + \Delta P_{inj}); x_e]$ for the intermodule instability, and in coordinates $[(\Delta P_{in\,red}; x_e)]$ for the reduced channels' stability. The data prove the possibility of simulating self-oscillating processes in a multichannel steam generator module with upstream and downstream hydraulic pipelines with local resistances reduced to those in single-tube models.

Of certain interest for practical applications is the possible creation of a model of the full-scale closed loop with small energy requirements for studies of the system-wide (loop-wide) instability. Let us consider a solution of this problem based on the model of the natural or forced coolant circulation loop having

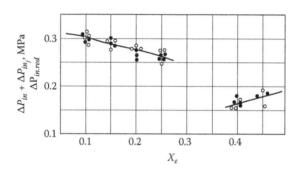

FIGURE 2.8
System-wide stability boundary in a two-channel steam generator; ●: the steam generator; ○: the reduced channel.

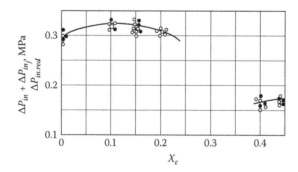

FIGURE 2.9
System-wide stability boundary of a three-channel steam generator; ●: the steam generator; ○: the reduced channel.

a steam generator and a condenser. Such loops are frequently employed by the nuclear power plant's autonomous emergency heat removal systems.

It is assumed that the steam generator and condenser of the loop in question consist respectively of identical steam-generating and -condensing channels with internal steam condensation. There is no interchannel instability in both steam generator and condenser. If the condenser provides external condensation, it may be generally regarded as the constant pressure point, thus simplifying the simulation problem.

The matrix transfer function "input vector of disturbances in the channel-output vector of disturbances" is identical for a single heated channel and for a system of identical parallel channels that is similar to the single channel (if the interchannel instability is absent). Then, the vector of disturbances will be

$$\left\{ \delta \bar{G} = \frac{\delta G_j}{G_j}, \delta P, \delta i \right\}.$$

Evidently, it is valid for identical channels, both steam-generating and -condensing ones.

So, a real multichannel steam generator composed of identical parallel channels is replaced by a single full-scale channel for the purpose of building a model of the loop for stability boundary investigations. The heat transfer process in the condenser proceeds less intensively as compared to the steam generator; therefore, the number of condenser channels is much bigger than that of the steam generator. In this case, the model loop condenser is a system of full-scale condensing tubes, the number of which is selected to meet the condition of equal flow rates through the steam-generating channel and condenser in the steady-state regime.

Further, to ensure coincidence between the dynamic properties of the model loop and those of the full-scale one (for coincidence of stability boundaries), the matrix transfer functions of respective liquid and steam loop sections are required to coincide. Since the sections are nonheated and frequently the working fluid is a single-phase one (liquid or steam), the conditions are not difficult to satisfy in an approximate manner.

The elements of the preceding matrix transfer functions are the solutions of the linearized and Laplace-transformed continuity, energy, and momentum equations. All disturbed parameters are nondimensional.

If the full-scale liquid and steam pipelines are replaced by models with the same length and identical space orientation, then the approximate simulation needs only the following conditions to be fulfilled:

$$\Delta P_M = \Delta P_{fs}; \quad d_M^2 = \frac{1}{n} d_{fs}^2; \quad \left[\frac{(A\rho C_P)_w}{(A\rho C_P)_c} \right]_M = \left[\frac{(A\rho C_P)_w}{(A\rho C_P)_c} \right]_{fs},$$

where w is the wall, C is the coolant, fs is the full scale, M is the model, and n is the number of channels in the full-scale steam generator.

To ensure identity of pressure drops in the model and full-scale pipelines, lumped resistance may be additionally introduced.

To verify the method of the loop processes' simulation, comparative experiments were conducted to investigate self-oscillating modes in two loops: the full-scale and model ones. Figures 2.10 and 2.11 present the obtained values of the relative flow rate fluctuation amplitudes for the full-scale and model loops depending on the ratio of pressure drops in the liquid and steam sections of the loop $\Delta P_L/\Delta P_{TP}$. The results show satisfactory agreement for relative amplitudes in the model and full-scale loops for both small and large fluctuation amplitudes.

If experimental thermal-hydraulic instability investigations experience significant problems with such coolant as, say, potassium or mercury, these may be replaced by a simulant (e.g., water). Let us consider specifics of simulating the coil-type potassium steam generator by a water–steam one. Geometrically, both steam generators are identical. The steam

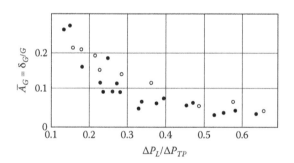

FIGURE 2.10
Development of loop flow rate fluctuations. $X_{e.SG} = 0.8$; $G_1 = 7$ g/s; $\Delta P_e = 0.2$ MPa; ●: model loop; ○: full-scale loop.

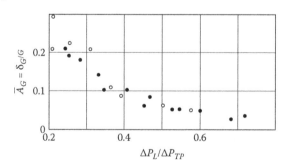

FIGURE 2.11
Development of loop flow rate fluctuations. $X_{e.SG} = 1.0$; $G_1 = 7$ g/s; $\Delta P_e = 0.18$ MPa; ●: model loop; ○: full-scale loop.

generator stability boundary is determined "in the small" by the linearized system of equations of continuity, energy, and momentum, and by boundary conditions. For the coincidence of stability boundaries of the full-scale and model steam generators, respective coefficients of the previous equations written for potassium and water flows should also coincide.

The numerical analysis of steam generator stability showed that the main stability-defining coefficients are as follows:

$$\frac{\partial \bar{\rho}}{\partial \bar{i}}; \quad \frac{\partial \bar{i}}{\partial z}; \quad \frac{G}{\Delta P}\frac{\partial \tau_w}{\partial G}; \quad \frac{1}{\Delta P}\frac{\partial \tau_w}{\partial \bar{i}},$$

where $\bar{i} = [i(z) - i_{in}]/\Delta i$; $\bar{\rho} = \rho(z)/\rho_L$; $\bar{Z} = z/H$; ΔP, Δi are the steam generator pressure drops and enthalpy, respectively; and τ_w is the friction shear stress.

Consequently, it suffices for stability processes' simulation that the following equalities between the model and full-scale coolant parameters are met:

$$\left(\frac{\partial \bar{\rho}}{\partial \bar{i}}\right)_M = \left(\frac{\partial \bar{\rho}}{\partial \bar{i}}\right)_{fs}; \quad \left(\frac{\partial \bar{i}}{\partial \bar{z}}\right)_M = \left(\frac{\partial \bar{i}}{\partial \bar{z}}\right)_{fs} \tag{2.13}$$

$$\left(\frac{G}{\Delta P} \cdot \frac{\partial \tau_w}{\partial G}\right)_M = \left(\frac{G}{\Delta P} \cdot \frac{\partial \tau_w}{\partial G}\right)_{fs}; \quad \left(\frac{1}{\Delta P} \cdot \frac{\partial \tau_w}{\partial \bar{i}}\right)_M = \left(\frac{1}{\Delta P} \cdot \frac{\partial \tau_w}{\partial \bar{i}}\right)_{fs}.$$

The first two conditions of (2.13) will be fulfilled if ρ_L/ρ_G, $\alpha(z)$, $i(z)$ coincide, from which it follows that the nonboiling lengths for both working fluids shall be identical. This is the basis for selecting initial temperature of the simulated working fluid. In our case, $(i_{in})_M$ is selected by superposing $\bar{\rho}_{fs}$, $\bar{\rho}_M$ the curves.

Investigations of the liquid metal flow hydraulic friction showed the friction laws for plain tubes and local resistances obey those for nonmetallic liquids [102]. Also, the results of a comparative experimental investigation of steam–metal and steam–water mixture flows are available. Hydrodynamic properties of such mixtures are shown to be practically the same. Hydraulic friction in coil-type channels obeys the same dependences for both potassium and water. Calculations using the formulas for T indicate that the third and fourth equalities of (2.13) are satisfied approximately with a water flow rate that is several tens of percentages higher than that of potassium.

The suggested technique was verified by experimental comparative tests of a liquid sodium-heated potassium steam generator and the glycerin-heated water steam generator identical in size and configuration. (Experimental results on liquid metals were obtained with participation of E. D. Fedorovich,

V. D. Khudyakov, and I. S. Kudryavtsev.) Parameters of the simulated working fluids were chosen using the preceding technique.

Figure 2.12 presents experimental data as dependencies of the inlet flow rate fluctuation amplitude (A_G) on the inlet pressure drop (ΔP_{in}) for both steam generators. Good quantitative agreement between the results is

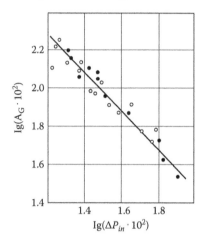

FIGURE 2.12
Dependence of the flow rate fluctuations' amplitude on the inlet resistance in an isolated coiled channel. $G_{ci.1} = 3$ g/s; $P_c = 0.15$ MPa; $G_{H2O} = 4.1$ g/s; $P_{H2O} = 0.205$ MPa; ●: potassium; ○: water.

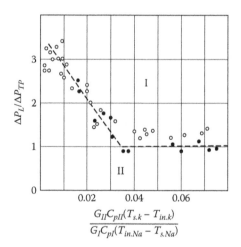

FIGURE 2.13
Stability boundary of an isolated coiled channel. ●: large-scale steam generator, single-tube section; ○: steam generator employing a model working fluid, single-tube section; I: stable region; II: unstable region. G_I, G_{II}, C_{P1}, C_{P2} are flow rates and heat capacities of the heating coolant and heated working fluid, respectively; T_{in}: inlet temperature; T_s: saturation temperature; ΔP_L, ΔP_{TP}: pressure drops across the liquid and two-phase steam generator sections.

obvious. Figure 2.13 shows the results concerning stability boundaries of the compared steam generators. The inlet flow rate fluctuation amplitude of $A_{GBn} = 0.2\,G_0$ was accepted as the stability boundary.

These results speak in favor of applying the suggested simulation method in the study of such complex problems as flow thermal-hydraulic instability.

One more essential issue in the simulation of thermal-hydraulic instability [9] in water coolant systems requires attention. When extrapolating the test facility results to the full-scale conditions, it is necessary to remember about geometry differences influencing pressure distribution and flow pattern, the impossibility always to ensure equal heat flux density distribution along the height, thermal inertia of the heating element, kind of heating, etc. Therefore, either a mathematical model should be used for estimating the quantitative effect of differing parameters on the experimentally obtained stability boundary or, knowing qualitative effect of differing parameters on the stability boundary, set them when designing test facilities to avoid as far as possible an increase in system stability. This should ensure conservative estimation of the full-scale plant flow stability boundaries.

3

Simplified Correlations for Determining the Two-Phase Flow Thermal-Hydraulic Oscillatory Stability Boundary

3.1 Introduction

Numerous experimental and numerical investigations performed in Russia and abroad made it possible to reveal clear regularities in the influence of regime and design parameters on the oscillatory stability boundary of a flow. Based on these results, simplified methods for the preliminary determination of the thermal-hydraulic stability boundary for a two-phase flow of the density-wave type in a channel with the constant pressure drop have been developed.

The methods of Lokshin, Peterson, and Shvarts [12] and Saha, Ishii, and Zuber [13] employing algebraic relationships and nomograms for the stability boundary determination have found the widest application and recognition. Although they are very simple, these methods nevertheless possess a number of important constraints that are frequently ignored when performing numerical evaluations. The basic constraints are as follows:

No account is made of the effect of heat flux distribution along the channel height, thermal inertia of heating elements and heat transfer surfaces, and hydrodynamic interaction of parallel channels.

The channel power input is assumed to be constant over time and no account is made of such feedbacks as the heat transport delay in convective heating and energy release change due to density effects of reactivity with nuclear heating, which may substantially distort the stability region boundary.

Regularities of steam quality distribution along the height in Lokshin et al. [12] and Saha et al. [13] are based on the tubular geometry, thereby disregarding the effect of channel complex geometry on the interphase slip in the two-phase flow and change of the pressure drop gravity component and friction pressure drop.

These methods are inapplicable for determining the oscillatory flow stability boundary at low exit qualities.

The method of Lokshin et al. [12] can be used for the determination of the stability boundary of only a steam–water two-phase flow in horizontal and vertical channels, while the techniques of Saha et al. [13] are applicable for evaluating stability of any two-phase flow, but in vertical channels only.

Therefore, the flow stability boundary determined by the preceding methods is approximate (frequently with a large margin) and is applicable for the initial rough estimates and selection of the techniques of flow stabilization at the design stage. The final determination of the two-phase stability boundary for a particular design is based on more accurate methods (e.g., direct numerical solution using a distributed dynamic model).

Recent work [65–70, etc.] report the development of a novel method for the flow stability boundary determination, which is free of a number of constraints typical of the methods of Lokshin et al. [12] and Saha et al. [13]. However, this method is not free from some constraints, described later.

Simple empirical correlations obtained in Unal [43,44] allow channel power to be established in relation to the stability boundary of density waves. However, these correlations are not universal and are applicable to specific conditions only—namely, to steam generators with tubes of $H/d_h > 1{,}200$ and $x > 1$. Therefore, due to the limited application range, the correlations will not be considered in further analysis.

Now, let us discuss the previously mentioned rough methods in more detail.

3.2 The CKTI Method

This method has been developed at the I. I. Polzunov Central Boiler and Turbine Research Institute (CKTI) and adopted as a standard tool for hydraulic design of boiler units in determination of the stability boundary for horizontal and vertical tubes [12].

It is based on numerical calculations of flow self-oscillations in a channel using a distributed nonlinear dynamic model and the results of comprehensive comparison of predictions with extensive experimental data [25–27,42]. The use of the method for boiler operating conditions is justified by either complete absence or weak expression of the majority of the previously mentioned constraints. In comparison with the method of Lokshin et al. [12], this one offers a more correct accounting for the effect of the outlet throttling and the transformed resistance coefficient for the nonheated outlet section on the flow stability boundary, which is of importance for steam generators. Also, the ranges of pressure, resistance coefficients, and inlet subcooling variation used in the nomogram have been widened.

The calculation procedure is as follows. For horizontal tubes, the boundary mass velocity at which flow oscillations start is derived from the formula

$$\left(\rho W\right)^h_{Bn} = 4.63 \cdot 10^{-3} \left(\rho W\right)_{o,p} \frac{\bar{q} \cdot H}{d_h},$$ (3.1)

where $\bar{q}$ is the average heat flux density on the channel heated surface, MW/m^2; d_h and H are the equivalent hydraulic diameter and the heated section length of the channel, respectively, m; $(\rho W)^h_{Bn}$ is the boundary mass flow rate in the given horizontal channel, kg/m^{2}s; and $(\rho W)_{o,p}$ is the boundary mass flow rate in a horizontal channel with a fixed heated length, equivalent diameter, and heat flux density obtained from the nomogram in Figure 3.1, kg/m$_2$s.

The $(\rho W)_{o,p}$ value depends on the hydraulic resistance coefficient (ξ), coolant subcooling at the channel inlet Δi_{in} (kJ/kg), and pressure P(MPa) and is found from the formula

$$\left(\rho W\right)_{o,p} = \left(\rho W\right)_0 K_P,$$ (3.2)

where $(\rho W)_0$ is the boundary mass velocity in a horizontal channel with a fixed heated length, equivalent diameter, and heat flux density at P = 10 MPa, derived (with preset ξ and Δi_{in}) from the right-hand part of the nomogram in

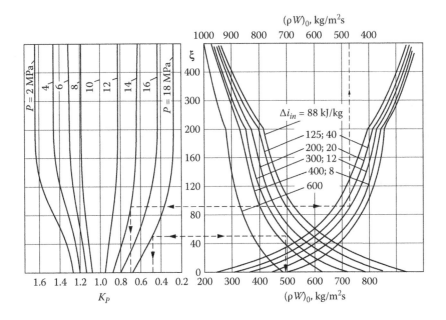

FIGURE 3.1
Nomogram for determining mass velocity at the stability boundary in a horizontal channel [12].

Figure 3.1; and K_p is the correction factor for pressure, defined (with preset ξ) from the left-hand part of the nomogram in Figure 3.1.

The hydraulic resistance coefficient ξ is a parameter in the nomogram in Figure 3.1 and equal to

$$\xi = \xi_{in} - \xi_{e}, \tag{3.3}$$

where $\xi_{in} = (\Delta P_{in} + \Delta P_{in.t} + \Delta P_{in.up.})/0.5\,\rho_L W^2_{in}$ is the reduced hydraulic resistance coefficient related to the channel inlet velocity, incorporating inlet resistance ΔP_{in}, inlet resistance of the throttling orifice $\Delta P_{in.t}$, and friction in the upstream nonheated section $\Delta P_{in.up}$; and $\xi_e = (\Delta P_e + \Delta P_{e.t} + \Delta P_{e.non})/0.5\,\rho_L W^2_{in}$ is the reduced outlet resistance coefficient related to the channel input velocity, incorporating outlet resistance ΔP_e, outlet resistance of the throttling orifice $\Delta P_{e.t}$, and friction in the nonheated outlet section $\Delta P_{e.non}$.

All hydraulic resistance coefficients for the economizer and evaporating sections are determined under conditions of the single-phase isothermal flow. At $\xi \geq 0$, $(\rho W)_0$ and K_p values are found directly from the nomogram in Figure 3.1.

If $\xi < 0$, the value of K_p is found from the left-hand part of the nomogram in Figure 3.1 at $\xi = |\xi|$, while $(\rho W)_o$ is derived from

$$(\rho W)_0 = 2(\rho W)_0^{\xi=0} - (\rho W)_0^{\xi=|\xi|}, \tag{3.4}$$

where $(\rho W)_0^{\xi=0}$ and $(\rho W)_0^{\xi=|\xi|}$ are determined from the right-hand part of the nomogram in Figure 3.1 at $\xi = 0$ and $\xi = |\xi|$, respectively.

For vertical channels, the boundary mass velocity is obtained from

$$(\rho W)_{Bn}^{v} = C(\rho W)_{Bn}^{h}, \tag{3.5}$$

where $(\rho W)^h_{Bn}$ is the boundary mass velocity in a similar horizontal channel, defined from (3.1), kg/m²s; and C is the coefficient dependent on the inlet subcooling Δi_{in} and pressure P. This coefficient is obtained from the curve in Figure 3.2 plotted for $\xi < 70$. At $\xi > 70$, C is overestimated and $(\rho W)^v_{Bn}$ is determined conservatively.

According to the method of Lokshin et al. [12], the channel flow is stable if mass velocity is larger than the boundary one; that is, $(\rho W) > (\rho W)_{Bn}$.

3.3 The Saha-Zuber Method

In Saha et al. [13], the introduction of two nondimensional parameters made it possible to derive a simplified dependence for determining the oscillatory stability boundary for two-phase flows in vertical channels. These are

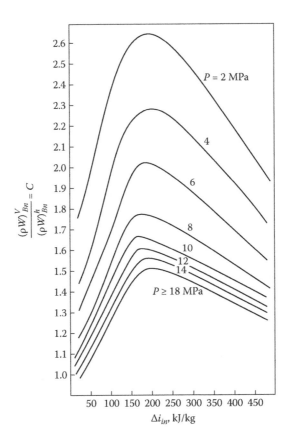

FIGURE 3.2
Coefficient for calculating mass velocity at the stability boundary in the vertical channel [12].

$$\text{the subcooling parameter, } N_{SUB} = \frac{(\rho_L - \rho_G)\Delta i_{in}}{\rho_G i_{LG}},$$

and

$$\text{the equilibrium frequency parameter, } N_{pch.eq} = \frac{10^3 \bar{q} \Pi H (\rho_L - \rho_G)}{A i_{LG} (\rho W) \rho_G},$$

where i_{LG} is the evaporation heat, kJ/kg; Π, A is the heated perimeter and channel cross-sectional area, respectively, m²; and ρ_L, ρ_G are densities of liquid and steam on the saturation line, respectively. Designations and dimensionality of other quantities were given previously.

For the equilibrium two-phase flow model at $N_{SUB} > 2$–3, the stability boundary is defined by

$$N_{pch.eq} - N_{SUB} = \frac{\xi_{in} + \xi_{TP} \, H\!/\!d_h + \xi_e}{1 + 0.5\left[0.5\xi_{TP} \, H\!/\!d_h + \xi_e\right]}, \tag{3.6}$$

where ξ_{in}, ξ_e are the channel inlet and outlet resistance coefficients, determined earlier, and ξ_{TP} is the two-phase flow friction coefficient.

It has been shown [13] that in the region of small subcooling, dependence (3.6) yields a stability boundary that strongly differs from the experimentally obtained one. Therefore, a refined method for determining the flow stability boundary in the region of small subcooling has been developed in Saha et al. [13], taking the nonequilibrium two-phase flow into account.

However, practical application of this method requires the knowledge of the equilibrium frequency parameter N^0_{pch} corresponding to zero subcooling with due account for the nonequilibrium boiling. The N^0_{pch} value was not determined in Saha et al. [13] and in the general case it shall be calculated using more complex techniques (e.g., frequency response methods or the direct numerical solution of nonlinear dynamic equations for steam-generating channels).

This circumstance deprives the modified method of its principal benefit—that is, the opportunity of using simple algebraic relations for determining the flow stability boundary. Therefore, within the framework of a simple algebraic relation of (3.6), the method of Saha et al. [13] may be applied only for an approximate determination of the flow stability boundary at high values of the inlet coolant subcooling ($N_{sub} > 2–3$). At low inlet subcooling, relation (3.6) defines the flow stability boundary with a large margin.

For a comparative determination of the flow stability boundary using these two techniques, let us transform (3.6) into the form of (3.1) and obtain

$$(\rho W)^V_{Bn} = 4 \cdot 10^3 \, \frac{\overline{q} H}{d_h} \frac{\rho_L - \rho_G}{\rho_G i_{LG}} \left[\frac{\xi_{in} + \xi_{TP} \, H\!/\!d_h + \xi_e}{1 + 0.5\left[0.5\xi_{TP} \, H\!/\!d_h + \xi_e\right]} + \frac{\rho_L - \rho_G}{\rho_G} \frac{\Delta i_{in}}{i_{LG}} \right]. \tag{3.7}$$

It is obvious from (3.1) and (3.7) that the effect of an average heat flux density on the predicted stability boundary is identical for both simplified techniques used. With other parameters in (3.1) unchanged, the effect of H/d_h on the stability boundary is of a linear nature, and with the constant $\overline{q} H/d_h$, the stability boundary predicted by the method of Lokshin et al. [12] is independent from H/d_h. Depending on (3.7), the effect of H/d_h on the stability boundary is of a more complex nature. Apart from the linear multiplier H/d_h before the brackets, H/d_h also enters the expression inside the brackets. However, an analysis of the expression inside the brackets in the denominator of (3.7), ($N_{SUB} > 2$), shows that at small values of ξ_{in} and ξ_e, as well as at large values of ξ_e and widely varying ξ_{in}, the stability boundary depends only insignificantly on H/d_h (by less than 20%) at a constant $\bar{q}H/d_h$. Similar results have been obtained theoretically for these conditions in Labuntsov and Mirzoyan [51]. At low ξ_e and high ξ_{in}, the value of H/d_h inside the brackets has a strong effect on the stability boundary at constant $\bar{q}H/d_h$.

Taking the described regularities into account, comparison of the numerically determined steam–water flow stability boundary using two techniques has been done for a vertical heated tube with the constant average heat flux density ($\bar{q}$ = 0.465 MW/m²), heated length (H = 2 m), equivalent diameter (d_h = 0.01 m) with widely varied inlet subcooling, system pressure, and the transformed channel inlet and outlet hydraulic resistance coefficients, and also for q = 0.0116 MW/m², H = 40 m, d_h = 0.01 m with ξ_{in} = 100, ξ_e = 5. Due to constraints of (3.7), the minimum inlet subcooling was determined as N_{SUB} = 2. The ranges of parameter variation and the mass velocities at the stability boundary predicted using methods of Lokshin et al. [12] and Saha et al. [13] are shown in Table 3.1.

Table 3.1 shows that the effect of pressure on the flow stability boundary is taken into account almost similarly in both simplified methods [12,13], and it is clearly shown in Figure 3.3. Close values of mass velocity at the stability boundary are predicted by these methods also with variation of channel inlet and outlet throttling within wide limits. This is illustrated by Figure 3.4 based on the data of Table 3.1 (neglecting ξ_{in} = 100; ξ_e = 5).

It is also seen from Table 3.1 that the conditions where there is a large inlet hydraulic resistance coefficient at low ξ_e (H/d_h = 4000) are an exception. In this case, the boundary mass velocity calculated using the method of Saha et al. [13] has a much lower value as compared to that predicted by Lokshin et al. [12]. For determining reliability of the stability boundary prediction using methods [12,13] at high ξ_{in} and low ξ_e, the results of calculations have been compared with experimental data [22] (see Table. 3.2). It is seen that the method of Saha et al. [13] significantly underestimates the stability boundary as compared to experimental data, while the method of Lokshin et al. [12] yields a conservative estimate, which guarantees system stability. The reason is that, at high inlet throttling, the method of Saha et al. [13] underestimates the boundary mass velocity almost in proportion to the increase in ξ_{in}, which is not real, since a superheating section equivalent to the exit resistance appears in the channel exit section at high ξ_{in} and low ξ_e, thereby reducing the positive effect of inlet throttling. This was also reported in Dykhuizen, Roy, and Calra [54].

In conclusion, it should be stated that

1. In comparison with the method of Lokshin et al. [12], the range of parameters for the stability boundary determination was broadened. The effect of inlet and exit nonheated sections and exit throttling on the stability boundary has been taken into account.

2. In the range of sufficiently high values of coolant inlet subcooling (N_{SUB} > 2) at moderate inlet throttling, the estimates of the two-phase

TABLE 3.1

Predicted Mass Velocity (ρW) Values at the Stability Boundary: Denominator and Numerator[a]

ξ_e	ξ_{in}	P = 3 MPA at Δi_{in}, KJ/kg				P = 5 MPA at Δi_{in}, KJ/kg				P = 10 MPA at Δi_{in}, KJ/kg			
		150	200	300	400	150	200	300	400	150	200	300	400
5	5	1460/1105	1220/1035	920/875	735/695	1150/895	1000/835	790/725	650/575	730/630	665/570	565/495	490/405
5	15	1160/985	1000/940	790/790	655/635	865/790	775/755	640/650	550/520	505/550	475/510	420/435	380/365
5	30	890/875	795/845	660/720	560/580	630/700	580/675	500/590	440/475	350/465	330/425	305/370	280/315
5	100	431/679	407/670	367/595	335/485	280/540	270/535	250/490	235/400	145/330	140/315	135/290	130/245
5	100*	1000/679	897/670	722/595	605/485	738/540	671/535	598/490	495/400	418/330	398/315	359/290	326/245
15	5	1580/1235	1305/1180	970/980	770/765	1280/995	1090/950	845/810	690/630	840/690	755/645	630/545	535/440
15	15	1390/1105	1170/1035	890/875	720/695	1080/895	945/835	750/725	625/630	670/630	615/570	530/495	460/405
15	30	1175/950	1015/905	797/765	655/620	875/760	785/725	645/625	550/510	515/520	480/485	425/420	380/355
15	100	685/695	628/685	537/585	470/500	465/535	440/525	390/460	355/395	250/325	240/310	225/280	215/235
30	5	1640/1395	1350/1305	990/1075	780/835	1350/1115	1140/1040	875/880	710/685	900/765	805/695	660/585	560/475
30	15	1515/1295	1260/1240	940/1000	750/790	1215/1045	1045/1000	815/825	670/655	780/720	705/675	595/555	510/455
30	30	1360/1105	1150/1035	880/875	710/695	1065/895	920/835	740/725	615/575	650/630	595/570	515/495	450/405
30	100	920/725	820/700	670/610	570/510	655/575	600/555	520/500	455/415	365/365	345/345	315/310	290/265

[a] Denominator: method from Lokshin, V. A., Peterson, D. F., Shvarts, A. L., eds. 1978. *Boiler Unit Hydraulic Design, Standard Method.* Moscow: Energia, 256 pp.; numerator: method from Saha, P., Ishii, M., Zuber, N. 1976. *Journal of Heat Transfer, Transactions of ASME* 98 (4): 616–622.

* Prediction for $H/d_h = 4000$ (q = 0.0116 MW/m²; H = 40 m; d_h = 0.01 m).

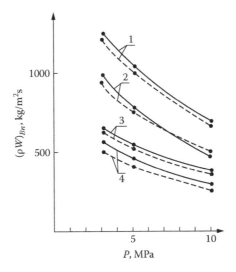

FIGURE 3.3
Dependence of mass velocity in the vertical channel at the stability boundary on pressure.
Solid line: calculation according to Lokshin, Peterson, and Shvarts [12]; dashed line: calculation
according to Saha, Ishii, and Zuber [13]. 1: $\xi_{in} = 15$; $\xi_e = 30$; $\Delta i_{in} = 200$ kJ/kg; 2: $\xi_{in} = 15$; $\xi_e = 5$; $\Delta i_{in} =$
200 kJ/kg; 3: $\xi = 15$; $\xi_e = 5$; $\Delta i_{in} = 400$ kJ/kg; 4: $\xi_{in} = 100$; $\xi_e = 30$; $\Delta i_{in} = 400$ kJ/kg.

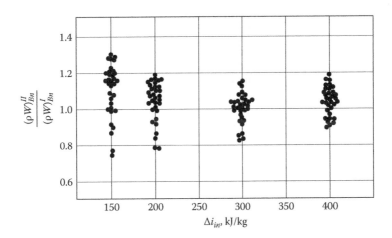

FIGURE 3.4
Comparison of the calculated boundary mass velocities in vertical channels at the stability
boundary using the methods of Lokshin et al. [12] (ρW) on the flow rate in the vertical steam-
generating channel (a) and corresponding regions of the interchannel flow rate instability for
two parallel identical channels (b) $(\rho W)^I_{Bn}$ [12] and $(\rho W)^{II}_{Bn}$ [13].

steam–water flow oscillatory instability in the vertical channel predicted using both methods are close in values.

3. In the range of low coolant inlet subcooling at high inlet throttling, as well as for horizontal channels, the method of Lokshin et al. [12] yields a better stability boundary approximation based on simple dependences than other methods.

4. Both methods are sufficiently approximate and are applicable for a preliminary rough estimation of the stability boundary (often rather overestimated) and also for selecting the ways of flow stabilizing at the design phase.

3.4 The Method of the Institute for Physics and Energetics (IPE)

Recently, a new, relatively simple method for determining the interchannel stability boundary has been developed at the Institute of Physics and Energetics (IPE) [65–70]. The method makes it possible to determine approximately the interchannel stability boundary for any two-phase steam–liquid flow in channels of vertical, horizontal, and any other orientation. This broadens the possibilities of stability boundary determination by relatively rough methods [12,13]. Also, the IPE method substantially differs from and supplements the approximate methods of Lokshin et al. [12] and Saha et al. [13] in determining the interchannel stability boundary at low exit qualities in vertical channels with the nonheated exit upflow section—as well as in taking the distribution of heat flux, quality, change of flow area, and presence of local resistances along the channel length into account. Basically, the IPE method is a sole simple method for the determination of the interchannel thermal-hydraulic stability boundary at low exit qualities.

However, the IPE method suffers from a number of constraints typical of the methods from Lokshin et al. [12] and Saha et al. [13] as it does not take into account

the thermal inertia of heat transfer surfaces

the hydrodynamic effect of thermally and hydrodynamically nonidentical parallel channels

the influence of thermal and transport delay in the case of convective heating

the effect of feedback on heat transfer dynamics in the case of nuclear heating, etc.

There is a rather wide region of parameters and boundary conditions where such constraints produce no appreciable influence on location of the interchannel stability boundary. At the same time, there also exists a range of parameters and boundary conditions where the neglect of such constraints may substantially deform the interchannel thermal-hydraulic stability boundary. (The examples demonstrating this influence are given in Sections 1.2–1.4 in Chapter 1.) That is why the IPE method cannot be considered as universal.

Application of the IPE method also poses a problem of accuracy in stability boundary determination. The constraints associated with this problem will be discussed herein.

The detailed theoretical and experimental substantiation of the IPE method, as well as its description and application procedure, are given in references 65–70. Here, it will be described in brief together with the procedure of application.

It should be noted that substantiation of assumptions made in the mathematical model and approbation of the method by a wide comparison of predictions and experimental data for determining the range of parameters and boundary conditions, wherein quantitative assessments by this method are valid, have not been completed yet. Therefore, so far, the method may be recommended for an approximate determination of the stability boundary and qualitative evaluation of the effect of different design and flow parameters on the location of the interchannel oscillatory stability boundary with the aim of optimizing the ways of improving stability of the system under investigation.

The criterion used in the IPE method for determining the interchannel fluctuation (ICP) boundaries is called the "D-criterion" in references 65–70 and has the form of

$$D = \frac{(d\Delta P_2/dG)}{(d\Delta P_1/dG)} = 1 \tag{3.8}$$

or

$$D = \frac{d\Delta P_2}{dG} - \frac{d\Delta P_1}{dG} = 0, \tag{3.9}$$

where

$$\Delta P_1 = \Delta P_{in.t} + \Delta P_{ec} + \Delta P_{in.up} \tag{3.10}$$

and

$$\Delta P_2 = \Delta P_{e.t} + \Delta P_{ev} + \Delta P_{e.non}. \tag{3.11}$$

$\Delta P_{in.t}$ and $\Delta P_{e.t}$ are the pressure drops at the inlet and outlet throttling orifice, respectively; $\Delta P_{in.up}$ and $\Delta P_{e.non}$ are the total pressure drops at the inlet and outlet of the channel heated section, taking into account the gravity component in the nonheated sections; and ΔP_{ec} and ΔP_{ev} are the total pressure drops in the economizer and evaporating sections, respectively, taking into account the gravity component.

Using the D-criterion from (3.9), the regime is stable with $d\Delta P_2/dG < d\Delta P_1/dG$ (D < 0) and is unstable if $d\Delta P_2/dG > d\Delta P_1/dG$ (D > 0).

To illustrate the IPE method application for the vertical channel conditions with the nonheated exit section at constant values of P, q, T_{in}, ξ_{in}, ξ_{e}, Figure 3.5 [66] presents qualitative behavior of channel pressure drop curves ΔP_C (G) and components of this drop ΔP_1 (G) and ΔP_2 (G). At points G_A, G_b, G_C, slopes

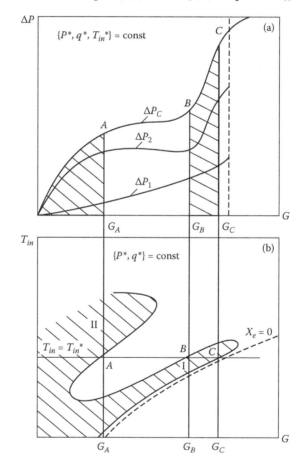

FIGURE 3.5
Pressure drop curves depending on the flow rate in the vertical steam-generating channel (a) and corresponding regions of the interchannel flow rate instability for two parallel identical channels (b) [66].

of ΔP_1 (G) and ΔP_2 (G) coincide (i.e., $d\Delta P_2/dG = d\Delta P_1/dG$ and D = 0). These points lie on the stability boundary. The regions $G < G_A$ and $G \in (G_B, G_C)$, where $d\Delta P_2/dG > d\Delta P_1/dG$ (D > 0), are the instability regions at preset parameters. It is seen that the instability at low exit qualities and the density-wave instability are on the positive stable branch of the hydraulic curve.

In Figure 3.5(b) [66], one can see an ICP map in the (G, T_{in}) plane showing stability boundaries constructed for a number of fixed values of T_{in} and constant other parameters in references 65–70.

Thus, construction of an ICP map requires curves $\Delta P_1(G)$ and $\Delta P_2(G)$ for the channel in question at fixed values of P, q, T_{in}, ξ_{in}, ξ_e. The analysis performed in Shvidchenko et al. [66] showed that predicted values of $\Delta P_1(G)$, $\Delta P_2(G)$, and their derivatives with respect to the flow rate differ greatly from those obtained experimentally. This leads to poor accuracy of instability region boundaries' determination using the D-criterion, when $\Delta P_1(G)$ and $\Delta P_2(G)$ are found numerically. Therefore, in order to increase accuracy, a method combining experimental and predictive techniques has been developed [65–70] for determining the previously mentioned pressure drops.

The method is based on experimental determination of the hydraulic characteristic (i.e., the pressure drop in headers $\Delta P_C(G)$) as well as the pressure drop at the inlet throttling orifice and across the nonheated inlet single-phase section ($\Delta P_{in.t}(G) + \Delta P_{in.up}(G)$) at different fixed values of T_{in}, for one of the investigated parallel identical channels with fixed values of P, q, ξ_{in}, ξ_e.

Then, $\Delta P_1(G)$ and $\Delta P_2(G)$ may be determined as follows:

$$\Delta P_1(G) = [\Delta P_{in.t}(G) + \Delta P_{in.up}(G)]_{exp} + \Delta P_{ec}(G) \qquad (3.12)$$

and

$$\Delta P_2(G) = \Delta P_{c.exp}(G) - \Delta P_1(G), \qquad (3.13)$$

where $\Delta P_{ec}(G)$ is the pressure drop across the economizer section, predicted on the basis of heat balance.

An example of the ICP boundaries' construction using the described method and applying the D-criterion is illustrated in Figure 3.6 [69].

For a regime to which application of the D-criterion is justified, the error in the ICP boundary determination by the preceding method is due to the error in experimental measurement of $\Delta P_{c.exp}(G)$ and $[\Delta P_{in.t}(G) + \Delta P_{in.up}(G)]_{exp}$ pressure drops (reaching 20% at low flow rates), the error in flow rate determination (ca. 5%), the error of ΔP_{ec} calculation, and the error of graphical determination of $d\Delta P_1/dG$ and $d\Delta P_2/dG$ derivatives, which may be rather large for flat derivatives.

The D-criterion (3.9) was developed in references 65–70 on the basis of P. A. Petrov's ideas [71]. First, Petrov developed the so-called Π-criterion for the stability boundary determination:

$$\Pi = \Delta P_1/\Delta P_2 = \text{const.}, \qquad (3.14)$$

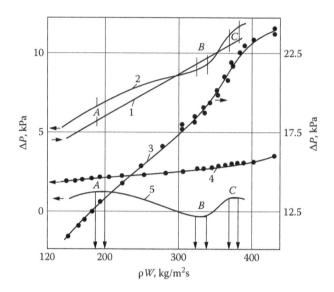

FIGURE 3.6
Construction of boundaries of the interchannel flow stability regions. P = 6 MPa, q = 0.375 MW/m²; ξ_{in} = 13; ξ_e = 1.6, T_{in} = 170°C; dark circle: experimental data; 1, 2: predicted and experimental pressure drop curves, ΔP_1 and ΔP_2, respectively; 3, 4: approximating curves of experimental channel pressure drops, P_c and ($\Delta P_{in.t} + \Delta P_{in\ up}$), respectively; 5: the ($\Delta P_2 - \Delta P_1$) curve used to determine the mass velocity range (ρW), where D = 0; A, B, C sections: errors of stability boundary determination.

which was further modified as follows:

$$\Pi \equiv \Delta P_1/\Delta P_2 = f\ (\rho W, p, q, d_h, \xi_{in}, \xi_e, T_{in}). \tag{3.15}$$

However, no universal regularities for the f function determination have been obtained so far. Therefore, Petrov's criterion is practically used as in (3.14). In this form, it has sometimes been applied to qualitatively evaluate the effect of parameters on the stability boundary, though in some cases it incorrectly reflects the effect even qualitatively.

Despite the fact that the D-criterion (3.9), like criterion (3.15), employs similar initial premises and requires knowledge of the static hydraulic characteristic only, the former does not coincide with Petrov's criterion. Indeed, the D-criterion (3.9) requires only slopes of ΔP_1 (G) and ΔP_2 (G) to be equal and imposes no limitations on the $\Delta P_1/\Delta P_2$ value that defines criterion (3.14). Since $d\Delta P_1/dG$ and $d\Delta P_2/dG$ can be equal at different $\Delta P_1/\Delta P_2$ ratios, the D-criterion has a more general nature.

Until now, there is no rigorous substantiation of the D-criterion applicability for the ICP boundaries determination and no parameter ranges have been specified where the made assumptions are appropriate. Therefore, references 65–70 present the results of experimental investigations on the

ICP boundary determination in a wide range of design and flow parameters' variation, aimed at substantiating applicability of the D-criterion for predicting the preceding boundary and evaluating accuracy of such predictions.

In these investigations, geometry of test channels, distance between headers, and heated/nonheated length ratios, ξ_{in}, ξ_e were varied. The ranges of parameter variation were as follows: P = 4.0–16 MPa; q = 0.3–0.75 MW/m²; pW = 110–600 kg/m²s; T_{in} = from 20°C to (T_S – 10°C); ξ_{in} = 13, 26, 52; ξ_e = 1.6, 5, 10; x_e < 1.

Characteristic examples of comparing the ICP boundary experimental maps based on the preceding range of parameters with the map obtained using the D-criterion are shown in Figure 3.7 [69].

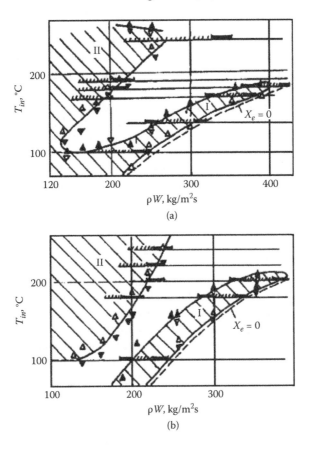

FIGURE 3.7
Comparison of the calculated and experimental interchannel flow stability boundaries [69]: (a) P = 6 MPa; q = 0.375 MW/m²; ξ_{in} = 13; ξ_e = 1.6; (b) P = 12 MPa; q = 0.5 MW/m²; ξ_{in} = 13; ξ_e = 1.6. ▲ and ▼: stable experimental regimes; △ and ▽: unstable experimental regimes; solid line: experimental boundary approximation; —accuracy of boundary calculation using the D-criterion (D = 0); — the region with D > 0 (calculated instability region); I: instability region at low exit qualities; II: density-wave instability region.

It follows from references 65–70 that in the preceding range of test parameter variation, the D-criterion yields a satisfactory qualitative description of the effect of design and flow parameters on the stability boundary. However, even in the previously given sufficiently limited range of parameter variation, the error of stability boundary determination using the D-criterion for some regimes is large. For example, it was noted in Shvidchenko et al. [66] that at a low flow rate and high q, a considerable disagreement was observed in the density-wave stability boundary determined experimentally and using the D-criterion. Under the same conditions, the stability boundary was described incorrectly also at low exit qualities, because it follows from, for example, Mitenkov et al. [7] that the lower boundary of this instability corresponds to $x_e < 0$, while with the D-criterion, it occurs only at $x_e \geq 0$. The error of the ICP boundary determination under such conditions is due, in the first place, to the fact that the D-criterion neglects surface boiling, which has an effect when calculating ΔP_{ec} and, hence, when determining ΔP_1 and ΔP_2.

It has been shown [13] that the effect of surface boiling on the stability boundary determination is the largest at low inlet coolant subcooling.

In Shvidchenko et al. [66], one can see that the error of the ICP boundary determination using the D-criterion increases markedly, if compared to the experimental boundary, under regimes with increased exit throttling ($\xi_e = 5; 10$).

Section 3.3 shows that the largest errors in the stability boundary determination using the methods from Lokshin et al. [12] and Saha et al. [13] are observed at high inlet throttling ($\xi_{in} > 100$) and in presence of long superheating sections. These cases are rather frequent in real once-through steam generators.

It should be noted that according to numerous publications, even the Petrov's criterion under such conditions fails to give a qualitatively correct reflection of the parameters' effect on the stability boundary. The D-criterion validity under these regimes, which are most difficult for the stability boundary prediction, has not been investigated until now. The range of parameters in the experimental investigations of references 65–70 was limited by $x_e < 1$ and by a sufficiently low inlet throttling ($\xi_{in} \leq 50$).

Thus, if applied to channels with the heat flux on the heat transfer surface constant over time, the IPE method yields a large error in prediction of the ICP boundary for the regimes with high specific heat flux and low coolant flow rate, as well as for the regimes with an increased exit throttling. Efficiency of the method has not been checked for the regimes with high inlet throttling ($\xi_{in} > 100$) and in the presence of a superheating section.

The performed analysis leads to the following conclusions:

1. The IPE method allows stability boundary prediction at low exit qualities, when predictions by the approximate methods of Lokshin et al. [12] and Saha et al. [13] fail.

2. In the cases when the hydraulic characteristic can be obtained experimentally, for the conditions of moderate inlet throttling, heat

flux constant over time, and the absence of a superheating section and exit throttling, the IPE method yields satisfactory accuracy and may be recommended for predicting interchannel stability boundaries. Particularly, it refers to the assemblies of heating surfaces with coolant upflow-downflow movement, for which approximate methods are inapplicable.

3. For the stability boundary preliminary evaluation while designing power equipment, when there are no experimental data on the hydraulic characteristic of the channel under investigation and the D-criterion may be only predicted, the IPE method has no advantage with respect to accuracy and is more complex for application as compared to the methods of Lokshin et al. [12] and Saha et al. [13].

4. Under conditions with exit throttling and high inlet throttling, in the presence of a superheating section, and at high surface heat fluxes, the validity and accuracy of the IPE method for stability boundary prediction have not been determined.

5. The IPE method, like those of Lokshin et al. [12] and Saha et al. [13], neglects the influence of heating surface thermal inertia, thermal and transport delays in the case of convective heating, and other dynamic characteristics of the processes in the channel under investigation. Therefore, when applied in conditions where such effects are strong, the approximate methods may suffer from large errors when predicting the interchannel stability boundary.

3.5 Determination of Oscillatory Stability Boundary at Supercritical Pressures

Deterioration of flow stability at supercritical pressure (SCP) was mentioned for the first time in Goldman [45]. In Russia, the oscillatory flow instability in parallel heated channels at SCP was experimentally discovered and investigated for the first time in Krasyakova and Glusker [46,47]. The bench-test experiments were conducted on full-scale models of boiler heating surface panels in loops with upflow and upflow-downflow coolant movement with steam and electrical heating. Water under supercritical pressure (P = 23–23.5 MPa) was used as the working fluid.

A theoretical study of flow oscillatory instability at SCP for conditions characteristic of utility boilers was made in Lokshin et al. [12] and Khabensky [48]. Flow fluctuations were studied in Khabensky by a direct numerical solution of the nonlinear distributed parameter dynamic model. Calculations were made using different values of pressure, inlet enthalpy, inner diameter, and heated length of the channel. Examples of channel inlet mass velocity

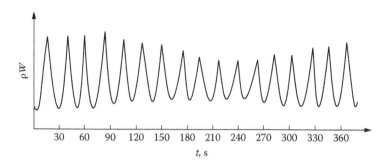

FIGURE 3.8
Fluctuations of mass velocity at the coil inlet with the upflow-downflow fluid motion in experimental investigations [46]. $\rho W = 300$ kg/m²s; $T_{in} = 350°C$; $q = 0.116$ MW/m²; $P = 23.5$ MPa.

fluctuations, experimentally obtained in Krasyakova and Glusker [46], are given in Figure 3.8.

Recently, experimental and predictive investigations of the oscillatory instability of the helium flow with supercritical parameters in heated channels have been reported in references 49–52. The reason is that under SCP, helium is used as a thermostating agent in various energetic, electronic computer, and research systems employing superconductivity. The appearance of low-frequency fluctuations in the SCP helium flow impairs efficiency of cooling systems and in some cases may lead to emergency situations [49].

The results of theoretical studies and experimental investigations of the coolant flow low-frequency oscillatory instability at SCP showed this type of thermal-hydraulic instability to have a nature and mechanism similar to those of the two-phase flow density-wave instability at subcritical pressure (see previous sections).

Irrespective of the fact that the medium is single phase at SCP, thermal-physical parameters of the medium, including density, were found to change sharply with enthalpy variation in the region of enthalpy that corresponds to the large heat capacities zone. Thus, at both sub- and supercritical pressures, there exists a delay in transfer of the flow rate disturbance along the channel length in the region of the maximum rate of the enthalpy-governed density change. Conditionally, it may be considered that enthalpy that corresponds to the beginning of strong dependence of density on enthalpy (i_m), signifies the boundary of the onset of pseudoboiling, which determines inlet subcooling $(\Delta i_{in} = i_m - i_{in})$ and conventional differentiation into the econo-mizer and evaporating-superheating sections. For the water coolant, enthalpy corresponding to initiation of pseudoboiling is determined by the dependence

$$i_m = 1.67 \cdot 10^3 + 42 \, (P - 23),\qquad(3.16)$$

where P is pressure, MPa; and i_m is the enthalpy at the onset of pseudoboiling, kJ/kg.

Numerical investigations [48] proved the experimentally discovered [53] flow fluctuations in a system of parallel heated tubes with vertical nonheated exit sections. These fluctuations appear when enthalpy of the medium at the channel exit reaches or somewhat exceeds the region of the onset of pseudoboiling ($i_e \geq i_m$). These fluctuations are equivalent to flow instability at low exit qualities at subcritical pressure and are caused by a change of the gravity pressure drop in the channel. Noticeable changes in coolant density in the channel exit cross section, which appear at low coolant flow rate disturbances, propagate along the height of the exit vertical nonheated section, and result in periodic coolant flow rate fluctuations due to oscillations of the gravity pressure drop.

The test section in Treshchev, Sukhov, and Shevchenko [53] was composed of two vertical tubes interconnected by headers. One of the tubes was heated by direct current, whereas the second, the nonheated one, served as a bypass. The ratio between flow rates through the heated channel and bypass was $G_c/G_{By} = 1/10$. The heated channel had an exit vertical nonheated section of variable height.

The periodic change of the heated channel exit flow temperature that occurred at exceeding some threshold power level by 1%–3% was accepted as the onset of self-oscillations.

The experimental results of Treshchev et al. [53], being in full compliance with the qualitative analysis of flow instability at low exit qualities and subcritical pressure offered in Section 1.2 (Chapter 1), show that

1. In this case, flow oscillations are not harmonic, though they have a sufficiently clear periodicity.

2. Inlet throttling somehow facilitates flow stabilization by decreasing the oscillation amplitude. However, in a sufficiently wide range of ξ_{in} variation, the effect of inlet throttling on the stability boundary (Figure 3.9) is insignificant. An exception is the abnormal change of the stability boundary at $\xi_{in} = 9$, which falls out of the general regularity.

3. The nonheated vertical exit section elongation results in a noticeable deterioration of flow stability (Figure 3.10).

However, for a medium with supercritical parameters, for which the rate of change of density change depending on enthalpy in the pseudoboiling zone is relatively small, the amplitude of fluctuations of the gravity pressure drop is a small fraction of the total channel pressure drop. Therefore, the amplitude of flow rate fluctuations with $i_e \geq i_m$ in both experimental and numerical investigations was small (<5% of the average channel flow rate) for the considered range of parameters.

Let us dwell on the density-wave flow instability.

Investigations described in references 46–52 show the effect of various design and flow parameters on the stability boundary at SCP to be similar to that of said parameters at subcritical pressure. For example, a change of heat

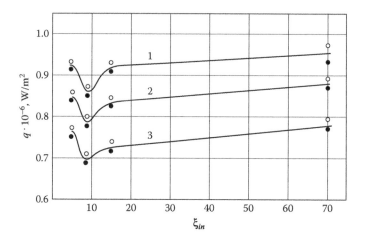

FIGURE 3.9
Surface heat flux density at the stability boundary depending on the hydraulic resistance coefficient at the channel inlet [53]. $\rho W = 1000$ kg/m²s; $T_{in} = 300°C$; O: unstable regimes; ●: stable regimes; 1: 26.5 MPa; 2: 24.3 MPa; 3: 22.5 MPa.

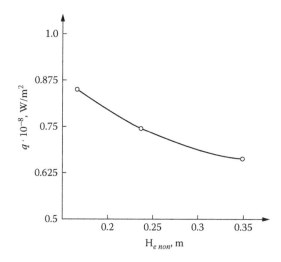

FIGURE 3.10
Surface heat flux density at the stability boundary depending on the unheated section height at the channel outlet: $\rho W = 750$ kg/m²s; $P = 22.5$ MPa; $T_{in} = 300°C$.

flux surface density and channel heated length leads to a directly proportional change, whereas a change of the channel diameter leads to a reversely proportional change of the boundary coolant mass velocity.

Similarly to the situation with the subcritical pressure, the coolant flow in a vertical tube is less stable than in an identical horizontal tube.

From references 48–52, one can see that a similar effect is produced on the stability boundary at both sub- and supercritical pressures by channel inlet and exit throttling, as well as heat flux surface density distribution along the channel length. The mechanism and nature of the preceding parameters' effect on the stability boundary are described in detail in Section 1.3 (Chapter 1) and thus not repeated here.

The effect of inlet enthalpy (or inlet subcooling) on the stability boundary is also similar to that described in Section 1.3 for the subcritical pressure. In this case, there were no flow oscillations with $i_{in} > i_m$. This may be explained by narrowing of the region with the enthalpy-governed intensive change of density in the considered channel. As a result, the delay time of the flow rate disturbance degenerates and the intensity of flow rate change along the channel length decreases. As Section 1.3 shows, this facilitates flow stabilization. At a subcritical pressure, similar conditions appear with the channel coolant inlet enthalpy of $i_{in} > i_s$.

Also, it was shown in Section 1.3 that the region of flow stability increases with the increasing pressure due to the weakening dependence of density of the medium in the two-phase area on the enthalpy and a smaller pressure drop fraction in the evaporating section in the total pressure drop in the channel because of the higher steam density. At SCP, the intensity of density dependence on enthalpy in the pseudoboiling area is minimal, thus providing the widest flow stability region as compared to subcritical pressure at other identical flow and design parameters. In this respect, Figure 3.11 is a clear illustration showing the results that Khabensky [48] obtained by solving the nonlinear distributed parameter mathematical model of the boundary mass velocity at the increase in pressure up to SCP at different inlet subcooling values. One can see from Figure 3.11 that, with the growing pressure, flow stability improves and transition to SCP does not change the monotonous increase of the stable region.

In this relation, the appearance of flow steady-state oscillations in the parallel-channel system at SCP requires that the pseudoboiling and superheating sections should be substantially increased by substantially reducing the coolant flow rate or increasing the heat flux surface density. In other words, the density-wave instability in this case manifests itself at a substantial superheating of the medium at the channel outlet. Numerical investigations for the water coolant and utility boilers [48] have established that flow fluctuations (without exit throttling) may appear only when the channel outlet enthalpy (i_e) is from 2.9 to 3.2 10^3 kJ/kg. If we take into account (see earlier discussion) that the initiation of fluctuations in a channel with the water coolant at SCP requires the inlet subcooling of $i_{in} < i_m$, then flow instability at low exit throttling manifests itself at $\Delta i_c > 1.5 \times 10^3$ kJ/kg. Since in modern boilers with supercritical parameters the enthalpy increment in an element does not exceed $\Delta i_c \leq 0.8 \cdot 10^3$ kJ/kg, as a rule, no flow fluctuations have been observed in them.

Taking into account that all regularities concerning the influence of parameters on the density-wave flow stability boundary are identical at

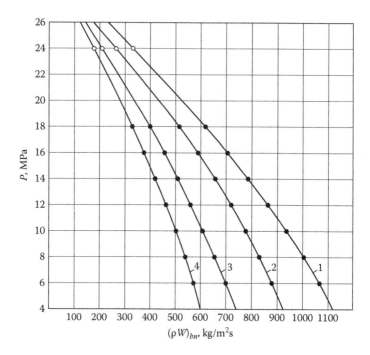

FIGURE 3.11
Dependence of boundary mass velocity on pressure in a horizontal tube with no inlet and outlet throttling. Dark circles: calculated points of stability boundary at subcritical pressure [48]; clear circles: same, at supercritical pressure [48]; 1: $\Delta i_{in} = 80$ kJ/kg; 2: 200 kJ/kg; 3: 400 kJ/kg; 4: 800 kJ/kg.

sub- and supercritical pressures, an approximate determination of the thermal-hydraulic stability boundary at SCP may be made with a small correction using the methods set forth in Section 1.3. All the constraints described there relate to supercritical parameters as well.

The procedures of the stability boundary determination at SCP using the methods of Lokshin et al. [12], Khabensky [48], and the modified method of Saha et al. [13] are given herein.

All designations and dimensionality of other parameters are the same as in Sections 3.2 and 3.3.

3.5.1 The CKTI Method

For horizontal channels, the boundary mass velocity at which flow fluctuations initiate is derived from

$$\left(\rho w\right)^h_{Bn} = 4.63 \cdot 10^{-3} \left(\rho w\right)_{o,p} \frac{\bar{q}H}{d_h}, \tag{3.17}$$

where $(\rho W)_{0,P} = (\rho W)_0 \, K_P$.

The value of $(\rho W)_0$ is determined at preset values of ξ and Δi_{in} ($\Delta i_{in} = i_m - i_{in}$; ξ and i_m are determined from (3.3) and (3.16), respectively) in the right-hand section of the nomogram in Figure 3.1.

The K_P coefficient accounting for pressure correction is derived from the curve in Figure 3.12.

For vertical channels, the boundary mass velocity is determined from

$$(\rho W)_{Bn}^{V} = C(\rho W)_{Bn}^{h},$$ (3.18)

where $(\rho W)_{Bn}^{h}$ is the boundary mass velocity for the identical horizontal channel; and C is the coefficient dependent on the inlet subcooling and obtained from the curve in Figure 3.2 at $P > 18$ MPa.

A comparison of the boundary mass velocity values ρW calculated by the methods of Lokshin et al. [12] and Khabensky [48] with the experimentally obtained ones [46] is given in Table 3.3 and shows satisfactory agreement. Systematic underestimation of the predicted boundary mass velocities as compared to the experimental ones may be due to small pressure drop fluctuations between headers caused by subcooling conditions in the heat exchanger of the test facility [46]. Pressure drop fluctuations could result in periodic flow rate oscillations in the stable region near the stability boundary.

Investigations of Khabensky [48] show that at high surface heat flux densities, the calculated boundary mass velocity may be underestimated because of the neglect of radial distribution of coolant density across the channel section in the mathematical model used.

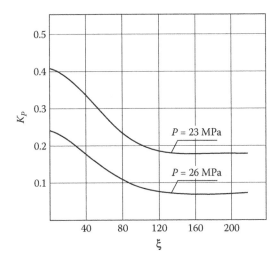

FIGURE 3.12
The coefficient accounting for the effect of pressure on the stability boundary at supercritical pressure [48].

TABLE 3.2

Comparison of Experimental and Predicted Boundary Mass Velocities

P, MPa	$\bar{q}$, MW/m²	ΔT_{in}, °C	H, m	d_h, m	ξ_{in}	ξ_e	$(\rho W)_{Bn}^{exp}$, kg/m²s	$(\rho W)_{Bn}^{I}$, kg/m²s	$(\rho W)_{Bn}^{II}$, kg/m²s
4.1	0.3085	105	5.25	0.0125	500	4	220	293	123
4.05	0.3	101	5.25	0.0125	520	4	220	278	117
4.05	0.26	100	5.25	0.0125	300	4	220	315	162
4.25	0.418	109	5.25	0.0125	500	4	298	396	161
4.25	0.449	100	5.25	0.0125	520	4	318	436	173
4.3	0.469	96	5.25	0.0125	520	4	335	445	175
4.2	0.465	96	5.25	0.0125	500	4	330	448	180

Note: I = method of Lokshin, V. A., Peterson, D. F., Shvarts, A. L., eds. 1978. *Boiler Unit Hydraulic Design, Standard Method.* Moscow: Energia, 256 pp. II = method of Saha, P., Ishii, M., Zuber, N. 1976. *Journal of Heat Transfer, Transactions of ASME* 98 (4): 616–622.

According to the methods of Lokshin et al. [12] and Khabensky [48], the stable region is determined by

$$(\rho W) > (\rho W)_{Bn}. \tag{3.19}$$

3.5.2 The Saha-Zuber Method

Using this method, the dependence for determining the boundary mass velocity by (3.7) has the form of

$$(\rho W)_{Bn}^{V} = 4 \cdot 10^3 \frac{\bar{q} \cdot H}{d_h} \frac{\rho_L - \rho_G}{\rho_G \cdot i_{LG}}$$

$$\times \left[\frac{\xi_{in} + \xi_{TP} H/d_h + \xi_e}{1 + 0.5\left[0.5\xi_{TP} H/d_h + \xi_e\right]} + \frac{(\rho_L - \rho_G)}{\rho_G} \times \frac{\Delta i_{in}}{i_{LG}} \right]^{-1}.$$

To obtain $(\rho W)^V_{Bn}$ at SCP using the preceding dependence, parameters ρ_L, ρ_G, and i_{LG} should be replaced with their equivalents in the SCP region.

This dependence may be modified to be applicable to the SCP region, taking into account two regularities. First, the effect of pressure on the flow stability boundary, as was shown in Section 3.3, is practically identical for both methods of Lokshin et al. [12] and Saha et al. [13] (Figure 3.3). Second, numerical investigations in Khabensky [48] have revealed the monotonous nature of the effect of pressure on the stability boundary at the transition from sub- to supercritical pressure (Figure 3.11). Therefore, with the boundary mass velocity determined at 22.5 MPa, its value may be approximately defined for the channel in question at P > 22.5 MPa, with other parameters being identical, by

$$(\rho W)_{Bn} = f(p) \cdot (\rho W)_{Bn}^{P=22.5}, \tag{3.20}$$

where $f(P)$ is the dependence determining the effect of pressure on the stability boundary at P > 22.5 MPa.

Numerical investigations of instability at SCP using the nonlinear mathematical model of Khabensky [48] yield the following approximate dependence:

$$f(P) = \left(\frac{P}{22.5} \right)^{-4}, \tag{3.21}$$

where P is the coolant pressure in the test channel at P > 22.5 MPa.

Taking (3.20) and (3.21) into consideration, dependence (3.7) may be presented as

$$(\rho W)_{Bn}^{V} = 4 \cdot 10^3 \frac{\overline{q} \cdot H}{d_h} \left(\frac{P}{22.5} \right)^{-4} \frac{\rho_L^* - \rho_G^*}{\rho_G^* \cdot i_{LG}^*}$$

$$\times \left[\frac{\xi_{in} + \xi_{TP} H/d_h + \xi_e}{1 + 0.5 \left[0.5\xi_{TP} H/d_h + \xi_e \right]} + \frac{(\rho_L^* - \rho_G^*)}{\rho_G^*} \times \frac{(i_L^* - \Delta i_{in})}{i_{LG}^*} \right]^{-1}, \tag{3.22}$$

where $\rho_L^*, \rho_G^*, i_{LG}^*$, and i_S^* are determined at P = 22.5 MPa.

TABLE 3.3

Comparison of Experimental Mass Velocities on the Stability Boundary with the Predicted Ones

P, MPa	$\overline{q}$, MW/m²	d_{in}, m	H, m	$i_{in} \times 10^3$, kJ/kg	$(\rho W)_{Bn'}^{exp}$ kg/m²s	$(\rho W)_{Bn'}^{I}$ kg/m²s	$(\rho W)_{Bn'}^{II}$ kg/m²s
23	0.0813	0.01	12.6	1.2	180	152	150
23	0.0695	0.01	12.6	1.2	150	130	129
23	0.116	0.01	12.6	1.59	300	308	255
23	0.0695	0.01	12.6	0.84	135	103	112
23.5	0.348	0.01	12.6	1.38	100	74	63
23.5	0.116	0.01	12.6	1.0	150	174	173
23.5	0.058	0.01	12.6	1.0	100	92	91

Source: Krasyakova, L. Yu, Glusker, B. N. 1965. *CKTI Proceedings*, Leningrad, 59:198–217.
Note: I = method of Khabensky, V. B. 1971. Application of Mathematical Methods and Computation Means in Heavy Machine Building Industry, NIIINFORMTYAZHMASH, Moscow, 15-71-3:17–28. II = modified method of Saha, P., Ishii, M., Zuber, N. 1976. Journal of Heat Transfer, Transactions of ASME 98 (4): 616–622.

The results of calculating boundary mass velocities at SCP using (3.22) and their comparison with those of Lokshin et al. [12], Khabensky [48], and experimental data from Krasyakova and Glusker [46] are given in Table 3.3.

To conclude, it should be noted that the described methods are rather rough and may be used for preliminary determination of the stability boundary at early design phases for selecting optimal design solutions. Further, regions of stable operation should be determined applying accurate methods (e.g., direct numerical solution of nonlinear dynamic equations).

4

Some Notes on the Oscillatory Flow Stability Boundary

4.1 Introduction

The lack of knowledge on two-phase thermal hydrodynamics of power equipment not only in unsteady- but also in steady-state conditions needs experimental verification of the adequacy of real processes' description by dynamic mathematical models. In particular, reliable prediction of coolant flow instability in full-scale systems is required.

Also, three factors can affect accuracy of a mathematical model and that of the stability boundary prediction. First, there is a need for substantial simplification of initial dynamic mathematical models and boundary conditions when dealing with practical instability predictions for complex geometry channels of real circulating loops. Second, frequency methods for determining the stability boundary in the first approximation demand linearization of the nonlinear mathematical model of two-phase flow thermal hydrodynamics. This, as will be shown later, is posing additional problems with establishing the real stability boundary in practice. Third, when studying the two-phase flow instability by direct numerical integration of nonlinear dynamic equations, the accuracy of solutions depends on the scheme of difference approximation of initial equations and boundary conditions, as well as on the method and stability of their numerical solution.

It has been noted already that reliability of theoretical predictions may be ensured only by comparing predictions and experimental data obtained on thermal-physical test facilities and on full-scale installations.

This, in turn, requires adequate criteria for the experimental stability boundary evaluation, as well as a clear idea of how predictions made by different techniques relate to the practical stability boundary. This knowledge is of practical importance for power equipment designers when determining a safety margin to ensure thermotechnical reliability of the equipment.

4.2 Experimental Determination of the Stability Boundary

Let us first consider the behavior of flow parameters (for instance, coolant flow rate) when the steam-generating channel operation loses stability. Generally, when approaching the stability boundary, flow parameters tend to change in small steps with a certain retention interval after each step.

In the case of the steam-generating channel stable operation, low-inertial flow meters commonly record a definite noise level available in the test loop. The noise may be due to the stochastic nature of boiling and change of flow patterns, produced by pump operation or power supply sources; vortexes formed around local obstacles; flow turbulence; disturbances caused by, say, mixing condenser operation; and other reasons. The average noise level may reach several percent of the nominal coolant flow rate.

The noise level may be over 10% of the nominal flow rate in isolated steam-generating channels with a plug flow at the outlet or with other two-phase flow specific features (e.g., in coil-type steam generators).

When approaching the stability boundary, the system disturbances may lead to an increased noise level in the channel flow rate due to resonance characteristics of the steam-generating channel, and transient oscillations in the channel with periodic disturbances may be accompanied by flow rate fluctuations occurring at a frequency close to that of the low-frequency thermal-hydraulic instability.

Near the oscillatory stability boundary, all the curves of self-oscillation development in the steam-generating channel may be roughly subdivided into two kinds.

1. A small stepwise increase in power (or in some other regime parameter) near the stability boundary leads to spontaneous flow rate swaying and establishment of high-amplitude self-oscillations. Sometimes, the amplitude may reach the level at which the channel may get destroyed in the case of electrical heating. This character of instability development is typical of steam-generating channels with linear properties at moderate flow rate fluctuation amplitudes. The oscillograms illustrating the development of flow rate self-oscillations show random noises observed in stable regimes. It should be noted that in this case the stability boundary can be recorded rather clearly if there are no large forced disturbances in the system, with the boundary being stable "in the small."

2. For a certain class of steam-generating channels and flow parameter regions, there exists a different mechanism of instability development. Let the input power increase, or channel input throttling decrease in a stepwise manner. In the latter case, the noise level may increase considerably, since dissipation losses decrease and the noise energy decays to a lesser extent. At the transition from the stable to

unstable region and in absence of large forced disturbances, the flow rate oscillogram exhibits regularity, which becomes apparent and clear with the further stepwise power rise. A slow increase of the regular flow rate oscillation amplitude corresponds to each step of power rise (with the reduced inlet throttling). There is no spontaneous acceleration of the process. In this case, the oscillation onset boundary is blurred and cannot be determined correctly without special measures (e.g., spectrum analyzer). Though the stability boundary is the boundary of stability in the small, it may fail to determine reliability of the steam-generating channel operation because of the low amplitude of parameter fluctuation in the instability region.

In the general case, it may be stated that self-oscillations are not obligatory monoharmonic and sometimes may have a more complex nature. In some cases, when the power to the channel is increased (or inlet throttling is decreased), the oscillogram exhibits the origin of the second and even third (which is rare) harmonics of the experimental flow rate. However, the preserved regularity is that each unstable channel regime has its own specific level of flow rate fluctuation amplitudes and set of harmonics. The development of complex oscillations may be illustrated by the results of flow stability investigations in the electrically heated coil-type channel [84]. The channel inlet and outlet constant pressure at a preset level was achieved by air cushions in the inlet and outlet headers. The channel was brought to the unstable regime by a stepwise decrease of inlet throttling using the control valve and by the appropriate feed vessel discharge in order to maintain constant average channel flow rate and pressure.

Figure 4.1 shows the oscillogram of the inlet flow rate change from Gerliga et al. [84] that illustrates the channel oscillation development. One can see that in region I with the inlet resistance ΔP_{in} of 0.3–0.15 MPa, small nondamping oscillations of pressure and flow rate at 3–10 Hz (depending on the coil diameter) were observed in all the tested channels. With the coil diameter of 56 mm, the frequency was 3–4 Hz and was found to increase up to 10 Hz with the coil diameter increasing up to 224 mm. The amplitude of flow rate fluctuations in this region amounted on the average to 7%–9%

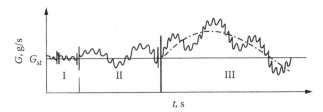

FIGURE 4.1
Flow rate fluctuations' development in the coiled channel. I: ΔP_{in} = 0.3–0.15 MPa; II: ΔP_{in} = 0.15–0.03 MPa; III: ΔP_{in} < 0.03 MPa.

of the steady-state value (G_o); with throttling decreasing down to $\Delta P_{in} = 0.15$ MPa, it increased to about 12% of G_o at the constant frequency. The cause of such oscillations is probably the changing structure of the two-phase flow in coils. The cycle of oscillations is substantially smaller than the time it takes the liquid particle to pass through the heated channel and does not correspond to the frequency of density-wave instability.

In region II, the flow rate and pressure fluctuations appeared at 1–1.3 Hz in all coiled channels at the decrease of the inlet resistance ($\Delta P_{in} < 0.15$ MPa). It was accompanied by growth of the oscillation amplitude at the unchanged frequency, finally reaching 60% of G_o with $\Delta P_{in} = 0.03$–0.05 MPa.

The nature of such oscillations is not fully clear and may be due to specifics of the test facility, apart from the density waves mechanism.

Finally, with $\Delta P_{in} < 0.03$ MPa in region III, the lowest (~0.3–0.1 Hz) frequency flow rate oscillations appear in the coils and previous oscillations become superimposed by the new ones. The amplitude of the 0.3–0.1 Hz oscillations increases at a higher rate and, at $\Delta P_{in} = 0.015$–0.02 MPa, it exceeds the steady-state value. The nature of these oscillations most of all corresponds to the mechanism of density wave low-frequency oscillations. At said frequencies, oscillations result in considerable periodic deformations of the coiled channel, which was found to contract and stretch axially as a spring (up to 6 cm at 0.3–0.1 Hz) and simultaneously twists and untwists in a tangential direction. These effects may lead to destruction of the channel. The provided example shows a rather complex structure of flow oscillations with superposition of oscillations of different types.

Let us consider criteria underlying the experimental determination of the flow stability boundary, as well as problems associated with the unambiguous prediction of the stability boundary in experimental investigations.

References 2, 8, 9, 13, 15, 32, and 55, as well as others, offer the criteria that are most widely used for determining the experimental boundary of the oscillatory instability initiation:

1. Small steady-state oscillations of the inlet flow rate are initiated at a typical frequency.

2. The inlet flow rate oscillation amplitude is extrapolated up to zero value at the amplitude change, depending on the input power.

3. There is a certain level of amplitudes of the inlet flow rate steady-state oscillations. In different published sources, this level corresponds to 5%, 10%, 20%, and 30% of the average steady-state flow rate in a channel or loop.

4. Flow rate oscillations with an increasing amplitude are initiated.

5. There is an appearance of a bend on the curve representing the dependence of the flow rate oscillation amplitude on the input power that characterizes a transition from slow to fast growth of the oscillation amplitude when power is increasing. An example of the stability

boundary determination using the preceding criterion [13] is shown in Figure 4.2. A typical oscillogram of the experimental channel inlet flow rate oscillations at different fixed power levels [13] is shown in Figure 4.3.

6. Periodic burnouts are initiated in the channel with coolant flow oscillations.

More rarely, such statistical methods as, for example, the least squares method for determining the line separating the regions with stable and unstable regimes, or the flow noise dispersion analysis, are applied for determining the stability boundary. An experimental method that is being developed for determining the flow stability boundary by flow parameter noises will be described later.

The preceding criteria clearly show that the stability boundary determined on their basis may differ in a rather wide range. Even the procedure of criteria selection is based on different approaches. So, the first and second criteria have been selected stemming from the initiation of system fluctuations. This condition relates also to small stable amplitudes of the inlet flow rate fluctuations (about 5% of the average value) in the third criterion.

The fourth, fifth, and, especially, sixth criteria of the stability boundary determination have been selected to meet safety and thermal reliability of power equipment. In part, the preceding condition relates to large

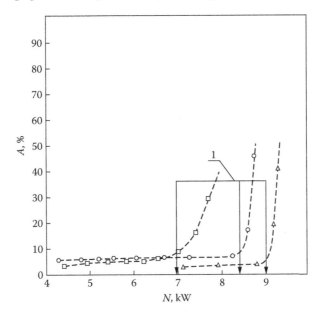

FIGURE 4.2
Dependence of the inlet flow rate fluctuations' relative amplitude on the channel power [13]. 1: stability boundary; $\bigcirc$: $\Delta i_{sub} = 13.7$ kJ/kg; $\square$: $\Delta i_{sub} = 24.4$ kJ/kg; $\triangle$: $\Delta i_{sub} = 51.2$ kJ/kg.

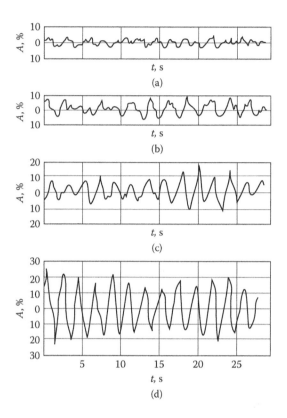

FIGURE 4.3
Coolant inlet flow rate fluctuations at different power levels in the test section [13]. Δi_{sub} = 24.4 kJ/kg; (a) N = 4.4 kW; (b) N = 5.85 kW; (c) N = 6.825 kW; (d) N = 7.1 kW.

amplitudes of steady-state fluctuations in the third criterion. It is the so-called "dangerous" stability boundary from thermal reliability considerations.

Sometimes, experimental determination of the stability boundary applying some of the previously listed criteria is confronted with difficulties. For example, the presence of constant disturbances in full-scale or test facilities— namely, natural noise due to coolant boiling caused by stochastic nature of boiling and change of flow regimes, disturbances introduced by circulating pumps or occurring in steam–water mixing heat exchangers upstream from the test section, periodic switching-on of automatic control systems, etc.—may cause the slowly converging flow rate fluctuations with a typical frequency in the stable region, which is sufficiently remote from the stability boundary. It should be noted that the initial amplitude of these fluctuations may be found to rise when approaching the stability boundary. Figure 4.3 may be offered as an example of the beginning of flow fluctuations' development at low power levels. Therefore, stability boundary determination for real systems using the first criterion may yield a substantial error.

Application of the second criteria may lead to uncertainty in stability boundary determination. Apart from the disadvantage of the first criterion, extrapolation to zero of the inlet flow rate fluctuation amplitude is frequently difficult due to the nonlinear dependence of the flow rate fluctuation amplitude growth on power (see, for example, Figure 4.2).

Determination of the stability boundary by the third criterion is a simple procedure, but the criterion itself is sufficiently conventional. Indeed, there is no clear justification of the flow rate fluctuation amplitude boundary value, which is determined by different investigators arbitrarily. Secondly, depending on the nonlinear system properties, the growth rate of the flow rate fluctuation amplitude may be different when moving off the boundary into the instability region. Therefore, for different circulating systems in the instability region, the distance from the stability boundary into the instability region at the same relative flow rate fluctuation amplitude may be different within the range of parameters in question. This complicates comparison of the experimentally determined stability boundary with that predicted using a linear model.

The fourth criterion, based on the determination of the onset of fluctuations with the increasing amplitude, is conventional, since oscillatory instability of two-phase flows in power equipment shows, as a rule, a limited amplitude growth due to the nonlinear feedbacks. Evidently, the fourth criterion in this case should be understood as a condition promoting a sharp growth of the flow rate steady-state fluctuation amplitude following either a small power change or reduction of channel throttling. Figure 4.4 [22] illustrates a sharp rise of the fluctuation amplitude at a small increase of power.

Practically, the fourth and fifth criteria have much in common; the latter has the purpose of determining the stability boundary by the point where the curve bends from slow to rapid rise of the flow rate fluctuation amplitude with the increasing power (see Figure 4.2). These criteria also have a number of drawbacks.

First, not all two-phase coolant systems in the instability region demonstrate a sharp change in the rate of the fluctuation amplitude growth when power is increasing. Experiments show that rather frequently the amplitude is found to rise smoothly (type B). Second, as one can see from Figure 4.2, a sudden bend on the curve showing the fluctuation amplitude change as a function of power may happen at a large flow rate fluctuation amplitude (about 10% of the average flow rate), which is already dangerous for thermal reliability of power-generating equipment.

Third, when the flow rate fluctuation amplitude is high, the flow instability development becomes influenced by dynamic characteristics of the model test facility, which may substantially differ from those of a full-scale plant. Therefore, even in the case of complete similarity between the investigated experimental and full-scale channels, the results of the stability boundary determination on the model test facility may substantially differ not only quantitatively, but also qualitatively, if, for instance, there is no constant

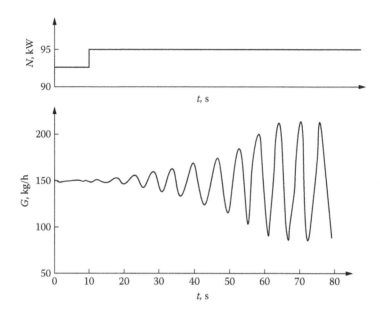

FIGURE 4.4
Coolant flow fluctuations' development at the test channel inlet with small power increments [22].

pressure drop in the parallel-channel system, no constant inlet enthalpy, or when circulating loops of test facilities incorporate compressible volumes (missing in the full-scale plants) to balance fluctuations of pressure and flow rate.

Fourth, quite frequently, a different parameter instead of power (e.g., inlet enthalpy) is varied in the experiments devoted to stability boundary determination. The growth rate of the inlet flow rate fluctuation amplitude may be different depending on the varied parameter. For example, it was previously shown (see Figure 1.21 in Chapter 1) that when the inlet coolant subcooling is increasing, the flow rate fluctuation amplitude, due to the ambiguous effect of subcooling on the stability boundary, may first grow and then decrease. This effect may contribute additional uncertainty to finding the stability boundary using the fourth and fifth criteria. It should be noted that the last three constraints may be manifested when determining the stability boundary on the basis of the third criterion with a high boundary amplitude of the flow rate fluctuation.

The sixth criterion, which makes the onset of periodic burnouts a condition for determining the stability boundary, is unacceptable. Indeed, periodic burnouts appear in the region of the developed instability at very high flow rate fluctuation amplitudes and lead to a rapid destruction of the heated channel. For instance, it has been shown [9] that a periodic burnout in the experimental channel was developing at a power amounting to

40% of the steady-state critical one, when the flow rate fluctuation amplitude reached high values in presence of the oscillatory instability. A typical example of the effect of instability of the critical heat flux (CHF) is given in Figure 4.5 [63], where CHFs are shown in dependence on the exit bulk quality at identical flow parameters in the nonfluctuating regime provided by strong inlet throttling, and in the fluctuating regime.

At low exit bulk qualities ($x_e \approx 0.05$) in the fluctuating regime, the minimal thermal load was observed to be over two times lower than that in the nonfluctuating regime under otherwise equal conditions. However, thermal reliability of the channel may be impaired at lower amplitudes of flow rate fluctuations due to the long-acting periodic thermal stresses or due to other reasons. Then, the condition of the onset of periodic burnouts may not be the minimum "dangerous" flow stability boundary.

Thus, a brief survey of the existing approaches to the evaluation of the experimentally determined two-phase flow stability boundary shows that

1. Experimentalists employ two approaches when determining the stability boundary: (a) the boundary is defined as a condition of the onset of steady-state flow fluctuations with a typical frequency— that is, the boundary of asymptotic stability ("in the small") of the steady-state regime, and (b) as a condition of the initiation of flow fluctuations, which are dangerous for the system thermal

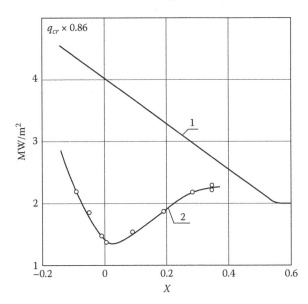

FIGURE 4.5

Comparison of CHFs in fluctuating and nonfluctuating regimes of the electrically heated tube [63]. $\Delta_{in} = 18$ mm with P = 9.1 MPa and $\rho W = 750$ kg/m²s; l: nonfluctuating regime; 2: fluctuating regime.

reliability, the so-called practical or "dangerous" stability boundary. The indicator of the dangerous stability boundary is specific for
each concrete system and is not determined by a preset fluctuation
amplitude. The practical stability boundary may be determined for
a concrete system in the following way: Depending on the required
channel life at the known fluctuation frequency, the permissible
fluctuation amplitude is obtained and then is used to determine the
practical stability boundary in the range of flow parameters.

2. Accuracy of the experimental determination of the onset of flow
   steady-state fluctuations with a typical frequency due to the noise
   present in the two-phase flow systems and due to the system
   nonlinearity may, in some cases, be sufficiently low and differ from
   the theoretically obtained asymptotic stability boundary.

3. When comparing stability boundaries determined numerically with
   those obtained experimentally by various authors, the knowledge
   that is required in addition to the design and flow parameters of
   a thermophysical test facility refers to (a) criteria of determining
   the stability boundary under given experimental conditions,
   (b) the habit of the coolant flow rate fluctuation amplitude with flow
   parameters' variation in the unstable region, and (c) the technique
   used for conducting experimental investigations on the stability
   boundary determination.

4.3 Experimental Determination of Thermal-Hydraulic Stability Boundaries of a Flow Using Operating Parameter Noise

A promising approach to experimental determination of the flow stability
boundary is being developed by one of the authors of references 100 and 101.
It is based on the use of the operating parameter noise for the purpose. This
can facilitate solving two problems:

1. The requirement of the steam-generating channel dynamic isolation
   prevents the test facility meant for investigating static problems
   to be used for stability investigations. The constraints may be
   removed if the stability boundary study is limited to that in the
   small, neglecting the specifics of instability development and self-
   oscillation amplitudes. The low-frequency noises of inlet coolant
   flow rate and pressure drop in the channel connected to the test
   facility without hydrodynamic pressure isolation make it possible
   experimentally to find a numerical value of the left-hand part of

the characteristic equation of the channel, depending on frequency. Further, application of the generally accepted techniques permits determination of the steam generator stability boundary in the small and solution of problems related to the stability margin.

2. Inferior reliability of the steam-generating channel thermal-hydraulic stability boundary determination by numerical methods is, in particular, due to approximate correlations for friction, steam void fraction, and heat transfer. Insufficiently reliable predictions are also due to the fact that, in the process of power-generating equipment operation, heat flux distribution along the channel length changes, as well as the boiling and friction conditions in the two-phase sections because of the uncontrolled thicknesses, structure, and composition of deposits. These peculiarities of real operation cannot be fully accounted for either in predictive or in physical models of the test channels, which above all fail to model dynamic behavior of the full-scale loop. Also, the experimental steam-generating channels are difficult to model due to a lower power in the model channel (as compared to the real one) and its changed geometry. Finally, there may be no full confidence in correctness of the extrapolation of both numerical and experimental results to the full-scale steam-generating channel. In such a situation, it is very tempting to possess a diagnostic method ensuring effective monitoring of the thermal-hydraulic stability margin during the current steady-state regime in the process of normal operation of the full-scale plant.

Now, let us consider a steam-generating channel with the inlet local pressure drop. Theoretically, the channel may be brought to the stability boundary by changing the inlet pressure drop. The boundary inlet pressure drop $\Delta P_{in.Bn}$ will correspond to the boundary state of the channel. Formally, it may even be negative. The difference in the real and boundary values of this pressure drop should be regarded as the stability margin. From the equations of motion that describe the steam-generating channel coolant state using the commonly accepted assumptions, upon linearization and Laplace transformation at zero initial conditions and taking the relationships

$$P_1 - P'_1 = \Delta P_{in}\,(G_{in}),$$

$$i_{in} = \text{const.}$$

into account,

$$\delta \tilde{P}_1 - \delta \tilde{P}_2 = \left[\frac{2\Delta P_{in}}{G} + \Pi(s)\right]\delta \tilde{G}_{in} \tag{4.1}$$

may be derived for the dynamically isolated steam-generating channel, where $\delta\tilde{x}$ denotes the Laplace-transformed noises of respective parameters x, G is the steady-state channel coolant flow rate, and G_{in} is the inlet channel unsteady flow rate. It has been taken into account that, for the quadratic inlet resistance

$$\frac{d\Delta P_{in}}{dG} = 2\frac{\Delta P_{in}}{G},$$

the expression

$$\frac{\delta P_1 - \delta P_2}{\delta G_{in}} = \frac{2\Delta P_{in}}{G} + \Pi(S)$$

represents the "channel flow rate-pressure drop" transfer function; S is the Laplace transformation parameter. The characteristic equation for analyzing the interchannel instability ($\delta P_1(t) = 0$, $\delta P_2(t) = 0$) looks as follows:

$$\Delta P_{in} + \frac{G}{2}\Pi(S) = 0. \tag{4.2}$$

Relationship (4.1) may also be obtained from the following simple considerations. The channel, located downstream relative to the inlet resistance, may be considered as a linear "black box" with the inlet δG_{in} and outlet $\delta\tilde{P}_1 - \delta\tilde{P}_2$. The relation between the inlet and outlet will be expressed via the transfer function

$$\delta\tilde{P}_1 - \delta\tilde{P}_2 = \Pi(S)\cdot\delta\tilde{G}_{in}. \tag{4.3}$$

From (4.3) and the linearized relationship for the inlet pressure drop

$$\delta\tilde{P}_1 - \delta\tilde{P}_1' = \frac{2\Delta P_{in}}{G}\delta\tilde{G}_{in},$$

expression (4.1) follows.

It is evident that the left-hand part of the characteristic equation coincides with the channel flow rate-pressure drop transfer function within the accuracy of the multiplier G/2:

$$\overline{\Pi}(S) = K\frac{\delta\tilde{P}_1 - \delta\tilde{P}_2}{\delta\tilde{G}_{in}},$$

where K = G/2.

The characteristic function,

$$\overline{\Pi}(S) = \Delta P_{in} + K\Pi(S),$$

upon the $S = j\omega$ substitution, may be obtained empirically by spectral analysis of the operation parameter noise [99]:

$$\overline{\Pi}(j\omega) = K \frac{S_{G\Delta P}(j\omega)}{S_{GG}(\omega)}, \qquad (4.4)$$

where $S_{G\Delta P}(j\omega)$ is the reciprocal spectral density of noise power $\delta G_{in}(t)$ and $\delta \Delta P(t) = \delta P_1(t) - \delta P_2(t)$; $S_{GG}(\omega)$ is the self-spectral density of the noise power $\delta G_{in}(t)$. These noises are, in fact, low-frequency stochastic fluctuations of flow parameters around their steady-state values. At the stability boundary, we have

$$\mathrm{Re}\,[\overline{\Pi}(j\omega_0)] = 0;$$

$$\mathrm{Im}\,[\overline{\Pi}(j\omega_0)] = 0. \qquad (4.5)$$

In the steady-state condition,

$$\mathrm{Re}\,[\overline{\Pi}(j\omega_0)] \neq 0, \qquad (4.6)$$

which ensures the nonzero margin for the steam-generating channel thermal-hydraulic stability. The drop of ΔP_{in} down to $\Delta P_{in.Bn.}$ corresponds to Re $[\overline{\Pi}(j\omega_0)]$ tending to zero. Practically, the margin is determined on the basis of the real axis section from the origin of coordinates of the complex plane $(0, j_0)$ up to the point of the axis crossing with the hodograph of the characteristic equation left-hand part.

Technical feasibility of the suggested method has been proved in laboratory conditions. The test facility represented a vertical closed loop containing a steam generator and condenser in the lower and upper parts, respectively. The steam generator was modeled, when required, either by two parallel, identical thin-walled joule-heated tubes, or by a single austenite Cr-Ni tube. The condenser was produced in the form of a multichannel construction of the "pipe within a pipe" type. Distilled water was used as the coolant. To ensure forced or combined coolant circulation, a pump could be connected to the loop. The quick-response tachometer with a tangential turbine was used as the primary flow transducer at the steam generator inlet, while an induction pickup of the DDI-20 type was used as the channel pressure drop transducer in combination with the IVP-2 converter. The outputs of both measuring channels were connected to the inputs of the deduction device based on the integral operational amplifier KI40UD8. The online logging of the flow parameter measured values was done using the NO67 analogous multichannel magntitograph.

The experimental technique was as follows. Using the results of preliminary tests of the two-channel steam generator, the boundary of the interchannel thermal-hydraulic stability areas was found to coincide with the stability boundary of a single dynamically isolated steam-generating channel. The experiments were conducted under different flow conditions at pressures ranging up to 10 MPa, mass velocities from 100 to 1500 kg/m²s, subcooling below the saturation temperature at the steam-generating channel inlet from 0 up to 150 K, and specific heat fluxes of up to 0.8 MW/m². Then, the single channel (without dynamic isolation, which is a prerequisite to the thermal-hydraulic instability onset in a steam-generating channel) was successively brought to different steady-state conditions of the stably operating loop. In each of the investigated steady-state conditions, the flow rate and pressure drop noises, after their filtration in the 0.01–2 Hz frequency range (covering the region of resonance frequencies ω_0, determined by the coolant transport through the steam-generating channel heated section) were recorded and subjected to spectral analysis according to (4.4), using the fast Fourier transformation algorithm [99]. Based on the results of the spectral analysis, the hodograph of the characteristic function (the left-hand part of the characteristic equation; see Figure 4.6) was constructed.

With a relative error of 12.5% of the spectral analysis, its frequency resolution was 0.156 Hz. Considering that, for frequencies a priori above 2 Hz the coherence function of the analyzed noises is close to zero as a rule, the high-frequency part of the hodograph may be unreliable. Therefore, the 0–2 Hz segment of the hodograph, which contains the basic resonance frequency,

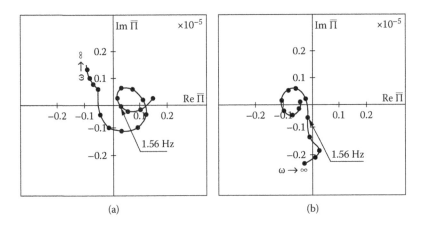

(a)   (b)

FIGURE 4.6
Hodograph of the $\Pi(jw)$ function for the steam-generating channel. (a) $T_{in} = 225°C$, $P = 6$ MPa; $q = 0.7$ MW/m²; ρW: 225 kg/m²s; (b) $T_{in} = 130°C$, $P = 6$ MPa, $q = 0.7$ MW/m²; $\rho W = 225$ kg/m²s.

should be controlled for the analysis of the low-frequency thermal-hydraulic stability. The steady-state operating conditions of the single dynamically nonisolated channel, under which the hodograph Π (jω) crossed the origin of coordinates of the complex plane, were classified as boundary (with respect to the dynamically isolated steam-generating channels) and marked as benchmarks in the "mass velocity-subcooling" plane of parameters.

Figure 4.7 illustrates the thermal-hydraulic stability boundary (shaded toward the stable steady-state conditions) determined according to the experimental results for a two-channel steam generator with a specific heat flux of 0.7 MW/m² and system pressure of 6 MPa.

The half-dark circles denote boundary points derived from the spectral analysis results for the flow rate and pressure drop noise across the single steam-generating channel and satisfying (4.5). The dark circles designate the conditions where the hodograph did not cover the origin of coordinates $(0, j_0)$, as is shown in Figure 4.6(b); this corresponds to stable operation of the two-channel steam generator. Light circles stand for the conditions where the hodograph covered the origin of coordinates $(0, j_0)$, as is shown in Figure 4.6(a); this corresponds to unstable operation of the two-channel steam generator. An opportunity of increasing accuracy in determination of thermal-hydraulic stability boundaries for a steam-generating channel using the operating parameter noises is associated with reducing the instrumental error of measuring devices used for the purpose, while the random error resulting from spectral calculations using the finite segments of noise realization may be reduced by selecting a larger number of averaging points.

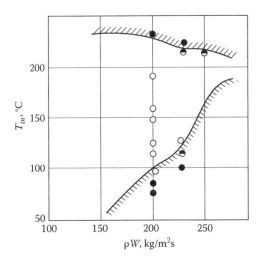

FIGURE 4.7
Experimental stability boundaries of a steam-generating channel; P = 6 MPa; q = 0.7 MW/m²; ●: stable steady-state regimes; ○: unstable steady-state regimes; ◐ – stability boundary regimes.

Application of the method in real operating conditions of a nuclear power plant presents no principal difficulties. However, it requires several additional conditions to be taken into account. For example, in contrast to the laboratory conditions, the diagnosed reactor core channel is influenced not only by the inlet flow rate and pressure drop disturbances, but also by additional ones such as low-frequency noise power $\delta N(t)$, which, according to the dynamic coupling "inlet flow rate—steam void fraction—reactivity—power" is related to the $\delta G(t)$ noise. In this case, the process channel as the object of diagnostics has two correlated inputs and one output. In contrast to the "one input—one exit" model of a steam-generating channel, the unity-approaching partial (but not general) coherence function should be used as the index of internal consistency between the noise measurements of flow parameters. The left-hand part of the characteristic equation for the analysis of the interchannel thermal-hydraulic instability may be obtained using noises of the three previously mentioned flow parameters. Considering the power noise and that of the channel coolant flow rate as correlated, in the general case, it may be written [100] that

$$S_{G\Delta P}(j\omega) = \Pi_{G\Delta P}(j\omega)\, S_{GG}(j\omega) + \Pi_{N\Delta P}(j\omega)\, S_{GN}(j\omega)$$

and

$$S_{N\Delta P}(j\omega) = \Pi_{G\Delta P}(j\omega)\, S_{NG}(j\omega) + \Pi_{N\Delta P}(j\omega)\, S_{NN}(j\omega).$$

From the system of equations, the expression for the unknown quantity is

$$\overline{\Pi}(j\omega) = \frac{G_{in}}{2}\, \frac{S_G\Delta P(j\omega)\cdot S_{NN}(\omega) - S_{GN}(j\omega)\cdot S_N\Delta P(j\omega)}{S_{GG}(\omega)\cdot S_{NN}(\omega) - S_{GN}(j\omega)\cdot S_{NG}(j\omega)}, \qquad (4.7)$$

where $S_{G\Delta P}(j\omega)$, $S_{NG}(j\omega)$, $S_{GN}(j\omega)$, and $S_{N\Delta P}(j\omega)$ are reciprocal spectral densities of power of respective pairs of operation parameter noises, while $S_{GG}(\omega)$ and $S_{NN}(\omega)$ are self-spectral power densities of respective noises.

Expression (4.7) may be used for determining the stability margin for the steam-generating channel within a parallel-channel system operating in the stable steady-state regime according to (4.6). With the presence of several correlated inlet disturbances of the diagnosed object, the expression for $\overline{\Pi}(j\omega)$ becomes somewhat more complicated. The working channels' thermal-hydraulic stability margin can be controlled in real time by combining the application of software for spectral analysis of the measured operation parameter noise realization with the use of devices for the diagnostic signal power spectral densities determination.

The described method for diagnosing thermal-hydraulic stability of the dynamically isolated steam-generating channels may probably serve as a basis for constructing diagnostic systems for both laboratory and industrial application.

4.4 The First Approximation Stability Investigation

Let us briefly review the basic numerical methods for determining the stability boundary and their inherent peculiarities and constraints.

It has been stated in the Introduction that two approaches are applied in theoretical investigations of thermal-hydraulic instability. The first approach deals with studying the behavior of the solution of a system of differential equations without search for exact or approximate solution. It is based on the utilization of results of the qualitative theory of differential equations and consists of the determination of critical parameters at which the disturbance path (operation process parameter variation over time in presence of disturbance) starts deviating from the undisturbed path when the disturbance is removed. In the context of thermal-hydraulic instability, the steady-state regime with parameters determined by the steady-state thermal-hydraulic equation is understood as the undisturbed flow.

The theory of stability created by A. M. Lyapunov is under constant development and at present is used (especially as it refers to the linear theory methods) in various branches of science and engineering.

The second approach deals with solving a system of nonstationary nonlinear partial differential equations with partial derivatives with given initial and boundary conditions. Investigations of stability (or, to be more precise, of the solution self-oscillating behavior) by means of direct numerical solution of the equations describing the nonstationary flow thermal hydrodynamics in the components of power-generating equipment acquire a wider application now due to the rapid development of computer techniques [8,9,11,15,29,54, etc.]. In many investigations of the two-phase thermal-hydraulic instability, this method is known as the method of analysis of a finite difference scheme for the nonlinear model.

Let us first consider the method of stability investigation in the first approximation. Some generally known definitions and notions of the stability theory are given in references 2, 3, 4, and 56.

Let parameter deviations from the steady-state condition δx_1 be described by a dynamic system of differential equations,

$$\frac{d\delta x_1}{dt} = \chi (t, \delta x_1,\ldots, \delta x_n); \quad (1 = 1,\ldots, n). \tag{4.8}$$

In terms of A. M. Lyapunov, such an equation is called the equation of disturbed motion.

Definition 1. The undisturbed motion is stable, if for each positive number ε, notwithstanding how small it is, there exists another positive number $\eta(\varepsilon)$, at which

$$|\delta x_1(t_0)| \leq \eta \qquad (4.9)$$

is met for all disturbed motions with $t = t_0$ at the initial moment, and

$$|\delta x_1(t)| < \varepsilon \qquad (4.10)$$

is satisfied for all $t > t_0$. ∎

Definition 2. If the undisturbed motion is stable and if the number η may be chosen so small that for all disturbed motions satisfying (4.9) the conditions

$$\lim_{t \to \infty} \delta x_1(t) = 0 \qquad (4.11)$$

will be met, then the undisturbed motion is called asymptotically stable. ∎

If region (4.9) of initial disturbances $\delta x_1(t)$, for which the asymptotic stability condition (4.11) is met, is not evaluated, then the asymptotic stability in the small is considered.

Let Q be a certain, a priori set region of changing variables δx_1 wherein the values of $\delta x_1(t_0)$ of the initial disturbances may lie.

Definition 3. The undisturbed motion is called asymptotically stable "in the large" (within the Q region), if it is stable and condition (4.11) is fulfilled for all $\delta x_1(t_0)$ within the Q region.

In particular, the Q region may be determined by the inequality

$$|\delta x_1(t_0)| \leq K$$

where K is the preset number. ∎

Definition 4. The undisturbed motion is called asymptotically stable "on the whole" if this motion is stable and condition (4.11) is fulfilled for any initial disturbances $\delta x_1(t_0)$, no matter how large they would be. ∎

The mathematical model for describing unsteady thermal-hydraulic processes in a heated channel with the two-phase coolant is in fact an essentially nonlinear system of partial differential equations. As a rule, stability investigations of the nonlinear system in the large and, even more so, on the whole are rather difficult and frequently represent an unsolvable problem. Therefore, for solving the majority of practical problems, investigations are limited to those of stability in the small (i.e., to those based on linear approximation of the initial nonlinear dynamic system). Such a method of stability investigation is known as the first approximation stability theory. The engineering experience of using the method of the two-phase flow stability boundary determination in the small with regard to thermal power plants shows that it is practicable for the analysis of qualitative effects of parameters on the stability boundary, and it is quite sufficient to provide quantitative estimates of the stability boundary for simple thermal-hydraulic schemes.

The first approximation methods of stability investigation are described in detail in references 2–4, 56, and 57. Here, in compliance with Morozov and Gerliga [2] and Goryachenko [3], the general principle and the procedure of an investigation for the assessment of possible errors introduced by assumptions and the reliability of comparing the predicted and real stability boundaries are given.

Let dynamics of the steam-generating channel be described by a system of nonlinear differential partial equations as deviations from the equilibrium

$$\frac{\partial \delta x_1}{\partial t} = F_1\left(z, \delta x_1, ..., \delta x_n, \frac{\partial \delta x_1}{\partial z}, ..., \frac{\partial \delta x_n}{\partial z}\right), \tag{4.12}$$

where $\delta x_1 = x_1 - x_{1o}$ ($1 = 1,..., n$). $\delta x_1 = 0$ corresponds to the undisturbed process.

For investigating stability of the solution for the system of equation (4.12), its right-hand parts are divided into linear and nonlinear components and written as

$$\frac{\partial \delta x}{\partial t} = A(z)\delta x + B(z)\frac{\partial \delta x}{\partial z} + \phi, \tag{4.13}$$

where $A(z)$ and $B(z)$ are square matrices of coefficients and ϕ is the column matrix of nonlinearities.

The linearized system of equations

$$\frac{\partial \delta x}{\partial t} = A(z)\delta x + B(z)\frac{\partial \delta x}{\partial z} \tag{4.14}$$

is called the first approximation system of equations and is further used for the evaluation of stability in the small of the initial nonlinear system (4.12). The boundary conditions are also subject to such linearization. It is supposed that all nonlinearities incorporated into the mathematical model can be linearized. However, as Morozov and Gerliga [2], linearization is impossible for the description of some thermal-hydraulics (e.g., for shock waves arising at passing of the two-phase flow parameters over the local sound velocity, etc.).

Solution of the linear system (4.14) with appropriate boundary conditions may be represented as

$$\delta x(t,z) = \sum_{k=1}^{\infty} e^{S_k t} U_k(z),$$ (4.15)

where $U_k(z)$ for the fixed z is the constant dependent on the boundary and initial conditions (the latter being the function of initial disturbances), while S_k is the roots of characteristic polynomial complex in the general case

$$S_k = \lambda_k + j\omega_k.$$ (4.16)

The values of S_k are defined by parameters of the steady-state system.

Then, the solution of (4.15) may be

$$\delta x(t,z) = \sum_{k=1}^{\infty} U_k(z) \cdot e^{\lambda_k t} (\cos \omega_k t + j \sin \omega_k t).$$ (4.17)

As one can see from (4.17), the asymptotic stability (or instability) of the linear system is determined by the sign of the real part of the characteristic polynomial roots. If all $\lambda_k < 0$, then the system (4.14) is asymptotically stable; if at least one $\lambda_k > 0$, it is unstable. It follows from (4.17) for the linear system that if a system is stable with small initial disturbances, then it will be stable at any large disturbances (i.e., on the whole). However, it is not true for the initial nonlinear system. When the linear system is asymptotically stable (4.14) on the whole, theory guarantees asymptotic stability of the initial nonlinear system (4.12) only in the small and does not guarantee it to be stable in the large. This is because only at small initial disturbances is behavior of the nonlinear system found to differ insignificantly from its linear approximation.

The subsequent investigations into the first approximation system stability envisage determination of the sign of the real part of the roots of characteristic polynomial of the linear system. When doing this, it is not mandatory to solve the characteristic equation and determine its roots. There are different methods providing necessary and sufficient conditions for determining the region of parameters where all roots of the characteristic polynomial have negative real parts. Usually, the system of linear equations and boundary conditions are Laplace transformed with zero initial conditions, thus obtaining the characteristic function $\Pi(S)$. Actually, it is equivalent to substituting a partial solution $e^{S_k t}$ into the obtained system of linear equations. For the case of asymptotic stability in the small, it is necessary and sufficient that all the roots of the characteristic equation $\Pi(S) = 0$ be located in the left-hand half-plane S that does not incorporate the imaginary axis.

To determine the region of parameters where stability of the linear system is ensured, a special method of separating the areas of stability—the D-division method—is widely used. It is described in detail in references 2, 3, and 57, so we will not dwell on it. The procedure of its application is well established and causes no fundamental difficulties when investigating stability of steam-generating channels. It should be noted that this method permits outlining of the region of parameters where all roots of the characteristic equation have negative real parts, thereby determining stability boundaries of the linear system.

Now, let us consider how the thus obtained system stability boundary determines practical stability of steam-generating channels of test facilities or real equipment.

If the stability boundary is of the "A" type, then with an adequate mathematical description of the process, the stability boundary in the small (predicted using either a nonlinear or linearized model) is generally close to the experimentally derived boundary with smooth variation of parameters and with no forced disturbances in the test loop.

As has been stated before, the investigation of stability in the small does not permit determination of the region of permissible initial disturbances at which the system is asymptotically stable. The maximum permissible initial disturbances may be smaller than those present in real boiling systems. In such cases, the real system, though theoretically linearly stable within the considered range, is practically unstable. This circumstance may substantially deteriorate the range of stable operation of the system. Because of great difficulties with theoretical substantiation of behavior of the dynamic system near the boundary and determination of the type of the stability area boundary (safe or unsafe), quite frequently the only criterion of checking conformity of the predicted linear stability boundary to the real one is an experimental investigation of thermal-hydraulic flow stability on thermophysical test facilities. This being the case, it is desirable to model not only the steam-generating channel and the boundary conditions therein, but also the disturbances that may present in the real system, as close to the real systems as possible.

On the other hand, some systems that, according to the linear theory, are within the instability region near its boundary may be considered as practically stable because the maximum amplitude of fluctuations happens to be too small ("B" type). Such instability is very difficult to identify experimentally on thermophysical test facilities because the level of parameter fluctuation may happen to be lower than natural noises always present in a real system. This may complicate comparison of theoretical and experimental stability boundaries and lead to incorrect conclusions on the accuracy of predictions based on the linear model.

A simple example [56] illustrates this. With regard to this work, let us consider the following equation:

$$\frac{dx}{dt} = \lambda^2 x - x^3 . \tag{4.18}$$

The root of the characteristics equation of the first approximation is λ^2; that is, it has the positive real part and the undisturbed motion is unstable. However, it follows from Malkin [56] that with any initial disturbances of x, its value tends over time either to $+\lambda$ or to $-\lambda$. If λ is very small, the disturbed motion will practically tend to the equilibrium state, and the system will be stable.

On the other hand, according to Malkin [56], if the equation of motion is written as

$$\frac{dx}{dt} = -\lambda^2 x + x^3, \tag{4.19}$$

the root of the characteristic equation will be negative, and the undisturbed motion will be asymptotically stable, according to Lyapunov. In this case, as it follows from Malkin [56], if the value of λ is very small, then the motion may be considered as practically unstable since with initial disturbance above $\lambda \times \lim_{t \to \infty} x \to \pm\infty$. It directly follows from (4.19), as with $|x| > \lambda$, the inequality $x(dx/dt) > 0$ holds true.

Though the offered example may be not typical for the known types of thermal-hydraulic two-phase flow instabilities, it cannot be excluded that such a pattern of the process development may be encountered in new thermal-hydraulic circuits.

Now, let us dwell on another aspect of practical system instability, which is not directly associated with Lyapunov's stability or instability. As was noted in Morozov and Gerliga [2], such dynamic characteristics as transient oscillations, the time of their decay, and the steam-generating channel resonance characteristics are of great importance for safe operation of power-generating systems. With unfavorable dynamic characteristics, the steady-state stability and even stability on the whole, as was noted in Vdovin [58], does not guarantee safe operation, since small disturbances present in the system may cause impermissibly large, slowly converging flow rate fluctuations in the steam-generating channel. The resonance characteristics of the channel become most pronounced when approaching the stability boundary. In boiling water reactor (BWR) conditions, periodic flow rate oscillation packages were observed in the core channels when the feed water relay control system was actuated. Under these conditions, the steady-state regime is not practically stable irrespective of its stability according to Lyapunov.

As an illustration, here is an example [58] where characteristics of a BWR with a large positive steam coefficient of reactivity are analyzed. The existence of such a feedback in combination with heat transfer inertia in the core may cause impermissible power surge and activation of emergency protection in case of small disturbances of input variables. It may result in sufficiently rigid requirements to permissible disturbances (or to the rate of their introduction) with regard to both coolant temperature and velocity at the core inlet.

With some limiting assumptions (upon introduction of a disturbance, the inlet flow rate and enthalpy are assumed to be constant), a diagram was plotted in Vdovin [58] and is offered here in Figure 4.8. It shows the boundaries of permissible and impermissible (leading to the emergency protection activation) disturbances depending on density of the reactivity coefficient γ_ρ. One can see from Figure 4.8 that when $\gamma_\rho = 10\beta$ (for the VK-50 reactor, as is noted in Vdovin [58], in some test conditions $\gamma_\rho \approx 30\beta$), the steady-state regime loses its practical stability, since the permissible overall disturbance ($\delta G/G - \delta i_{in}/\Delta I$) shall not exceed 1.8% for the conditions described in Vdovin [58]. In this case, δG and δi_{in} are the inlet flow rate and enthalpy disturbances, respectively, while G and ΔI are the steady-state values of flow rate and enthalpy drop in the core. It should be noted that in the given example the equilibrium state is stable on the whole according to Lyapunov (i.e., at any large feedback coefficients).

Thus, a brief analysis of the first approximation method for stability investigation shows that

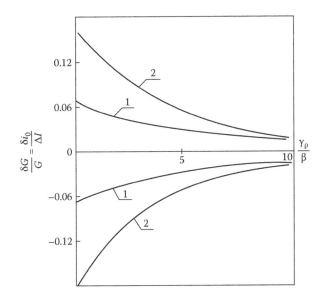

FIGURE 4.8
Boundary of permissible (1) and impermissible (2) disturbances.

1. The linear stability calculation permits conducting illustrative qualitative and quantitative analyses of the effect of various design and flow parameters on the stability boundary.

2. This stability analysis yields a quantitative assessment of the stability boundary that is close to the real one only for the nonlinear systems with the so-called "soft excitation," for which stability in the small also determines stability in the large. However, the correspondence between the stability boundary determined by the linear analysis and that of a real full-scale plant described by the investigated nonlinear system can be practically proved by comparison of the predicted boundary with that obtained experimentally on the thermophysical test facility capable of simulating the real system with real disturbances.

3. The linear approximation stability analysis fails to determine the nature of parameter variation in the unstable region. At the same time, for some nonlinear systems in the unstable region near the boundary, deviation of parameters from the steady-state values may be small, and the system can be practically stable. Under such conditions, exact experimental determination of the stability boundary on the thermophysical test facility, as has been shown previously, becomes problematic because of natural noises in the system, and it complicates the comparison of the predicted and experimental boundaries.

4. Unfavorable dynamic characteristics of the nonlinear systems may create dangerous operating conditions of power plants even with the existence of stability on the whole, according to Lyapunov, because in this case small disturbances may lead to the appearance of slowly converging parameter fluctuations with a large initial amplitude, thus making the system practically unstable.

These constraints suggest that the quantitative data on the stability boundary obtained on the basis of linear approximation should be treated with care.

4.5 Stability Investigations Based on Direct Numerical Solution of the Unsteady System of Nonlinear Equations

When an adequate mathematical model and numerical solution are available, stability investigations, based on the direct numerical integration of the unsteady nonlinear system of partial differential equations that describes thermal-hydraulics of the two-phase flow in the components of power equipment, have a number of significant advantages as compared to the methods of search for the boundary using the linear theory, and even to respective experimental investigations on model test and full-scale facilities.

In contrast to the linear methods of boundary determination, direct numerical solution of the nonlinear system allows modeling of real thermophysical test facilities and full-scale plants with the inherent system disturbances and feedbacks. Additionally, a numerical solution, being in fact a numerical experiment, permits determination of not only the stability boundary but also the nature of development of local and integral parameter fluctuations in the unstable region, as well as behavior of parameters in the stable region near the boundary influenced by the permanent system disturbances. This provides a more exact quantification of the asymptotic stability boundary, determination of the "dangerous" or "practical" stability boundary, and a more reliable comparison of predictions and experimental results to validate adequacy of the mathematical model.

In comparison with the experimental investigation, a numerical solution, as has been stated (e.g., in Patankar [62]),

helps to decrease the cost of research, especially when studying stability in complex systems

makes it possible to perform a large number of calculations varying some design and flow parameters in a wide range in ideal conditions and keeping other parameters unchanged, for the sake of selecting optimal conditions of increasing system stability (as it is a problem to make on the experimental facilities)

yields detailed information on spatial and time variation of all basic parameters, including those difficult to measure (as it is practically impossible to register by sensors during experiments)

provides knowledge on the time-dependent behavior of parameters in the entire range of solutions and a possibility of purposeful variation of design and flow parameters, and of initial and boundary conditions, discriminating ideally between the influence of different factors and thus revealing the physical mechanism of instability and the regularities in the effect of basic parameters on the stability boundary

permits modeling instability conditions of the real plant being designed, which is difficult under conditions of a model thermophysical test facility

However, despite these considerable advantages of the stability investigation techniques by the direct numerical solution, they have not found wide application yet in engineering practice, though a number of recent works describes computer codes and the results of investigation of thermal-hydraulic instability by this method, while dynamic investigations of transients and emergency situations in power plants have become customary in the engineering practice.

Apart from conventional difficulties associated with the numerical solution of a nonlinear system of differential equations of the two-phase flow, which will be considered later, additional problems of creating an adequate mathematical model for investigating thermal-hydraulic instability and its numerical realization on the computer are aggravated by the fact that parameters defining the system stability boundary are critical ones; in their region, the solution of the system of nonlinear equations may bifurcate or change the qualitative structure of the phase portrait. This leads to the situation when the solution near and at the stability boundary is very sensitive to small variations of the boundary and initial conditions, small system disturbances, type of solution on the approach to the boundary, and the oscillatory nature of parameters' variation in the unstable region. Therefore, the accuracy of stability boundary determination and parameters' variation in the unstable region using a numerical solution largely depends on:

1. An adequate description of thermal-hydraulic parameters' distribution in components of the considered plants
2. Substantiation of the choice of nonlinear correlations for closing the system of equations, which should reflect both static and dynamic characteristics of thermal-hydraulic processes
3. Selection of a scheme for the difference approximation of the initial equations and a method of their numerical solution, etc.

It should be noted that these requirements are more rigid than in the case of investigating transients and emergency situations by dynamic mathematical models in the stable region of parameters.

It should be noted that the existing mathematical theory of numerical solution of the nonlinear partial differential equations, as is stated in Roache [60], fails to provide a rigorous investigation of numerical stability, evaluation of the error, and convergence of the solution. Also, the numerical solution of thermal hydrodynamics becomes more difficult because of complex geometry, strong nonlinearity, substantial variation of liquid properties, and phase changes., As a rule, formalized principles cannot be used for building algorithms and computer codes for the numerical solution of complex thermal-hydrodynamic problems, and each successfully solved concrete problem is mostly due to the calculator's skill. Therefore, a numerical solution often must

be based on experimental results of thermal-physical investigations

rely on intuition and heuristics

use physical peculiarities of the processes to simplify equations and upgrade numerical methods and solution algorithms

employ various algorithmic "tricks"

As a rule, all this not only reduces computation time several times or essentially increases accuracy of solution, but sometimes determines the possibility of solving a concrete problem numerically. Let us illustrate some basic difficulties and constraints encountered when the thermal-hydraulic two-phase flow instability is investigated by the direct numerical solution of a nonlinear dynamic mathematical model.

The difference of the numerical solution of the mathematical model from characteristics of the real physical process may be determined by the following factors:

1. Validity of the mathematical model used (i.e., by the degree of its correspondence to the real thermal-hydraulic processes under investigation). Equally, this factor determines reliability of other methods for stability investigations (e.g., the methods of the linear theory)

2. Validity of building a discrete analog of the initial system of equations and correspondence of the discrete analog solution to that of the initial system

3. Selection of the optimal stable method of numerical solution of the obtained system of equations and optimal approximation of nonlinearities

4. Appearance of various effects of the algorithmic nature caused by discrete algorithm for describing continuous processes instead of physical peculiarities of the process

Now, let us briefly consider each of these factors, requiring careful substantiation in order to minimize the difference of the numerical solution from real parameters of the nonstationary thermal-hydraulic process.

The following analysis is of a qualitative nature, just to illustrate the difficulties and constraints to be overcome in the numerical solution of the problem of stability boundary determination and two-phase flow parameters' variation in the unstable region.

Validity of the initial mathematical model is the decisive factor of successful numerical solution of the flow stability problem. If such a model is inadequate to the process under investigation, no numerical method will yield true results.

The two-phase flow thermohydrodynamics obeys the basic laws of conservation used in single-phase hydromechanics. However, such equations are very complex in the case of two-phase flows and the inner forces determining phase interaction are not strictly defined. Therefore, the presently used mathematical models are unfit for the description of local parameters.

These difficulties are caused, as was mentioned in Wallis [72], by the infinite number of the interphase interface forms and two-phase flow regimes, flow prehistory effect on the interface structure, stability of the collective interaction of interphase surfaces and other, very often sudden factors. Theoretically, it is impossible to describe the behavior of two-phase system parameters not only for unsteady-, but also for most steady-state conditions. It should be noted [73] that now there are no complete experimental data on velocity fields, gas content, and turbulent characteristics for gas–liquid flows. Only individual problems (e.g., growth of a single bubble on the heated surface, film boiling, annular flow, etc.) may be solved by approximate theoretical analysis. Therefore, modification of the existing and development of new mathematical models of the two-phase flow are largely based on experimental investigations. During these investigations, flow regimes, their boundaries, and the averaged two-phase distribution of phases and velocities typical of the regimes in question are determined. Also, dependence of quality, friction, and heat transfer on the mean integral two-phase flow characteristics such as mass-averaged velocity, enthalpy, pressure, heat flux, channel geometry, etc. is studied.

Though significant success has been achieved in experimental and predictive two-phase flow investigations (e.g., in those devoted to flow regime boundaries [74] and other thermal-hydraulic characteristics determination), no reliable quantitative relations have been revealed yet. This is due to the fact that the flow regime boundary is separated by intermediate zones where physical processes are unknown, stability depends on many uncontrolled factors, and the experimental setting of the regimes is insufficiently reliable [75].

The dependencies defining dynamic and thermal nonequilibrium, as well as heat transfer and hydraulics, are related to the flow regime, and the form of these dependencies is different for each regime. For some processes, the structural form of dependencies is determined on the basis of simple physical hypotheses, while the constants and power exponents are refined in experimental investigations. For other processes, the dependencies represent empirical correlations valid only for the conditions of their derivation and not based on physical notions of the process.

These constraints of the theory necessitated the development of a scheme of transforming general local two-phase flow equations with indefinite functions of internal forces (which mainly reflect the interphase interaction) into a practically applicable form. Such a technique is described, for instance, in references 76–79. Here, one can find only a simplified description of these transformations as an illustration of the limitedness and, in some cases [72], even "unreliability" of the initial system of two-phase flow equations.

To make them practically applicable, the local equations of conservation for the two-phase flow are averaged with respect to a certain time interval (shorter than the characteristic time of the unsteady process) to eliminate the instantaneous oscillation components of basic parameters. Then, they are averaged with respect to the channel cross section to express the unknown functions of the interphase surface interactions in the integral form. Thus, one-dimensional spatial nonstationary partial differential equations expressing the laws of mass conservation, momentum, and energy are obtained for each phase.

Such a system of equations is known as the separated flow model (or two-fluid or heterogeneous flow model). The time- and cross-section averaged functions of the interphase interaction and interaction on the channel inner surfaces, as was stated earlier, cannot be found theoretically. However, they can be determined through experimental investigations of each flow regime. The equation of the separated flow model incorporates these correlations as additional empirical terms or various coefficients. For example, the empirical correlation for the interphase slip, or the relation obtained on the basis of the flow drift model, is used to determine the dynamic nonequilibrium.

Thus, one can see that in a two-phase flow each regime is described by different equations characterized by additional terms and coefficients derived experimentally. It is impossible to describe with sufficient accuracy, say, bubble and annular flow regimes by the same mathematical model.

The thus obtained mathematical model provides a basis for investigating unsteady thermal-hydraulic processes, including instability, in steam-generating channels and power plants employing two-phase coolant systems. These equations, together with that for the channel wall thermal conductivity, meet the conservation laws and, with some accuracy, determine the real temperatures, density, velocity, and flow section of each phase in both equilibrium and nonequilibrium conditions. The accuracy of these parameters' description depends on the following constraints:

1. Some members of the initial system of equations have either semi-empirical or empirical origin. These dependencies have been derived for steady-state conditions. They describe static characteristics well, but fail to reflect fully dynamic properties of the system. The example offered in Section 1.4 (Figure 1.26) in Chapter 1 proves it, as it shows that a description of surface boiling applying two different techniques yields different results in determination of the flow stability boundary by the numerical solution, though for the steady-state conditions these techniques yield similar results.

2. Under the unsteady conditions in the steam-generating channel in fixed coordinates with respect to the channel length, change of flow regimes may be observed in different periods of time, starting with the single-phase liquid flow with the subsequent change of all patterns of the two-phase flow and terminating in the steam flow. In this case, different equations describing flow regimes have to be used for some channel sections where change of flow regimes takes place. The differences in members of such equations may change even the type (hyperbolic, elliptic) of the system of equations. When a numerical solution for the unsteady conditions in question is sought, there arises an extremely complex problem of joining these equations, which is due to the change of the equation

type, function discontinuity, jumps in derivatives, and the presence of nonlinearities of different natures.

3. The empirical correlations may not only differ because of changes in two-phase flow regimes, but may also be of different form even for the same flow regime, depending on the channel geometry. Therefore, a computer code for numerical investigation of the two-phase flow instability in channels with specified geometries is not universal and often cannot be used without essential correction and checking for investigating instability in channels with different geometry.

Therefore, the full-scale use of the stratified flow model, which is an approximate model itself, is associated with considerable difficulties. Usually, the initial system of equations is subjected to further simplifications in order to solve particular unsteady problems. These simplifications are determined by the nature of the studied regime, range of parameters, intensity of disturbing influences, and other factors. The most frequently used simplifications are as follows:

1. The neglect of thermal nonequilibrium (for instance, the model does not describe nonequilibrium boiling)

2. The neglect of dynamic nonequilibrium (the relative velocity of phases is either disregarded or assumed constant)

3. The transformation (though with some loss of accuracy) of empirical or semi-empirical correlations incorporated into equations, their representation as a single dependence on continuous parameters (e.g., real quality, mass velocity, pressure), excluding their association with the flow regime

The latter simplification makes it possible to avoid the possible equation type change and function discontinuity when the channel flow regimes are changed.

When equations are simplified, two opposite tendencies arise. On the one hand, it is desirable to make the model as simple as possible to facilitate numerical solution and interpretation of the results obtained; on the other hand, the model should be as accurate as possible. Thus, the main constraint in simplification is the required accuracy.

Figure 4.9 [80] illustrates qualitative relation between accuracy and simplicity. The curve L confines from above and right a "domain" of all possible mathematical models. However, of practical interest are only the models designated M_1, M_2, and M_3 and located in the shadowed area where the models are computable and offer acceptable accuracy.

It is an extremely difficult problem to find the optimal relation between simplicity and accuracy, especially at an early stage of investigations when

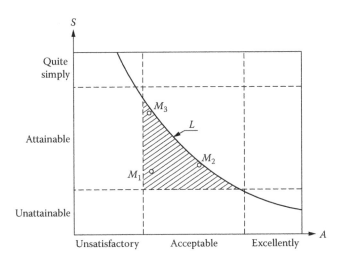

FIGURE 4.9
Selection of the optimal mathematical model [80]; S: simplicity; A: adequacy.

the dominating processes are not yet clear. In such a case, quite often one should rely on personal intuition, previous experience, and qualitative understanding of the processes.

No methodological prescriptions for simplifying and checking the adequacy of a mathematical model exist; nevertheless, some general recommendations commonly used in practice may be given:

1. It is necessary to develop specific mathematical models capable of solving concrete urgent problems instead of universal ones, applicable for description of all details of a real process. Depending on the investigation objective, specific models may be designed for describing separate processes, individual components, or a plant as a whole; and concrete operating modes and difference in the degree of detailing, accuracy, speed, and area of application. It is very important in view of the fact that investigators usually tend to construct a model that takes into account all details of a real process—often secondary ones—relying on the computer's ability to solve all real complexities. Such an approach, as is stated in Shannon [64], is erroneous not only because of extremely increased numerical solution difficulties and computation time, but also largely because of the danger of missing basic interrelations and dominating processes in a variety of secondary details.

2. A mathematical model of a steam-generating channel, especially for the reactor core conditions, is a complex system describing different processes. It is important to meet a number of conditions in order to select the model components and their interactions correctly. In this

respect, the principle of equal accuracy of mathematical description [61] may be mentioned. According to it, all the modeled interrelated physical processes should be described in approximately equal detail. A more detailed description of one process or element may lead not only to increased computation time and complicated program, but also to a lower accuracy of the mathematical model. This statement may be illustrated by an example [61] where a numerical solution of the accident induced by a sudden steam extraction in the single loop PWR was obtained. Figure 4.10 [61] shows numerical solutions of the accident based on different models. Curve 1 (Figure 4.10) was plotted on the basis of the point description of kinetics and one-dimensional thermal hydraulics, disregarding nonequilibrium boiling. Refining of the model by using spatial (one-dimensional) kinetic equations has yielded a solution (curve 2) that provides a worse description of the experiment (curve 4) than a more rough model (curve 1). The nonequilibrium boiling taken into account in the model in combination with spatial description of neutron kinetics gives a solution (curve 3) that practically coincides with the experiment. Such a behavior of solutions has a clear physical explanation, which is offered in Malkin and Filin [61] and illustrates the principle of equal accuracy.

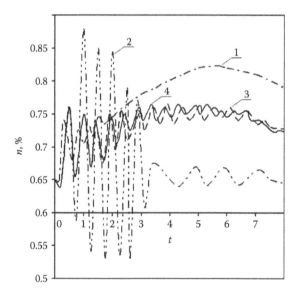

FIGURE 4.10

Results of the neutron flux power (n) calculation in emergency conditions with steam extraction disturbance [61]. 1: Point kinetics and thermal hydraulics model without surface boiling; 2: spatial kinetics and thermal hydraulics model without surface boiling; 3: spatial kinetics and thermal hydraulics model with surface boiling; 4: reactor experiment.

3. Validation of a mathematical model requires the results of the model-based numerical solution to be compared with experimental results obtained on a thermophysical test facility or a real plant. It should be mentioned that complexities of numerical solution of a detailed model may cause difficulties in validating adequacy of the initial differential mathematical model, which is additionally aggravated by the problems of accuracy and validity of its numerical solution. Therefore, it is desirable, especially at early stages of investigations, to check adequacy of the maximally simplified mathematical model, whose numerical solution presents no principal difficulties. Such a model should describe the test regimes sufficiently well. Afterward, the model may be refined and numerical and experimental studies closely coordinated. One should bear in mind [64] that the extent of understanding a phenomenon is inversely proportional to the number of unknowns. If one follows the natural striving to a more detailed model, the latter will become so complex that no interpretation of the results it yields will be possible.

4.5.1 Construction of the Discrete Analog of the Initial System of Differential Equations

Since the rules and methods of constructing discrete analogs for hydrodynamics and heat-mass exchange problems are clearly and completely presented in, for example, references 59, 60, and 62, this section offers some typical examples illustrating the difficulties that appear when a numerical solution is sought for the initial system of differential equations, as well as the caution the obtained results should be treated with. In the following, the basic concepts and typical examples from Roache [60] and Patankar [62] are explained in a simplified manner.

The initial equations describing thermal hydraulics express the conservation laws and they may be reduced to the general form [62] of

$$\frac{\partial}{\partial t}(\rho\phi) + div(\rho\bar{u}\phi) = div(\Gamma\nabla\phi) + S, \tag{4.20}$$

where ϕ is the variable, ρ is density, Γ is the diffusion factor, and S is the source term.

The form of Γ and S depends on the sense of variable ϕ; for instance, for the energy conservation equation, S is the volumetric rate of heat release, while Γ is the thermal conductivity coefficient. Enthalpy, temperature, velocity component, or turbulent kinetic energy may be the variable.

In the general case, the differential equation (4.20) includes four components—that is, the nonstationary, convective, diffusion, and source term. Some terms—say, the source, or convective, or the diffusion term—may be absent. For the steady-state problem, the nonstationary term is not used. If the diffusion term is not expressed in the gradient form, it may be related to the source term.

When $\phi \equiv 1$, $\Gamma = 0$, and $S = 0$, (4.20) becomes the continuity equation.

In the general case, the coordinates x, y, and z and time t are independent variables. For the problem of numerical investigation of two-phase flow self-oscillations, the differential equations are used in **one-dimensional** approximation (i.e., the variables depend on one spatial coordinate).

For numerical solution, each differential equation is transformed into the algebraic form, which is called the discrete analog of the initial equation, whereupon the obtained equations are solved.

It is useful to consider construction of the discrete analogs in the form of algebraic equations and the algorithms of their solution as two separate problems and to neglect the interaction effect for the sake of clarity.

These two problems are related to the notions of approximation and iterative convergences. The discrete analog correctly approximates the initial equations, if with $\Delta x \to 0$, $\Delta t \to 0$, the algebraic equation tends to the solution of the partial differential equation. The iterative convergence is the way to solve the finite-difference algebraic equation in the process of iterations. The condition of convergence is $\phi_{K+1} \approx \phi_K$.

If the numerical solution is stable, the fulfillment of these conditions for the two problems in question yields the solution of the initial problem.

Let us dwell on the methods of equation discretization.

Substitution of the initial equations by the discrete analog supposes the replacement of continuous information contained in the exact solution with the values of ϕ variable in the finite number of points in the region of independent variables called mesh nodes.

The algebraic equations of the discrete analog links values of the dependent variable ϕ in a certain group of adjacent mesh nodes.

When constructing these equations, the nature of ϕ variations between the nodes should be specified using some considerations or assumptions. The existence of a multiplicity of discrete analogs is the consequence of different assumptions concerning the type of the dependent variable profile (linear, fixed, exponential, etc.) and ways of analog constructing.

Some assumptions used when constructing the discrete analog, as is shown in Roache [60] and Patankar [62] and illustrated later, may change not only quantitative accuracy, but also qualitative behavior of the initial system. For instance, some kinds of discrete analogs introduce the effects (e.g., viscosity effect) absent in the initial equation or lead to the nonphysical change of parameters.

A discrete analog may be obtained using (a) the Taylor series expansion, (b) the variation method, (c) the weighted residuals method, (d) the control volume method, etc.

Apart from mathematical difficulties, the variation method suffers from a constraint that it is not inapplicable to all differential equations of interest and has not found wide application in treating problems of thermal hydrodynamics.

The essence of the weighted residuals method is in minimization of the difference between the exact and approximate solution specified by a polynomial. For minimization, a sequence of weighted functions is selected, and their forms define different versions of the method, each having its own name. The simplest and most frequently used weighted functions are W = 1 and W = 0. This variant of the method is known as the control volume method.

Let us consider in more detail the method that is most frequently used to deal with problems of thermal hydrodynamics, which employs the Taylor series expansion and the control volume method.

The essence of the method employing the Taylor series expansion is in approximation of derivatives in the differential equation by the finite number of the Taylor series terms. In this case, the central difference scheme of the finite-difference equation is obtained. When the difference analog is derived by the Taylor series expansion, it may seem that the condition of the approximation convergence is met automatically. But in reality it is not always so. The method yields good results when the dependence of ϕ on x is close to the polynomial. For example, the Taylor series expansion gives good approximation for the thermal conductivity equation. However, for the case of, say, exponential dependence of ϕ on x, as is the case with the convective-term energy equation, application of this method yields erroneous results. This is due to the fact that the truncated Taylor series approximates the exponent with insufficient exactness.

To illustrate this, here is an example [60]. Let us consider a one-dimensional steady-state energy equation in the dimensionless form with a convective term constant coefficient,

$$\frac{\partial \overline{\phi}}{\partial \overline{x}} = \frac{1}{Re} \frac{\partial^2 \overline{\phi}}{\partial \overline{x}^2}, \tag{4.21}$$

with the following boundary conditions:

$$\overline{\phi}(0) = 0; \quad \overline{\phi}(1) = 1; \quad 0 \le \overline{\phi} \le 1; \quad 0 \le \overline{x} \le 1. \tag{4.22}$$

Re is the Reynolds number.

Using the Taylor series expansion, the central finite difference scheme is obtained:

$$\frac{\overline{\phi}_{i+1} - \overline{\phi}_{i-1}}{2\Delta \overline{x}} = \frac{1}{Re} \frac{\overline{\phi}_{i+1} - 2\overline{\phi}_i + \overline{\phi}_{i-1}}{\Delta \overline{x}^2}. \tag{4.23}$$

When the $\Delta \overline{x} = 0.1$ step size is chosen, the boundary conditions in the discrete form are $\overline{\phi}_1 = 0; \overline{\phi}_{11} = 1$.

Figure 4.11 shows the exact solution of the initial equation (4.21) when Re = 1 and Re = 100, while Figure 4.12 offers the numerical solution of the finite difference equation (4.23) by the sweep method with the same Re numbers. One can see that the numerical solution with Re = 1 is smooth and coincides with the exact solution, while with Re = 100, typical sawtooth oscillations are formed. To be more exact, the solution of the initial equation is smooth with the latter Re value; however, it has a very large derivative. The sawtooth oscillations in this case, as is noted in Roache [60], are not caused by the iterative process instability and nonlinearity (the latter often blamed for the occurring difficulties), but in fact represent the exact steady-state solution of (4.23). Thus, in presence of the exponential growth of the

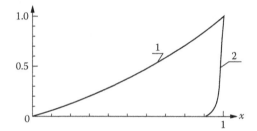

FIGURE 4.11
Exact solution of (4.21). 1: Re = 1; 2: Re = 100.

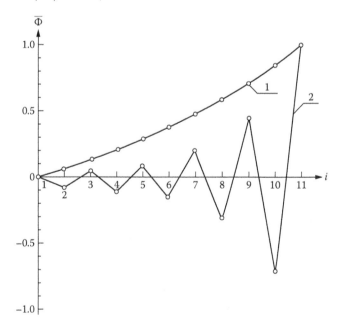

FIGURE 4.12
Solution of (4.23) with central differences. $\Delta\bar{x}$ = 0.1; ϕ = 1; u = const.; 1: Re = 1; 2: Re = 100.

derivative, its approximation by the truncated Taylor series may yield a discrete analog that does not correspond to the initial equation. For an unsteady equation, this effect is observed at each time step. Such problems belong to the class of the so-called rigid type problems, the theory of which is intensively developed at present and solving of which is confronted with great difficulties.

To eliminate oscillations, various artificial techniques for the wall boundary conditions are used: either the time steps are considerably reduced with respect to the spatial coordinate or the step size is varied with respect to the spatial coordinate. The radical way—switching to another finite-difference scheme—will be discussed later.

The control volume method is simple to interpret physically and its essence is as follows. The prediction area is subdivided into several nonintersecting control volumes comprising a node. To obtain a discrete analog, the distribution function of the dependent variable ϕ between the nodes is specified and the initial differential equation is integrated over each control volume.

One of the main features of the control volume method is in conservation, under certain conditions to be described, of mass integral quantities, momentum, and energy in any control volume and, hence, in the entire prediction area. This case satisfies the requirement of conservatism that consists in identical conservation in the finite-difference scheme of exact integral balances laid down in the initial equations for microvolumes. Thus, the numerical solution, even with the coarse mesh used, should satisfy the exact integral balances.

Like any other finite-difference technique, the control volume method yields a solution in the form of the variable ϕ values only in the nodes.

Prior to considering special conditions to be fulfilled when using the control volume method to achieve conservatism, it should be remembered that for the integration of equations by the control volume, the variable profile between the nodes should be specified. It is of importance that, for different variables, different profiles can be used and there is no need in using the same profile for all terms of the equation. The previously mentioned freedom in selecting the profiles of ϕ between the nodes results in numerous methods of constructing discrete analogs of initial equations. However, special requirements that ensure conservatism limit the range of potential profiles of ϕ when the control volume method is applied. These special conditions are as follows [62]:

1. A numerical solution should be physically plausible.
2. The solution should keep the integral balance.

The first condition means that the discrete analog solution should qualitatively coincide with the exact solution and may be easily checked for simple cases [62]. For example, in the case of a thermal conductivity problem without heat sources, the node temperature should not be beyond the limits of temperatures in the adjacent nodes. Such qualitative tests should be applied when constructing a discrete analog, and it is indicated that a researcher may achieve optimal results with the help of the numerical solution if physics of the studied processes is sensed well.

The second condition says that heat fluxes, mass flow rates, and momentum flows should correctly reflect the balance with appropriate sources and sinks for any number of nodes. The condition is satisfied with the equality of flows between the adjacent control volumes on the surface separating these volumes. An example of this condition violation is the construction of a discrete analog for the energy equation with convective and diffusion terms (4.21) using the linear profile between the nodes. The control volume method in the case of linear profile of ϕ leads to a discrete analog with the central differences (4.23). That is, in this case, the discrete analog coincides with that obtained by the Taylor series expansion for the variable and, with large Re numbers, yields a physically implausible solution (Figure 4.11). Thus, the ϕ variable linear profile between the nodes is unacceptable for the convective term, though it may be applied to the diffusion one.

The previously mentioned conditions are met if the value of ϕ for the convective term on the boundary of the control volume is equal to that in the adjacent node on the side opposite to the direction of mass velocity (i.e., with the stepwise variable profile). Such a difference analog is called the upwind difference scheme and is successfully used for the convection problems' numerical solution. Such a profile has a clear physical sense for this problem, since at a high mass velocity, when convection substantially exceeds diffusion, the heat flux through the control volume boundary is completely determined by the convective component in the direction of the coolant flow and equals zero in the opposite direction. With the linear profile of ϕ between the nodes, the flow between the adjacent volumes at the boundary of the control volume is directed both ways irrespective of the direction and value of mass velocity. This contradicts physics of the process. It should also be noted that the upwind difference scheme is very convenient in the case of coolant reversal, since the change of coolant direction is automatically taken into account in the difference scheme and requires no artificial techniques often used for the purpose—say, nodes' renumbering.

Here is another example of possible violation of the numerical solution physical plausibility caused by the form of the initial equations presentation. Numerical solution of energy equations has physical sense if the preset velocity and density fields satisfy the continuity equations. The flow field is frequently not really preset, but rather is calculated iteratively from the momentum equation; thus, with the use of the finite-difference analogs of energy and continuity equations and their simultaneous solution at intermediate iterations prior to the final convergence, the flow fields may not meet the continuity equation and physically erroneous solutions may be obtained. Therefore, it is desirable to rearrange the energy equation using the continuity equation. Such a form of the energy equation is known as conservative (or divergent) and, when building a discrete analog, it permits keeping the integral nature of the flow in the finite-difference scheme and the flow continuity at the control volume boundaries.

Conservatism of the difference scheme is not obligatorily connected with the increasing accuracy of the scheme. For some types of equations, the nonconservative scheme may yield more exact results. However, the conservative scheme ensures a physically plausible difference analog.

It has been stated already that determination of the actual velocity field meets great difficulties. The essence of these difficulties and possible ways of eliminating them are briefly described next in accordance with Patankar [62].

The velocity field components are described by the momentum equation, and the difficulties of calculating the velocity field are due to the unknown pressure field. There is no explicit equation for determining the pressure field; the pressure gradient is only a part of the source term in the momentum equation (in the one-dimensional case, the term is dP/dx).

When constructing a discrete analog of the momentum equation—for example, for the one-dimensional case (Figure 4.13, where W, P, and E are nodes, and ω and e are the control volume boundaries)—the integration over the control volume yields the $P_\omega - P_e$ difference, which, in fact, is the pressure force applied to the control volume. By assigning the pressure piecewise-linear profile between the nodes, we obtain

$$P_\omega - P_e = \frac{P_W - P_E}{2}. \tag{4.24}$$

Thus, the pressure difference is not taken from the adjacent but rather from the alternate nodes (i.e., from a coarser mesh). However, this is not the main disadvantage. If the pressure in the nodes has the wavy character, as is shown in Figure 4.14, then the obtained difference approximation will show zero ($P_W - P_E = 0$) pressure drop at each node P (i.e., the wavy pressure field in the momentum equation is treated as uniform).

Similar difficulties arise when a discrete analog of the continuity equation is constructed. For the steady-state flow with the linear-piecewise velocity profile, the discrete analog has the $U_E - U_W = 0$ form (for nodes, designation, see Figure 4.13). That is, velocities in the discrete analog are equal in the alternate instead of the adjacent points. Therefore, the discrete analog may be consistent with the nonphysical wavy velocity field similar to that in Figure 4.14.

These difficulties should be eliminated before the numerical method has been formulated to obviate the nonphysical solution in the problems incorporating velocity and pressure components.

These problems are overcome thanks to a technique according to which velocity components are determined using a mesh that is different from that

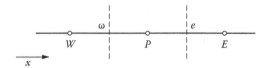

FIGURE 4.13
Three successive nodal points (the dashed area is the control volume) [62].

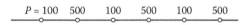

FIGURE 4.14
A wavy pressure field [62].

used for other variables, the so-called staggered mesh. In this case, the nodes for all variables, with exception of velocity, are within the control volume, while the nodes for velocity are on the control volume boundary. The staggered mesh offers the following advantages:

1. Mass flow rate through the control volume boundary is determined without interpolation of velocity components.
2. The discrete analog of the continuity equation incorporates the difference of the velocity components in adjacent nodes located inside the control volume, and therefore the wavy velocity field will not satisfy the continuity equation.
3. The pressure difference between the adjacent nodes located inside the control volume determines the velocity components located between them; therefore, the wavy velocity fields will not be treated as uniform and cannot be used as possible solutions.

The staggered mesh somewhat complicates the program logic and requires a larger RAM capacity, but it facilitates obtaining a physical solution. The TRAC code [82], currently the most advanced one for predicting unsteady two-phase flow thermal hydraulics in reactor systems, employs this mesh structure concept, according to which such variables as pressure, internal energy, and steam void fraction have been obtained for the mesh cell center, while the flow variable—for the cell boundaries.

Another way of eliminating these difficulties is to present the momentum equation in the integral form using the Mayer-Rose method. However, it ignores sound effects and in this case only the low-frequency two-phase flow oscillations can be investigated using the integral model.

Let us dwell on one more aspect in constructing the difference analog, which is the presentation of source term S in equation (4.20). Often, this term is a function of the dependent variable ϕ. Since the solution of the discrete analog is generally found by the method of solving a system of linear algebraic equations, then S should be presented as a linear function of ϕ:

$$S = S_c + S_p \, \phi, \qquad (4.25)$$

where S_c and S_p are constants.

In such an approximation, it is assumed that the ϕ variable value within the entire control volume equals that at the node ϕ_p. The character of solution depends on the type of (4.25) approximation, and even a nonphysical solution may be obtained. To avoid the latter, the Sp coefficient should be nonpositive. Even if S is linearly dependent on ϕ and S_p is positive in reality, while the calculation process is iterative, then sometimes S_p should be artificially made zero or negative for the sake of convergence at each step.

The nonphysical solution possibility may be easily illustrated by an example [62]. Let the discrete analog for a one-dimensional thermal conductivity equation with a volumetric source of heat release be an algebraic equation,

$$a_p T_P = a_E T_E + a_W T_W + S \Delta x, \qquad (4.26)$$

where T_E, T_W are temperatures at the adjacent nodes, and a_E, a_p, and a_W are constant coefficients.

Let us represent S as the linear function of T (4.25) and introduce into (4.26):

$$(a_p - S_p \, \Delta x) \, T_P = a_E \, T_E + a_W \, T_W + S_c \Delta x. \qquad (4.27)$$

If S_p is positive and has a sufficiently large value, it may turn out that $a_p - S_p \Delta x \leq 0$. With $a_p - S_p \Delta x = 0$, no solution of (4.27) exists, while with $a - S_p \Delta x < 0$, the solution is nonphysical, as the increasing temperature at adjacent nodes T_W and T_E causes a decrease in T_p temperature at the intermediate node.

An unsteady term in the equation leads to the appearance of the unidirectional time coordinate t and the solution is sought by time traversing from the preset initial distribution of the dependent variable; that is, the values of ϕ preset for the nodes at the time moment t shall be used for determining values of ϕ at the time moment $t + \Delta t$:

$$\phi(t) \rightarrow \phi_P^0, \phi_E^0, \phi_w^0 ; \quad \phi(t + \Delta t) \rightarrow \phi_P', \phi_E', \phi_w'$$

When constructing a discrete analog of the initial equation, it should be integrated over the control volume and time interval Δt.

It is assumed that the variables ϕ_p, ϕ_E, and ϕ_W change within the t to $(t + \Delta t)$ interval as follows:

$$\int_t^{t+\Delta t} \phi \, dt = \left[f \phi' + (1 - f) \phi^0 \right] \Delta t , \qquad (4.28)$$

where f is the weight coefficient with values from 0 to 1. Then, the difference analog without the source term will be

$$a_p \phi'_p = a_E \left[f \phi'_E + (1-f) \phi^0_E \right] + a_W \left[f \phi'_W + (1-f) \phi^0_W \right]$$

$$+ \left[a^0_p - (1-f) a_W - (1-f) a_W \right] \phi^0_p. \tag{4.29}$$

For example, for the one-dimensional thermal conductivity equation without the source term, with the constant thermal conductivity coefficient K, and with the control volume boundary located in between the nodes, the coefficients in (4.29) have the form of

$$a_E = \frac{2K}{\Delta x}; \quad a_W = \frac{2K}{\Delta x}; \quad a^0_p = \frac{\rho c_p \Delta x}{\Delta t}; \quad a_p = f a_E + f a_W + a^0_p. \tag{4.30}$$

When the weight function value f is fixed, the discrete analog yields different finite-difference schemes. For instance, the explicit scheme is obtained for f = 0, the Crank-Nicolson scheme for f = 0.5, and the completely implicit scheme for f = 1.

The explicit scheme implies that the $\phi_p{}^0$ value from the previous step is maintained within the entire time step with exception for (·) (t + Δt). Thus, to obtain a new value of ϕ_p', it is necessary to know values of ϕ at nodes of the previous step (i.e., it is determined explicitly through the known values at the previous step).

The implicit scheme means that ϕ_p sharply changes from $\phi_p{}^0$ to ϕ_p' at a time moment t and then remains constantly equal to ϕ_p' throughout the entire time step. In the case of the Crank-Nicolson scheme, ϕ_p varies linearly during the time step from $\phi_p{}^0$ up to ϕ_p'. Let us consider qualitative behavior of the solution for each of the schemes.

For the explicit scheme (f = 0) from (4.29), we have

$$a_p \phi'_p = a_E \phi^0_E + a_W \phi^0_W + (a^0_p - a_E - a_W) \phi^0_p. \tag{4.31}$$

From algebraic equation (4.31) one can see that the physically plausible solution can be obtained when the coefficient is positive with $\phi_p{}^0$, i.e., with $a_p{}^0 > a_E + a_w$. If this condition is violated, the increased variable ϕ in the point under consideration and at the adjacent nodes of the previous step may cause ϕ_p' to decrease.

For example, for a thermal conductivity equation with coefficients as in (4.30), the positive coefficient with $T_p{}^0$ yields

$$\Delta t < \rho c_p (\Delta x)^2 / 2K. \tag{4.32}$$

That is, time and space coordinate steps should be matched.

For the semi-implicit Crank-Nicolson scheme (f = 0.5) from (4.29), we have

$$a_p \phi'_p = 0.5a_E(\phi'_E + \phi^0_E) + 0.5a_W(\phi'_W + \phi^0_W) + (a^0_p - 0.5a_E - 0.5a_W)\phi^0_p \qquad (4.33)$$

This scheme yields a physically plausible solution when $a_p > 0.5(a_E + a_w)$. For example, for a thermal conductivity equation with coefficients from (4.30), the positive coefficient with $T_p{}^0$ gives

$$\Delta t < \rho c_p (\Delta x)^2 / K. \qquad (4.34)$$

Thus, the time and space coordinate steps should be matched in order to obtain a physically plausible solution using the Crank-Nicolson scheme. The semi-implicit scheme differs from the explicit one, first, by less rigid requirements imposed by the physical plausibility condition on the time step selection. Second, the presence of values of variables at the adjacent nodes for t + Δt in the right-hand part of (4.33) yields a solution that oscillates around the true solution with the damping amplitude.

For the completely implicit scheme (f = 1) from (4.29), we have

$$a_p \phi'_p = a_E \phi'_E + a_W \phi'_W + a^0_p \phi^0_p . \qquad (4.35)$$

It is obvious from (4.35) that all coefficients in the algebraic equation are positive and the completely implicit scheme should be absolutely stable at any combination of the time and space steps.

However, for practical use of these schemes for the numerical investigation of dynamic and especially self-oscillating processes of two-phase flows in power-generating equipment, everything is not as simple as it may seem. Let us consider the problems in brief.

First, the offered considerations on stability of the explicit and completely implicit schemes are valid for linear equations only. Even the completely implicit scheme may be unstable for the equations with strong nonlinearities.

Second, the selection of a large time step even for the stable completely implicit scheme leads to a considerable loss of accuracy when it relates to physical processes with rapidly changing parameters and especially for oscillatory processes of large frequency and amplitude. The use of a small time step in multi-iteration at each step may result in the fact that the completely implicit scheme becomes less efficient economically as compared to the explicit one due to large computation time.

Third, a theorem was proved [81] that for any (stable or unstable) differential system, there exists such a value of time step, $\Delta \bar{t}$, that the completely implicit analog of the initial system is asymptotically stable at the $\Delta t > \Delta \bar{t}$ step. Thus, if the differential system is unstable, its discrete analog may be asymptotically stable when the integration step $\Delta t > \Delta \bar{t}$ has been accordingly selected, and these two systems describe qualitatively different processes.

This can be illustrated by an example [83] that considers a vessel with a horizontal tube with saturated steam–water mixture at P = 6.9 MPa and with quality of 0.8 maintained constant. At the initial time moment, these systems are connected to the vessel under constant pressure (P) of 5.17 MPa. The system of unsteady equations for the two-phase flow may be brought to the nonhyperbolic form if some simplifying assumptions (say, pressure equality in steam and liquid phases) are made. Solution of the problem with initial conditions for such equations is unstable (though the physical process is stable) and its discrete analog, constructed using the completely implicit scheme and solved on the basis of stable iteration, is also unstable. However, if the time step is increased and goes above a certain value, the solution becomes stable. However, fails to correspond to the solution of the modified initial equation and loses physical sense (Figure 4.15), since pressure in the pipeline cannot be below that in the vessels it is connecting.

This reasoning leads to the fact that the explicit or semi-implicit schemes are usually used in mathematical modeling employing the computer-based numerical solution of dynamic and especially self-oscillatory regimes in two-phase flows.

In conclusion, it should be noted that special conditions permitting the construction of the optimal discrete analog using the control volume technique are met by adhering to four simple rules [62]:

Rule 1. The control volume boundary flows should be matched (i.e., the presentation of the flows across the boundary common for two adjacent control volumes should be identical in formulation of discrete analogs for such volumes).

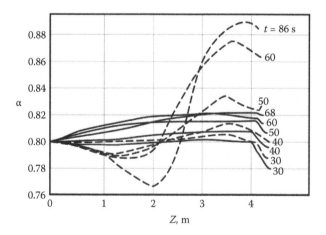

FIGURE 4.15
Calculation of steam quality variation with a leaky pipe [83]. Dashed lines: nonhyperbolic equation system; solid lines: hyperbolic equation system.

Rule 2. The coefficients of the discrete analog dependent variables should be positive. Physical sense of this rule may be illustrated by positive coefficients a_p, a_e, and a_w in equation (4.26) without the source term, in which positivity of a_e and a_w requires a_p also to be positive so that the rise of temperatures at the adjacent nodes T_e and T_w would lead to the temperature rise at the middle node T_p.

Rule 3. When the source term is linearized as $S = S_c + S_p T_p$, the coefficient S_p should be negative or zero. Physical sense of this rule was illustrated by equation (4.27) and is linked to the observation of Rule 2.

Rule 4. For the cases when a differential equation is satisfied also with the addition of a constant value to the dependent variable, it is necessary for the coefficient of the variable at the considered node to be equal to the sum of the coefficients of variables at adjacent nodes—that is,

$$a_p = \sum a_{nb} \qquad (4.36)$$

The relationship (4.36) preserves this property of the differential equation in the discrete analog. If the source term depends on ϕ, the rule is inapplicable.

It should be noted that these rules help not only to construct the discrete analog correctly, but also to ensure or optimize the iteration convergence (i.e., to select an optimal numerical method that would ensure fast convergence when solving the discrete analog).

For the two-phase flows, the preceding difficulties and problems reveal themselves to a greater extent in numerical investigation of self-oscillatory regimes. First, this is because of the increased nonlinearities in the equations due to the enthalpy-influenced sharp variation of the two-phase density and other thermal-physical coolant parameters, as well as to nonlinear empirical correlations. Second, periodical switching from one set of nonlinear empirical dependencies to another when the flow pattern changes is possible during the problem solving. This switching is frequently accompanied by function discontinuities and changes in the character of nonlinear correlations. Third, this is connected with the appearance and disappearance of single- and two-phase coolant sections in the channel and with the necessity of using artificial algorithmic techniques to eliminate the division by zero in this case.

In thermal-hydraulic instability investigations, these effects have been shown [6] to be capable of initiating nonphysical periodic fluctuations either due to calculations getting into an endless loop or due to the appearance of periodic disturbances during logic operations aimed at switching between sets of dependencies.

To eliminate or mitigate such effects, it is necessary, as was noted in Syu [82], to minimize logical transfers; eliminate, wherever possible, singularities and discontinuities when selecting correlation dependencies; use the continuous transfer between the correlations that describes different mechanisms of the physical process; and, where possible, strive to simplify nonlinear correlations at the stage of program debugging, even at the expense of losing some accuracy.

4.6 Conclusion

Thus, the overview provided in this chapter shows that

1. For predictive quantitative determination of the stability boundary, the "dangerous" or "practical" boundary included, most preferable is the direct numerical solution of the adequate nonlinear dynamic model of the system or plant under consideration.

2. Prediction of the stability boundary using the first approximation model is preferable for a qualitative investigation of the effect of design and flow parameters on the stability boundary, and for selecting the optimal way of increasing system stability. Also, acceptable quantitative results may be obtained if the system properties are close to the linear ones.

3. Adequacy of the description of real thermal-hydraulic processes by the nonlinear dynamic model and accuracy of numerical solution are verified by experimental investigations of thermal-hydraulic instability on a model thermophysical test facility with subsequent comparison of experimental results with predictions. In so doing, the conditions of (4.1) will be fulfilled.

4. If the design thermal-hydraulic parameters of the full-scale plant ensure just a small stability margin of the system as predicted by $[(\rho W)/(\rho W)_{Bn} < 1.3]$, then the full-scale plant commissioning test should include experimental determination of the stability boundary for all channels dangerous in this respect.

5

Static Instability

5.1 Basic Definitions

The static coolant flow instability in a heated channel, sometimes called the Ledinegg instability, is manifested as a spontaneous "drift" of the steady-state operating parameters upon reaching the stability boundary. At the onset of static instability, the channel parameters acquire a new steady state. However, if static instability is developing along with the decreasing coolant flow rate, then either burnout conditions or oscillatory instability may develop in the channel (or the interaction between oscillatory and static instabilities may start with fluctuations of the channel pressure drop). An opportunity of the onset of oscillatory instability at the initiation of static instability should be taken into account when analyzing the experimental results of flow stability in power equipment. Cases are known [2] in which static instability was confused with oscillatory instability in conditions when self-oscillations of the coolant flow developed immediately after the short-term flow rate reduction.

The experimentally registered behavior of the coolant flow rate in a steam-generating channel for three cases of static instability is shown in Figures 5.1, 5.2, and 5.3. The presented oscillograms have been recorded for a channel with the constant pressure drop and smooth variation of parameters on the approach to the stability boundary.

The reason for static thermal-hydraulic instability of coolant flow is the ambiguity of the static hydraulic curve of a heated channel or a group of channels. In the presence of this ambiguity, the dependence of the channel pressure drop (ΔP_c) on the channel coolant flow rate (G_c) in a certain region of flow rates has the negative slope, and several flow rate values (G_c) correspond to the preset channel pressure drop (Figure 5.4).

The hydraulic curve of a channel is understood as the dependence of pressure drop on the coolant flow rate $\Delta P_c = f(G_c)$ when inlet enthalpy and power supplied to the channel are constant.

As one can see from Figure 5.4, a higher pressure drop corresponds to a lower static flow rate in the (G_c^A, G_c^B) region in the $(\Delta P_c^A, \Delta P_c^B)$ range.

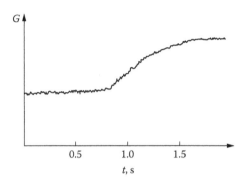

FIGURE 5.1
Static instability development in the coiled channel, followed by transition to a new steady-state regime.

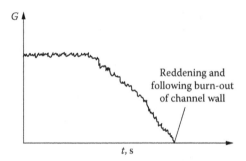

FIGURE 5.2
Static instability development in the coiled channel, followed by transition to the crisis regime.

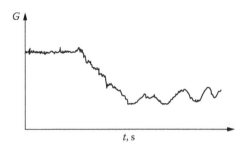

FIGURE 5.3
Static instability development, followed by transition to the self-oscillatory regime.

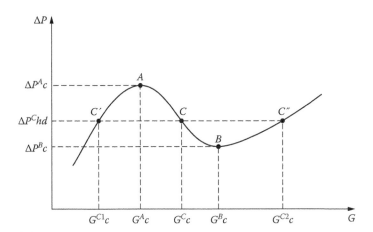

FIGURE 5.4
Hydraulic curve of a channel with an ambiguity region.

Therefore, stable steady-state channel operation in this region of flow rates is impossible if the channel pressure drop ΔP_{hd} is maintained constant.

Let ΔP^c_{hd} be a preset pressure drop across headers, independent of flow rate fluctuations in the given channel. If we admit that there exist conditions with G_c^c flow rate, then any random flow rate decrease will lead to an increase in channel pressure drop ($\Delta P_c > \Delta P^c_{hd}$), which, in its turn, will result in the further flow rate drop down to a stable G_c^{c1} value at a constant ΔP^c_{hd}. If the flow rate is increased randomly, the process takes a different direction from G_c^c: The channel pressure drop decreases and at ΔP^c_{hd} = const. causes the further flow rate rise up to a stable G_c^{c2} value. Thus, the flow rate regime within the G_c^A, G_c^B range cannot be realized with the preset pressure drop, since small disturbances, which are always present in the system, cause flow rate change to a new value that corresponds to one of the stable states (G_c^{c1} or G_c^{c2}), depending on the direction and kind of disturbance.

The boundaries of static stability as well as of oscillatory stability are generally determined for one out of three (or several) most typical thermal and hydrodynamic boundary conditions, which are considered in detail in Section 1.1 in Chapter 1.

The conditions for the static stability boundary determination for these three types of boundary conditions may be derived from the analysis of simple models.

Let us first consider the conditions of static stability of a circulation loop (the third type of boundary condition) for a simple two-phase system presented in Figure 5.5. Let the pressure difference between vessels $P_3 - P_1$ be constant. For the sake of simplicity, let us cancel the hydrostatic and friction pressure drops in the outer loop portion (with exception for the heated channel). Let us set forth the solution obtained in Yadigaroglu [15]. In accordance with Yadigaroglu, a relation,

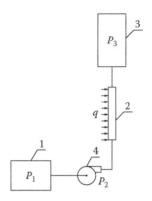

FIGURE 5.5
Simplified hydraulic loop scheme. 1 and 3: Constant pressure tanks; 2: heated channel; 4: circulating pump.

$$P_3 - P_1 = \Delta P_c - \Delta P_p - \left(\frac{H_1}{A_1} - \frac{H_c}{A_c} \right) \frac{dG}{dt} = \text{const.,} \qquad (5.1)$$

should be fulfilled for any moment in time. In this relation, ΔP_p is the pump head characteristic; ΔP_c is channel hydraulic characteristic in steady-state conditions; H_c and H_1 are the lengths of the loop outer section and heated channel, respectively; A_c and A_1 are the loop outer section and heated channel flow areas, respectively; and G is the coolant flow rate.

For small deviations from the steady-state conditions designated by zero subscript, (5.1) may be written as

$$\delta \Delta P_p(t) - \delta \Delta P_c(t) = \overline{H} \frac{d\delta G(t)}{dt}, \qquad (5.2)$$

where

$$\overline{H} = \frac{H_1}{A_1} - \frac{H_c}{A_c}.$$

Let us represent $\delta \Delta P_p(t)$ and $\delta \Delta P_c(t)$ as

$$\delta \Delta P_p(t) = \left. \frac{\partial \Delta P_p}{\partial G} \right|_0 \delta G(t) \qquad (5.3)$$

and

$$\delta \Delta P_c(t) = \left. \frac{\partial \Delta P_c}{\partial G} \right|_0 \delta G(t). \qquad (5.4)$$

By substituting these expressions into (5.2), we obtain the equation for the flow rate deviation from the steady-state condition:

$$\overline{H}\frac{d\delta G(t)}{dt} = A\delta G(t),$$ (5.5)

where

$$A = \frac{\partial \Delta P_p}{\partial G}\bigg|_0 - \frac{\partial \Delta P_c}{\partial G}\bigg|_0.$$

The solution of (5.5) for the initial deviation δG_o is

$$\delta G(t) = \delta G_0(t) \cdot e^{(A/\overline{H})t}.$$ (5.6)

One can see from (5.6) that system disturbances will increase with positive A (i.e., the system will be unstable), if the negative slope of channel hydraulic curve is smaller than that of the pump head curve

$$\frac{\partial \Delta P_p}{\partial G} > \frac{\partial \Delta P_c}{\partial G}.$$ (5.7)

In this case, the stability boundary will be defined by

$$\frac{\partial \Delta P_p}{\partial G} = \frac{\partial \Delta P_c}{\partial G}.$$ (5.8)

In practice, the channel pressure drop is usually much smaller than the system absolute pressure, and the loss of pressure in the channel, being insignificantly dependent on pressure variation along the path, is determined by the system absolute pressure. Therefore, partial derivatives in (5.7) and (5.8) may be substituted by total derivatives with respect to coolant flow rate.

Figure 5.6 presents the thermal-hydraulic curve of the heated channel with an ambiguity region and head curves of circulating pumps of different types. Obviously, a piston pump with an abrupt dropping head curve provides stable operation of a single channel in accordance with (5.7) even in the ambiguity region of the static hydraulic characteristic. Indeed, a random reduction of, for example, flow rate causes in this case a smaller increment in the channel pressure drop than the pump head increase, which results in the increased flow rate and system stabilization. However, if the condition of (5.7) is fulfilled, but the pump head characteristic is not very steep, resonance characteristics of the system may result in slowly decaying oscillations in the presence of disturbances. To ensure fast decay of disturbances and better

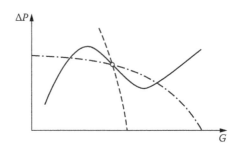

FIGURE 5.6
Hydraulic curve of the heated channel and pumps of different type [15]. Solid line = channel hydraulic curve; dashed-dot line = head curve of the centrifugal pump; dashed line = head curve of the piston pump.

control, close approaches to the static stability boundary should be avoided. According to Bailey [9], the following relation should be maintained:

$$\frac{\delta \Delta P}{\Delta P} \cdot \frac{G}{\delta G} \geq 0,3 + 0,6, \tag{5.9}$$

where ΔP and G are the channel (loop) pressure drop and flow rate at a working point, respectively, and $\delta(\Delta P)$ and δG are the pressure drop disturbance and the respective coolant flow rate static disturbance.

For the first type of boundary conditions corresponding to the isolated channel with a preset pressure drop, the preceding analysis (with no circulating pump) yields the following instability condition:

$$\frac{d\Delta P_c}{dG} < 0, \tag{5.10}$$

and the relation for determination of the stability boundary is

$$\frac{d\Delta P_c}{dG} = 0. \tag{5.11}$$

Thus, the static stability boundary coincides with the hydraulic curve extremum for an isolated heated channel with the constant pressure drop. The instability range encompasses the entire region of hydraulic curve ambiguity.

In the case of the second type of boundary condition, when the preset inlet enthalpy and the total flow rate in a system of parallel channels are preserved, the hydrodynamic interaction between parallel channels is observed in the instability area, and the pressure drop between the headers is not constant. Using the previous simplified analysis of stability and taking into account that, at the constant total flow rate in the component, flow rate variation in unstable channels leads to an equivalent change (with the opposite sign) of the flow rate in stable channels, we obtain the following condition of the heated channel static instability for the considered case:

$$-\frac{d\Delta P_{c1}}{dG} > K\frac{d\Delta P_{c2}}{dG},$$ (5.12)

and the condition for determining the stability boundary for the second type of boundary condition will be

$$-\frac{d\Delta P_{c1}}{dG} = K\frac{d\Delta P_{c2}}{dG},$$ (5.13)

where ΔP_{c1} is the static hydraulic characteristic of the hot channel with an ambiguous characteristic and ΔP_{c2} is the static hydraulic characteristic of the averaged channel from a group of stable parallel channels.

$K = \left|\dfrac{\Delta G_{c1}}{\Delta G_{c2}}\right|$ is the ratio between the averaged coolant flow rates in stable

and unstable channels at the preset total flow rate in the system of parallel channels.

The value of $K\cdot(d\,\Delta P_{c2}/dG)$ defines the derivative of pressure drop between headers with flow rate variation in the unstable channel. The conditions of stability boundary determination (5.11) and (5.13) show that the hydrodynamic influence of nonidentical parallel channels on each other, under otherwise equal conditions, leads to an increased static flow stability in parallel channels in comparison with the isolated channel with the constant pressure drop. Physically, the reason for stability improvement is that within a portion of the hot channel hydraulic curve with the negative slope, where $0 < -d\Delta P_{c1}/dG < K(d\Delta P_{c2}/dG)$, the flow maintains stability. This is because when the flow rate decreases in the hot channel, the flow rate increases in the averaged channel (in relation to K). According to the condition, the rise of pressure drop between the headers in this area is greater than in the hot channel, leading to an increase in flow rate in the latter channel and to system stabilization.

It can be seen from (5.13) that with a large number of stable channels, when $K \to 0$ and the flow rate change in the unstable channel has little influence on the pressure drop variation between the headers, the condition of (5.13) transits into that of (5.11); that is, the static stability boundary in the parallel-channel system approaches that of the isolated channel. With a rigid connection of two operating parallel heated channels or parallel steam generators, $K = 1$, and the maximum influence of the hydrodynamic interaction of components operating in parallel on the static stability boundary is observed. In the latter case, stable operation of a steam-generating component within the unstable region of the hydraulic curve is frequently observed in practice.

The static stability boundary for the three types of boundary conditions respectively defined by (5.11), (5.13), and (5.8) can be determined in practice as follows.

In simple cases with a large number of simplifying assumptions, a third-order algebraic equation of coolant flow rate may be derived for the isolated channel from the steady-state (stationary) equations for momentum, energy, and continuity. Using this algebraic equation, the flow rates corresponding to the hydraulic curve extremum—that is, satisfying the condition of (5.11)—may be obtained analytically.

However, the method has a limited application due to the following reasons: (1) It can be used for a limited number of sufficiently simple particular cases and therefore is not universal, and (2), a large number of simplifying assumptions used to obtain the analytical expression undermine accuracy of the static stability boundary determination.

At present, two methods are employed for the static stability boundary determination:

1. On the basis of the steady-state equations of momentum, energy, and continuity with closure correlations for the coefficients of friction, heat transfer, steam quality, etc., the hydraulic curve of a channel, component, or circulation loop is numerically predicted (mostly by computer codes) and, depending on boundary conditions (5.11), (5.13), or (5.8), the flow static stability boundary is determined by numerical differentiation. The nature of the parameters' change, depending on the law of control, heat and hydraulic irregularities in the component, changes in the nature of heat transfer in the case of convective heating and other factors, should be taken into account.

2. The second method is based on the application of the linear theory of stability—that is, on the investigation of stability by the first approximation [2]. To this end, the unsteady equations of conservation for the two-phase flow are linearized and a search is made for a region with no real root in the solution of the characteristic polynomial using qualitative methods of the theory of differential equations.

To optimize the design of power equipment and eliminate the appearance of static instability in the working range of parameters, it is necessary not only to determine static stability boundaries reliably, but also, first of all, to understand the physical mechanism of instability clearly and reveal the effect of different design and flow parameters on the static stability boundary.

The ambiguity region that appears on the hydraulic curve causes the onset of static instability. Its appearance depends on the working medium type, pressure loss components' ratio in the channel, nature of coolant flow (flow pattern), input power distribution along the channel, kind of heating, range of preset parameters, coolant flow direction, etc.

Let us first consider the shape of hydraulic curves for the boiling channel with constant heat supply, and qualitatively assess the effect of various design and flow parameters on the appearance of the ambiguity region. In this case, the reduction of the static channel flow instability is understood as a decrease of the negative and increase of the positive slope of the hydraulic curve, while the flow destabilization is understood as an increased negative and decreased positive slope of the hydraulic curve.

The three most typical situations when the channel hydraulic curve may include a region with the negative slope are as follows: (1) the appearance of sections with surface and volume boiling (with low bulk quality in the latter case) at the heated channel exit, (2) the presence of a long superheating section in the heated channel, and (3) the presence of a coolant downflow section in the heated channel or its portion. In some cases, the hydraulic curve ambiguity may be caused by the nonlinear dependence of thermophysical properties of the one-phase coolant (say, viscosity) on temperature, or by complex channel and circulation loop geometry.

5.2 Ambiguity of Hydraulic Curve due to Appearance of a Boiling Section at the Heated Channel Exit

When flow rates in the heated channel are high, the coolant is in a single-phase state and the pressure drop decreases almost in quadratic dependence with the decreasing flow rate. Starting from a certain flow rate level, first surface and then volume boiling appear at the heated channel exit. The friction pressure drop starts increasing in the boiling section because of a sharp reduction of coolant density. This increase may exceed the pressure drop reduction in the single-phase section and result in the overall growth of the channel pressure drop despite the decreased flow rate. With the further coolant flow rate reduction, the boiling section increases, while the zone of sharp density reduction shifts away from the exit. Therefore, the integral coolant density in the boiling section decreases not so sharply when the flow rate decreases. The growing pressure drop due to the reduced density fails to compensate for the quadratic reduction of the friction pressure drop along with the coolant flow rate decrease, and the overall pressure drop starts to decrease again.

For the sake of the initial qualitative analysis of the parameters' influence on the static stability boundary in this characteristic case, let us assume that the channel pressure drop is determined only by the pressure loss across the inlet and exit orifice plates.

The channel friction in this case is made up by pressure losses across the inlet and exit orifice plates, ΔP_{in} and ΔP_e, respectively:

$$\Delta P_c = \Delta P_{in} + \Delta P_e. \tag{5.14}$$

The first component depends only on the coolant flow rate because of the unchanged inlet density,

$$\Delta P_{in} = \xi_{in} \frac{G^2}{2A^2 \rho_L}, \tag{5.15}$$

where A is the channel flow passage area, G is the coolant flow rate, and ρ_L is the liquid density.

The second component of (5.14) is dependent on both the steam–liquid mixture flow rate and mixture density, which is the function of the two-phase flow enthalpy at the exit from the channel heated section:

$$\Delta P_e = \xi_e \frac{G^2}{2A^2 \rho(i_e, P)}. \tag{5.16}$$

Considering that $i_e = i_{in} + (\Pi H q / G)$ (where H is the length of the channel heated section, Π is the heated perimeter, and q is the heat flux density), we have

$$\Delta P_e = \frac{\xi_e}{2A^2} \frac{G^2}{\rho\left(i_{in} + \dfrac{\Pi \Pi q}{G}, P\right)}. \tag{5.17}$$

One can see from (5.17) that a reduced G causes simultaneous diminishing of the numerator and denominator. In some cases, the rate of coolant density reduction may be so intense that the pressure drop ΔP_e increases along with the decreasing flow rate.

According to the condition (5.10), the system stability in the small may be evaluated for the isolated channel with the preset pressure drop on the basis of the flow in the operating point (designated by "0"), depending on the sign of the complete channel pressure drop derivative:

$$\left.\frac{d\Delta P_c}{dG}\right|_{G_0} > 0 = \text{the channel process is statically stable;}$$

$$\left.\frac{d\Delta P_c}{dG}\right|_{G_0} < 0 = \text{the channel process is statically unstable.}$$

By differentiating (5.14), taking (5.17) into account, and neglecting $(\partial \rho / \partial G)$ and $(\partial \rho / \partial P)$, we obtain

$$\left.\frac{d\Delta P_c}{dG}\right|_{G_0} = \frac{2}{G_0}\left\{\Delta P_{in} + \Delta P_e\left[1+0.5\frac{i_e - i_{in}}{\rho_e}\left(\frac{\partial\rho}{\partial i}\right)_e\right]\right\}_{G_0}. \tag{5.18}$$

The derivative $(\partial\rho/\partial i)_e$ has the negative sign, and its qualitative distribution depending on enthalpy is shown in Figure 5.7. The maximum of the $(\partial\rho/\partial i)_e$ modulus is located in the area where the flow section averaged enthalpy is close to the saturation enthalpy (or, in the case of supercritical pressure fluid, to the enthalpy of maximum heat capacity area). It is seen from (5.18) that the expression in brackets may be negative, and with appropriate relation between ΔP_{in} and ΔP_e, the right part of (5.18) may turn out to be negative with the resultant static instability in the channel.

In the general case, there are no troubles with numerical determination of the derivative $(d\Delta P_c/dG)|_{G_0}$, with due account for the channel axial distribution of thermohydraulic parameters and pressure losses due to flow acceleration, friction, and gravity. However, even in the general case, the effect of ΔP_{in}, ΔP_e, $(\partial\rho/\partial i)_e$, $(i_e - i_{in})/\rho_e$ on flow stability is qualitatively the same as in (5.18), which is applied for the analysis of the effect of some parameters on the coolant flow static instability in the isolated steam-generating channel.

5.2.1 Effect of Local Hydraulic Resistance

It is obvious from (5.18) that the increase of inlet resistance ΔP_{in} (by means of inlet throttling, or nonheated upstream section) may yield a positive value of the right part of (5.18), thereby eliminating the ambiguity region on the hydraulic curve and stabilizing the system as regards the static stability. Physically, it can be clearly explained as follows. With the increasing ΔP_{in}, the share of the channel pressure drop increases in the single-phase section; therefore, the increased two-phase flow friction (whose contribution to the overall pressure drop reduces) at the channel exit becomes, with the

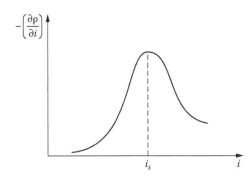

FIGURE 5.7
Dependence of the $(\partial\rho/\partial i)$ derivative on flow enthalpy.

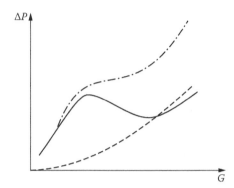

FIGURE 5.8
Effect of channel inlet throttling on flow static instability [7]. Solid line: hydraulic characteristic curve of the heated channel; dashed line: pressure drop across the inlet throttle; dashed-dotted line: hydraulic characteristic curve of the heated channel with the inlet throttle.

decreasing coolant flow, less than the pressure drop reduction value in the single-phase section. Figure 5.8 [9] exemplifies elimination of the ambiguity section on the hydraulic curve by additional inlet throttling. This approach is a major contributor to stabilization of statically unstable flows.

It is seen from (5.18) that an increase in the exit pressure drop ΔP_e when the expression in the brackets is negative destabilizes the flow in terms of static stability, since it increases the modulus of the negative term in the right part of the relation. The mechanism of such an effect of ΔP_e is as follows: When ΔP_e increases, the share of the pressure drop in the boiling section within the total channel pressure drop also increases, and the increased pressure drop in the boiling section along with the decreased flow rate fails to be compensated for by a lower pressure drop along the single-phase section. This leads to the appearance of the ambiguity region on the hydraulic curve or increases the negative slope of the curve. To the greatest extent, the desta-bilizing effect of exit throttling is manifested at $x_e \approx 0$, where the maximum of the derivative $(\partial \rho / \partial i)_e$ modulus is attained (see Figure 5.7). At $x_e > 0$, the destabilizing effect of exit throttling drops with increasing exit bulk quality and, as is shown in Morozov and Gerliga [2], ends at

$$x \approx \frac{i_G - i_{in}}{i_{LG}} - \frac{2}{\dfrac{\rho_L}{\rho_G} - 1}.$$

Therefore, the ambiguity region on the hydraulic curve appears when the boiling section forms in the exit section of the heated channel. When the boiling section becomes longer and, consequently, the exit bulk quality increases, the hydraulic curve, starting from a certain value, becomes unam-biguous with the positive slope.

Worth mentioning is an example [2] that shows the influence of the exit resistance on static stability. Let there exist a nonheated region at the heated channel exit. Let the ambiguity section on the hydraulic curve appear when surface boiling starts at the heated channel exit ($x_e < 0$). Let us suppose that steam bubbles that arrive from the region of surface boiling have enough time to condense before the exit throttle located in the nonheated region and the single-phase current flows through the throttle. For such conditions, it was found [2] that the increased exit throttling is a stabilizing factor as regards the static stability. Indeed, in this case, the exit throttling, like the inlet one, decreases the pressure drop in accordance with the quadratic equation, reduces the flow rate, and eliminates the ambiguity region, since the share of the pressure drop due to the single-phase flow friction becomes increased.

5.2.2 Pressure Effect

The qualitative effect of pressure on the flow static stability boundary in an isolated channel at a preset pressure drop can be traced from (5.18). The increasing pressure, other conditions being equal, stabilizes the flow in the heated channel for two reasons:

1. With identical bulk qualities, the two-phase flow density increases, thus decreasing the factor before the $(\partial\rho/\partial i)_e$ multiplier in (5.18) because of a lower pressure drop across the boiling section.
2. The $(\partial\rho/\partial i)_e$ derivative modulus decreases because of the lower rate of density change along with the increasing enthalpy at a higher pressure.

These two factors substantially diminish the modulus of the negative term in brackets in (5.18) by decreasing the negative slope of the hydraulic curve, or making it positive. The stabilizing effect of an increased pressure becomes especially apparent at small absolute pressures ($P < 4$ MPa), when the two-phase flow density reduces sharply along with the increasing pressure, and the $(\partial\rho/\partial i)_e$ modulus value is high. At low pressures, the destabilizing effect of exit throttling is also most noticeable.

5.2.3 Effect of Channel Inlet Coolant Subcooling

The amount of channel inlet coolant subcooling (i_e) has an ambiguous influence on the static stability boundary. Let us consider a case with a stable vertical steam-generating channel with preset flow rate, pressure, heat flux density, geometry, and throttling. With the increasing inlet coolant subcooling (at $i_e \geq i_s$), i_e will be approaching i_s and, as is clearly seen from Figure 5.7, the modulus of the $(\partial\rho/\partial i)_e$ derivative will be increasing intensively. The coefficient before $(\partial\rho/\partial i)_e$ in (5.18) reduces at the expense of the increasing

ρ_e at $i_e \rightarrow i_s$. However, the total negative value of the second term in the brackets of (5.18) decreases, as a rule, because of a more intensive decrease of $(\partial\rho/\partial i)_e$ and leads to the appearance of the negative slope of the hydraulic curve and to static instability. The further increase of inlet subcooling makes $i_e < i_s$ and, as is seen from Figure 5.7, the value of the modulus of the $(\partial\rho/\partial i)_e$ derivative decreases intensively and leads to flow stabilization from the point of view of static instability.

Most strongly, the described inlet subcooling influences static instability at low pressures (≤ 4 MPa), at which the $(\partial\rho/\partial i)_e$ derivative changes most intensively.

5.2.4 Effect of Heat Flux Density, Channel Length, and Equivalent Diameter

With the increasing heat flux and other design and flow parameters being constant (except for the mass velocity), the region of static instability shifts toward the higher flow rates region. As an illustration, Figure 5.9 shows the experimentally obtained hydraulic curves for the П-shaped panels under different heat loads [46]. The reason for this influence of the heat flux density is as follows: When q increases at low exit qualities, the surface boiling

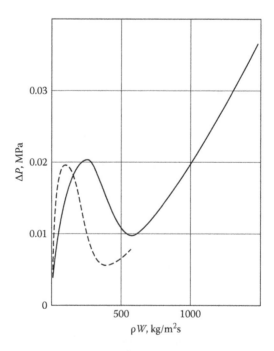

FIGURE 5.9
Experimental hydraulic curves for П-shaped tubes at P = 14.5 MPa [46]: solid line: q = 0.12 MW/m²; dashed line: q = 0.06 MW/m².

section increases and the coolant density gradient $(\partial\rho/\partial i)_{ev}$ in the channel exit heated section becomes larger due to a more rapid increment of mass quality per channel length unit. As one can see from Figure 5.9, this leads to a displacement of the static instability boundary to the higher flow rates region when q is increasing. An increase in coolant flow rate while keeping the outlet quality unchanged increases the slope of the hydraulic curve negative branch, since ΔP_{in} and ΔP_e are proportionate to G^2. Therefore, as one can see from (5.18), $d\Delta P_c/dG \sim G$. At the same time, when the heat flux density is increasing along with the increasing enthalpy-determined area where the $(\partial\rho/\partial i)_e$ modulus value is large, max $|(\partial\rho/\partial i)_e|$ may decrease with the increasing q (Figure 5.10) and, to a certain extent, reduce the destabilizing effect of the increased q.

The influence of a longer channel on static instability may be analyzed on the basis of respective heat flux density variation. Indeed, if the input power is assumed to be preset and constant, an increase of the channel heated length should be accompanied by a corresponding decrease in the heat flux density, which, as was shown before, decreases the hydraulic curve ambiguity area.

If the heated length increase is supposed to occur at constant other design and flow parameters, the flow may stabilize if the channel is within the region of the maximum variation of the $(\partial\rho/\partial i)$ gradient below the channel exit section; the flow may destabilize if the channel was previously in the stable region and coolant was in the single-phase state. For such conditions, an increase of the channel heated length leads to the static stability boundary displacement toward the region of higher flow rates and may have the stabilizing effect, if the nominal flow rate is less than that corresponding to the lower boundary of static stability at the channel exit when the nominal flow rate $x_e > 0$.

The effect of the increased channel equivalent diameter on static instability may be analyzed in a similar manner. At fixed values of P, ρW, i_e, i_{in}, H, and q, an increase in the channel equivalent diameter leads to the static instability

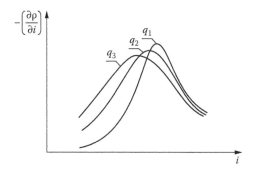

FIGURE 5.10
Dependence of the $(\partial\rho/\partial i)$ derivative on flow enthalpy at different heat flux densities: $q_1 < q_2 < q_3$.

boundaries' displacement into the lower mass velocities region (ρW), though it does not rule out a possibility of static instability appearance during the partial-load operation of the equipment.

5.2.5 Effect of Heating Distribution and Kind

For the considered conditions, in which the onset of static instability is associated with the appearance of surface and bulk (with low x_e) boiling at the heated channel exit section, the relative height distribution of heat flux density along the boiling section is essential for the static stability boundary.

As was shown previously, an increase of the relative heat flux density in the boiling section (i.e., in the channel exit section) broadens the region of static instability. The relative heat flux density height distribution along the single-phase coolant section produces no marked effect on channel flow static instability.

Hydraulic characteristics are reasonably easy to analyze in the case of joule or radiation heating of the heating surface, when the coolant fluid flow rate may be varied at constant heat input. However, the analysis becomes sufficiently more complicated when hydraulic characteristics are considered and flow static stability boundaries are determined for the convection-heated steam generator channels. For example, these are steam generators with liquid-metal, water, or gas heating fluid.

Despite the fact that a whole series of work has been dedicated to steam generator instability with convective heating [86–90], no clear notion of the flow behavior under these conditions has been obtained so far, and further research is required. In part, such an analysis will be provided in Section 5.3.

5.2.6 Effect of the Pressure Drop Gravity Component and Steam Slip Coefficient

Static instability with coolant downflow is considered in Section 5.4, so here we evaluate only the effect of the pressure drop gravity component and the steam slip coefficient on the static stability boundaries in the vertical channel with coolant upflow. At high coolant mass velocities, the steam void fraction is mostly determined by mass quality and increases with the decreasing mass velocity, thereby decreasing the pressure drop gravity component. Therefore, as was stated in Yadigaroglu [15], with the decreasing flow rate, the hydrostatic pressure in the vertical channel with coolant upflow decreases the negative slope of hydraulic curve (or increases the positive slope); it is, in fact, the very stabilizing factor as regards the static instability. At low mass velocities caused by steam slip, the steam void fraction changes not as sharply at the reduction of velocity as in the case with high mass velocities. Therefore, the pressure drop gravity component in these conditions may have a flatter dependence on mass velocity, and the stabilizing effect of the hydrostatic pressure decreases. This aspect of steam slip influence has a destabilizing effect on static instability. However, steam slip, causing a lower exit quality at a decreased mass velocity, decreases the rise of pressure drop due to friction in the boiling section, thereby increasing system static stability. The integral effect of steam slip on static instability from the previous oppositely acting effects depends on the ratio of change of hydrostatic pressure drop components and friction pressure drop in the boiling section.

In conclusion of this section it should be noted that, as a rule, the two-phase flow static instability at low exit qualities is observed in hydraulic systems with forced coolant circulation. The natural-circulation systems (e.g., natural-circulation boilers and single-loop NPPs (nuclear power plants) with boiling reactors and natural circulation) feature low coolant subcooling at the heated channel inlet and low circulation rate. As a rule, this ensures their operation in the entire working range of power on the left stable branch of hydraulic curve (to the left of point A in Figure 5.4). However, the onset of flow static instability cannot be ruled out for two-loop NPPs with boiling reactors and natural circulation or for natural-circulation boiler lighting up modes, when high coolant subcooling at the heated channel inlet and low exit qualities may appear at a definite combination of parameters—in particular, when the height of individual risers is small [7].

5.3 Hydraulic Characteristic Ambiguity in the Presence of a Superheating Section

5.3.1 Specifics of the Ambiguity Region Formation

A region of ambiguity on the hydraulic curve of the steam-generating channel may appear with the presence of a considerably long superheating section and high coolant subcooling at the channel inlet. Examples of predicted hydraulic curves for the coil-type steam generator with liquid-metal (Na) heating, with due account for the friction loss distribution along the channel length and different inlet temperatures of heating water coolant, are given in Figure 5.11.

Let us analyze the channel pressure drop components when considering the physical mechanism of appearance of an ambiguity region on the hydraulic curve under such conditions. First, the numerical analysis shows that the pressure drop due to acceleration and the gravity component of pressure drops are, with flow rate variation, of monotonous type with positive gradient (for the vertical channel with coolant upflow), while the hydraulic curve ambiguity is due to the friction pressure drop along the channel. To reveal the preceding behavior of the friction distributed along the channel, let us consider the predicted curves of variation of superheating, boiling, and economizer section lengths (Figure 5.12) and the curves of pressure drop

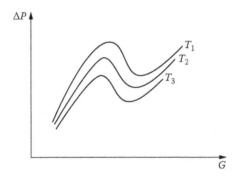

FIGURE 5.11
Hydraulic characteristic of a coil-type steam generator with liquid-metal heat-transfer medium (Na); $T_1 > T_2 > T_3$.

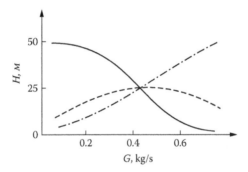

FIGURE 5.12
Channel section lengths variation at the flow rate change. ——— superheating section; ------ boiling section; _._._ economizer section.

variation in these sections (Figure 5.13) depending on the working fluid flow rate variation.

One can see from Figures 5.12. and 5.13 that in the case of the example in question, the nonlinear decrease of the superheating section length with the increasing flow rate, and the quadratic rise of pressure drop in the same section with the increasing coolant flow rate per length unit leads, in a certain range of flow rates, to the appearance of a maximum on the curve of pressure drop variation in the superheating section, depending on coolant flow rate. Such a manner of pressure drop change in the superheating section leads to the appearance of a region of ambiguity on the hydraulic curve.

The considered mechanism of an unstable region formation on the hydraulic curve somewhat differs from the previous one that formed at the origin of a boiling section at the heated channel exit. Therefore, the effect of parameters on static instability for these two mechanisms of ambiguity regions' formation may also differ.

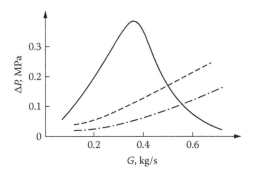

FIGURE 5.13
Pressure drop variation across different sections depending on the flow rate. ——— superheating section; ------- boiling section; _._._ economizer section.

For example, in the case of the considered mechanism of the instability region origin, both exit and inlet throttling may increase the system static stability. Without influencing the regularity of change of the relative lengths of the superheating, boiling, and economizer sections, increased inlet and exit throttling increases the share of the single-phase flow (liquid or steam) local resistance in the channel's overall pressure drop.

However, if density of the working fluid steam decreases noticeably along with the increasing temperature, then the pressure drop across the exit throttle may decrease less intensively with the flow rate reduction than in accordance with the square law, and the stabilizing effect of exit throttling may disappear.

If the dependence of steam density on temperature is strong for some coolants, exit throttling may cause flow destabilization as regards the static instability.

An increased pressure may destabilize the flow, as it decreases the share of friction pressure drop in the evaporating section in the total channel pressure drop due to a reduced evaporating section length because of a lower heat of evaporation. The mechanism of this effect of pressure will be considered in more detail in the next section. Static instability is much influenced by the kind of heating, especially by convective heating, some peculiarities of which are also considered in the next section.

5.3.2 Effect of the Kind of Heating

Steam generators with convective heating have found wide application in various branches of engineering. In the first place, this refers to nuclear power engineering where water, gas, and liquid–metal coolants are used. Steam generators with convective heating find application in space power plants, chemical and food industries, auxiliary systems of thermal power engineering, etc.

In presence of a superheating section and high inlet subcooling of the heated fluid in steam generators with convective heating, a serious problem arises as to ensuring static stability at nominal and especially partial-load operation. Predictive and experimental results show that convective heating may cause unfavorable conditions for the appearance of static instability, as compared to joule or radiation heating. This fact is of particular importance not only for identifying regularities that determine the effect of convective heating on flow static stability in steam-generating channels, but also for validating techniques of simulation of flow instability in real components of thermophysical test facilities.

When joule and radiation heating are used, the value and manner of distribution of the heat supplied to the heating fluid are independent of its flow rate. A specific feature of convective heating is in the fact that, depending on the heating fluid flow rate and inlet temperature, as well as its thermal-physical properties (heat capacity in particular), variation of the heating fluid flow rate may greatly change the heat input and its distribution along the steam-generating channel. Thus, the distribution of lengths of the economizer, evaporating, and superheating sections, which determine the nature of dependence of the channel pressure drop on the heated coolant flow rate, may vary considerably.

Figure 5.14 [86,87] presents thermal-hydraulic characteristic curves of a steam generator with convective heating employing different heat input methods. For a comparison, the figure also shows the hydraulic characteristic curve of a steam generator (curve 3, Figure 5.14) with a heat input independent from the typical working fluid flow rate (e.g., of radiation or joule heating). The calculations of the thermal-hydraulic characteristic of a steam generator with convective heating consider two cases. In the first case

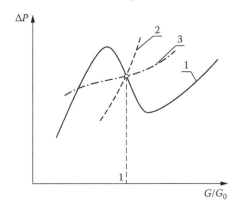

FIGURE 5.14
Hydraulic characteristic of a steam-generating component with differently specified heat exchange [86]. 1: Heat balance preservation; 2: constant heating surface temperature distribution along the component length; 3: constant heat flux density distribution along the component length.

(curve 2, Figure 5.14) with the varying working fluid flow rate, the heating medium temperature distribution along the heating surface was assumed constant and corresponding to the working point ($D/D_0 = 1$); that is, it was supposed that variation of the heated medium flow rate has no substantial effect on the temperature of the heating medium. Practically, this case may realize, for instance, when the heating medium flow rate is high ($G_0 \gg D_0$), or when the thermal-hydraulic characteristic of one of the steam generator tubes with crossflow of multiturn coils by a well mixed heating medium, the temperature of which is practically independent of thermal condition of the pipe in question. As one can see from Figure 5.14, the thermal-hydraulic characteristic in the vicinity of the working point is unambiguous and features a sharp rise of the curve.

In the second case (curve 1, Figure 5.14), a calculation of the thermal-hydraulic characteristic assumed that the working fluid flow rate variation causes, according to heat balance, a change of the heating medium temperature distribution along the heating surface, while the heating medium flow rate and inlet temperature are kept constant. In reality, such conditions can occur in a tube system of the parallel-connected steam generators, or in one of parallel tubes of a steam generator in the absence of the heating medium mixing. Under some conditions [86,87], the thermal-hydraulic characteristic shows an ambiguity region in the second case (curve 1, Figure 5.14).

Thus, Figure 5.14 indicates that, depending on the heat input distribution along the steam generator length, various deformations of the channel pressure drop variation curve are possible depending on the heated medium flow rate in the working point region. These deformations can span from the positive stable slope with a large gradient up to the unstable characteristic curve with an ambiguity region.

Let us analyze the influence of heat input distribution on the shape of the thermal-hydraulic characteristic curve and possible appearance of the ambiguity section.

From the previously described mechanism of an ambiguity region appearance on the thermal-hydraulic characteristic curve of a steam-generating channel with the superheating section, it follows that the main reason is the nonlinear variation of lengths of the superheating, evaporating, and economizer sections caused by varying the heated medium flow rate. Let us determine the regularities in variation of heat input to the steam-generating channel depending on the medium flow rate.

To analyze this regularity, let us consider the results of calculations performed for a coil-type steam generator in which water with the inlet temperature of 100°C, exit pressure (P_e) of 4 MPa, and working fluid flow rate at the nominal load of 60 t/h was used as the heated medium, and water with the inlet temperature of 360°C, inlet pressure (P_{in}) of 20 MPa, and flow rate at nominal load of 300 t/h was used as the heating medium. The results of calculations for three values of heating medium flow rate ($2G_0$, G_0, and $0.5G_0$) are presented in Figures 5.15, 5.16, 5.17, and 5.18.

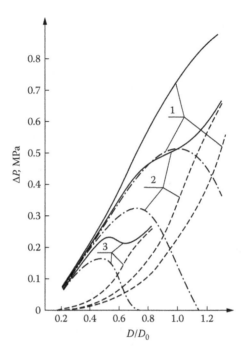

FIGURE 5.15
Pressure drop and its components' variation across the coil-type steam generator depending on the heated medium flow rate at different flow rates of the heating medium. 1: $G = 2G_0$; 2: $G = G_0$; 3: $G = 0.5G_0$; —— total pressure drop in the steam generator; _._._ friction pressure drop across the superheating section; ------- friction pressure drop across the evaporating section.

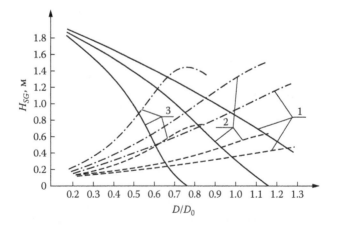

FIGURE 5.16
Variation of the economizer, evaporating and superheating sections lengths depending on the heated medium flow rate at different flow rates of the heating medium. 1 – $G = 2G_0$; 2 – $G = G_0$; 3 – $G = 0.5G_0$; —— superheating section; _._._ evaporating section; ------- economizer section.

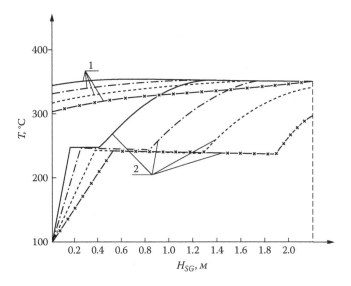

FIGURE 5.17
Distribution of the heating and heated media temperatures along the steam generator at $G = 2G_0$; 1 – heating medium; 2 – heated medium. ——— temperature distribution at D = 25 t/h; _._._ temperature distribution at D = 40 t/h; ------- temperature distribution at D = 60 t/h; -x-x- temperature distribution at D = 80 t/h.

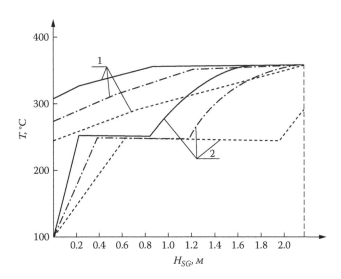

FIGURE 5.18
Distribution of the heating and heated media temperatures along the steam generator at $G = 0.5G_0$; 1: heating medium; 2: heated medium; ——— temperature distribution at D = 20 t/h; _._._ temperature distribution at D = 30 t/h; ------- temperature distribution at D = 40 t/h.

It can be seen from Figure 5.15 (curve 1) that when the heating medium flow rate is $2G_0$, the steam generator thermal-hydraulic characteristic curve has a stable positive slope with a large gradient in the entire range of the heated medium flow rate variation ($D/D_0 = 0.2/1.4$). At the nominal flow rate of the heating medium ($G = G_0$), the steam generator thermal-hydraulic characteristic curve maintains the stable positive slope, but has a small gradient in a certain range of heated medium flow rates (curve 2 of Figure 5.15). And, finally, at $G = 0.5G_0$, the thermal-hydraulic characteristic curve has an ambiguity region (curve 3, Figure 5.15.)

Let us consider the cause of the thermal-hydraulic characteristic deformation at different fixed flow rates of the heating medium with other parameters being constant. At a high flow rate of the heating fluid ($G = 2G_0$), the difference of its inlet and exit temperatures is small; with a large temperature difference between the heating medium and saturation temperature of the heated medium, a change in the heated medium flow rate has an insignificant effect on the temperature head between the heating and heated media (Figure 5.17). Thus, the heat load in the economizer and evaporating sections varies but little at the heated medium flow rate variation. It is clearly seen from Figure 5.16 (curve 1) that, in accordance with heat balance, it leads to an almost linear change of the lengths of the economizer, evaporating, and superheating sections when the heated medium flow rate is varied.

A simple model yields for the preceding case an approximate estimate of the condition stipulating the existence of an ambiguity region on the channel hydraulic characteristic. At the constant specific density of the heat flux, with neglect of pressure losses due to acceleration and gravity components, as well as of inlet and exit throttling, the channel pressure drop may be written as

$$\Delta P_C = \frac{\xi_{ec}}{2d_h A^2 \rho_L} \cdot \frac{i_s - i_{in}}{q\Pi} G^3 + \frac{\xi_{TP}}{2d_h A^2 \bar{\rho}_{TP}} \frac{i_{LG}}{q\Pi} G^3$$

$$+ \frac{\xi_{sh}}{2d_h A^2 \bar{\rho}_{sh}} \left[H_h - \frac{(i_s - i_{in}) + i_{LG}}{q\Pi} G \right] G^2,$$

(5.19)

where
  A is the channel flow area
  Π is the channel heated perimeter
  d_h is the channel equivalent diameter
  H_h is the channel heated length
  G is the coolant flow rate
  q is the heat flux specific density
  ξ_{ec}, ξ_{TP}, and ξ_{sh} are friction coefficients in the economizer, evaporating, and superheating sections, respectively
  i_{in}, i_s, and i_{LG} are the heated medium enthalpies at the inlet and at the saturation line, respectively, and the heat of evaporation

ρ_L, $\bar{\rho}_{TP}$, and $\bar{\rho}_{sh}$ are the average heated medium densities at the economizer, evaporating, and superheating sections, respectively

It may be assumed approximately that pressure losses due to friction in the economizer section are much less in comparison with ΔP_C. Firstly, such an assumption is close to reality, since in presence of a superheating section the share of the economizer section is much smaller than that of the evaporating section; secondly, the assumption makes the stability estimate more conservative, since the friction pressure drop in the economizer section proves to be a stabilizing factor as regards the static instability.

Since the bulk quality X changes from 0 to 1 in the evaporating section of a steam-generating channel with a superheating section, then, taking assumptions from Lokshin, Peterson, and Shvarts [12] into account and performing simple transformations, we can write (5.19) as

$$\Delta P_C = \frac{\xi_{ec}}{2 d_h A^2 \rho_L} \frac{\bar{\psi}}{4} \left(\frac{\rho_L}{\rho_G} \right) \frac{i_{LG}}{q\Pi} G^3 + \frac{\xi_{ec}}{2 d_h A^2 \rho_L} \left(\frac{\rho_L}{\rho_G} \right)$$

$$\times a \cdot \left[H_h - \frac{(i_s - i_{in}) + i_{LG}}{q\Pi} G \right] G^2,$$

(5.20)

where $\bar{\psi}$ is the correction for two-phase flow inhomogeneity in the evaporating section, derived from the nomogram in Lokshin et al. [12]

$$a = \left(\frac{\rho_G}{\bar{\rho}_{sh}} \right).$$

Let us differentiate (5.20) with respect to flow rate and, by determining

$$H_{ec} = \frac{i_s - i_{in}}{q\Pi} G;$$

(5.21)

and

$$H_{TP} = \frac{i_{LG}}{q\Pi} G$$

(5.22)

from the heat balance, we obtain

$$\frac{d\Delta P_C}{dG} = \frac{\xi_{ec}}{2 d_h A^2 \rho_L} \left(\frac{\rho_L}{\rho_G} \right) G \left[\frac{3}{4} \bar{\psi} H_{TP} + 2 a H_h - 3 a H_{ec} - 3 a H_{TP} \right].$$

(5.23)

The channel flow static stability requires the fulfillment of $\dfrac{d\Delta P_c}{dG} > 0$ or, from (5.23),

$$H_h - \tfrac{3}{2}H_{ec} - \left(\tfrac{3}{2} - \tfrac{3}{8}\dfrac{\bar{\Psi}}{a}\right)H_{TP} \geq 0 \qquad (5.24)$$

Taking into account the small value of 3/8 ψ/a, (5.24) may be written as:

$$H_h \geq \tfrac{3}{2}(H_{ec} + H_{TP}) \qquad (5.25)$$

or

$$H_{sh} \geq \tfrac{1}{2}(H_{ec} + H_{TP}) \qquad (5.26)$$

In order to analyze the influence of parameters on the static instability boundary in presence of the superheating section, (5.26) may be written for convenience as:

$$H_{sh} \geq \tfrac{1}{2}\left(\dfrac{i_s - i_{in}}{q\Pi} + \dfrac{i_{LG}}{q\Pi}\right)G \qquad (5.27)$$

For the results of the coil-type steam generator calculations at $G = 2G_0$ (curve 1) presented in Figures 5.15, 5.16, and 5.17, $H_{sh} = 1$ m, while $0.5\,(H_{ec} + H_{Tp}) = 0.6$ m. In accordance with the condition from (5.27), it ensures flow static stability.

From the condition of static stability (5.27) it may be seen that an increased inlet enthalpy of the heated medium, with all other parameters unchanged, leads to a smaller value of the right part of (5.27) and stabilizes the flow as regards the static instability. An increase of the heated medium pressure at the constant inlet subcooling leads to a lower value of the right part of (5.27) at the expense of the decreasing i_{LG} and stabilizes the flow in terms of static instability.

Let us consider the physical mechanism of said effect of inlet subcooling and pressure of the heated medium. As it is seen from Figure 5.15, the friction pressure drop component in the superheating section increases as the flow rate grows, reaches its maximum, and then starts decreasing and achieves zero at the superheating section degeneration. The friction pressure drop reduction in the superheating section after the maximum has been reached is explained by the fact that the rate of the superheating section decrease in this area exceeds the quadratic rise of the flow rate (i.e., $G^2 \cdot H_{sh}$ decreases with the increasing flow rate).

For the uniform heat flux density, it follows from (5.15, 5.16 and 5.17) that the maximum pressure loss due to friction across the superheating section is attained when the superheating section length equals approximately $0.4 \div 0.5\ H_h$. Actually, starting from this superheating section length, there is no excess heating surface, and an increase in flow rate leads to a lower channel exit steam temperature.

Evidently, the appearance of an ambiguity region on the hydraulic characteristic requires the rate of friction pressure loss in the superheating section at the decreasing flow rate to exceed the rate of pressure loss in the evaporating section. The friction pressure losses in the economizer section are negligible. With the increasing flow rate, that of the friction pressure loss decrease in the superheating section depends on the rate of superheating section shortening, which is determined by the increasing lengths of the economizer and evaporating sections; the rate of friction pressure loss increase in the evaporating section depends on the increase of the evaporating section length only. Therefore, an increase in the inlet enthalpy of the heated medium decreases the economizer section length and the economizer section length growth rate (and accordingly that of the superheating section length decrease) at the increasing flow rate, without affecting the rate of the evaporating section length change.

Depending on mass flow rate and pressure of the heated medium according to (5.26), variation of the coefficient ψ (a correction for the two-phase flow inhomogeneity) may also affect flow static instability. At $\psi > 1$, when the mass flow rates and pressures are low, the coefficient of H_{Tp} in (5.26) decreases, thus leading to a decrease of i_{LG}/qH in (5.27) and an increase in flow static stability in the presence of a superheating section. At $\psi < 1$ (high mass flow rates and pressures), the flow static stability decreases. This effect is also pretty obvious. At $\psi > 1$, the two-phase flow friction coefficient increases in the evaporating section and the rate of friction pressure loss increase in the evaporating section grows with the increasing flow rate, while the rate of friction pressure loss reduction in the superheating section remains unchanged. This either liquidates or decreases the negative slope section on the hydraulic characteristic curve. The opposite effect is observed at $\psi < 1$.

Now let us consider a case when the specific heat flux, being constant in the economizer and evaporating sections, is different in these sections (i.e., $q_{TP} \neq q_{ec}$). Then, by repeating the reasoning applied in the case of heat flux constancy along the channel, instead of (5.27) we arrive at the stability condition in the form of

$$H_{sh} \geq \frac{1}{2}\left(\frac{i_s - i_{in}}{q_{ec}\Pi} + \frac{i_{LG}}{q_{ec}\Pi}\frac{q_{ec}}{q_{TP}}\right)G \qquad (5.28)$$

At $q_{TP} > q_{ec}$, it is obvious from (5.28) that the two-phase flow static stability in the channel with a superheating section is worsening when the ratio between heat fluxes in the evaporating and economizer sections is increasing, i.e., when the heat flux is nonuniform and increases along the channel length.

The approximate qualitative analysis of flow static instability at constant heat input at the economizer and evaporating sections explains the appearance of an ambiguity region of the thermal-hydraulic characteristic resulting from the coil-type steam generator calculations at $G = 0.5G_0$ presented in Figures 5.15, 5.16, and 5.18 (curves 3). The difference between the heating medium inlet and exit temperatures increases under such conditions with the heated medium increasing flow rate (Figure 5.18), and this leads to a decreased temperature head between the heating and heated media in the economizer section. As is seen from Figure 5.18, this occurs more abruptly than in the evaporating section. The reduced temperature head leads to a higher rate of the economizer and evaporating sections lengthening and to a higher rate of the superheating section shortening (Figure 5.16, curve 3). The rate of the superheating section length reduction past the maximum of friction pressure loss in the superheating section exceeds that of the superheating section length growth. The preceding nonlinear change of the superheating and evaporating section lengths with the varying flow leads to a more intensive reduction of the friction pressure loss in the superheating section as compared to the increased friction pressure loss in the evaporating section (i.e., to the negative slope of the thermal-hydraulic characteristic).

Let us note another property of convective heating systems, considered in Dulevsky [189], which may influence the area of flow static instability. If thermal inertia of the heating medium is such that the time of its temperature field restructuring substantially exceeds that of coolant transport in the economizer and evaporating sections, then the static flow instability in the case of short-term random disturbances may be considered to have the constant temperature field of the heating medium in the working point. It has been shown previously that it improves stability of the system. In the case of long-term disturbances, this property of the heating medium may lead to a longer time of transition to a new stable state when the system finds itself in the static instability area.

5.3.3 Influence of Parallel-Channel Operation and Means of Controlling Parameters

Hydrodynamic interaction between parallel channels and the effect of the circulation loop cause a change in the conditions of static instability onset (conditions of Equations 5.8 and 5.13), if compared with those of static instability onset in the isolated channel with constant pressure drop (condition of Equation 5.11).

Let us use experimental investigations [87] to illustrate the effect of parallelism of channels (or steam generators) and the circulation loop on static instability of channel flow with convective heating.

In Belyakov, Breus, and Loginov [87], the test section consisted of two parallel steam generator modules. Each module represented a vertical tube 273×30 mm^2 and 2300 mm high, with a bundle of three-pass coils (190 mm in diameter; with tube diameter of 12×2 mm^2) arranged inside. Supercritical water with $P_{in} = 25$ MPa and $T_{in} = 375°C$ was used as the heating medium, whereas water at $T_{in} = 100°C$ and module inlet pressure $p_{in.h} = 10$ MPa was used as the heated medium. The flow rates of the heated media were regulated by the piston pump. In this case, the heating medium was passed through the heater prior to being fed into the intertube space of the steam generator module, while the heated medium was fed to the module coiled tubes via the throttle with a pressure drop of $\Delta P_{in} = 15$ MPa. The heating and heated media flows were countercurrent.

Stable operation of one module with three parallel coils and the second module out of operation was studied at the first stage of the investigations. The inlet heating medium parameters were kept constant, while the heated medium flow rate was reduced in small steps, with parameters kept constant for some time, from a high initial value ($D = 1.35D_0$) down to full stabilization.

An experimental thermal-hydraulic characteristic curve of the steam generator module for one of the regimes is presented in Figure 5.19 showing the minimum of this characteristic at a certain heated medium flow rate. This minimum corresponds to the appearance of a superheating section at the module coil's exit. The further decrease of the heated medium flow rate results in the module pressure drop growth and superheating section increase. With the continued flow rate reduction, the pressure drop reaches its maximum and starts to decrease. The point of the module pressure drop maximum corresponds to the superheated steam temperature that is close to the heating medium inlet temperature.

As one can see from Figure 5.19, a thermal-hydraulic characteristic curve has an ambiguity region and supposes the possibility of appearance of the loop and interchannel flow static instability. However, in the preceding series of experiments, there was no deterioration of flow static stability in any of the parallel coils or the circulation loop. The heated medium flow rates were stable also on the unstable branch of the thermal-hydraulic characteristic curve. Let us consider the causes of such a behavior of the flow rate.

Two reasons make loop static instability impossible in this system. First, intensive flow throttling past the pumps makes the hydraulic characteristic curve of the circulation loop unambiguous, despite the unstable thermal-hydraulic characteristic curve of the steam generator module. Second, the supply of the heated medium into the steam generator module is provided by the piston-type pump that has an almost vertical head characteristic (Figure 5.6) and ensures flow static stability even with the ambiguity section on the loop hydraulic characteristic curve (see the condition of Equation 5.8).

The absence of interchannel static instability is less evident in the given experimental series; on the contrary, there was stable operation of parallel

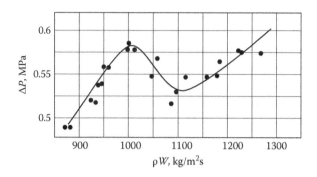

FIGURE 5.19
Experimental thermal-hydraulic characteristic curve of the steam generator module. (From Belyakov, I. I., Breus, V. I., and Loginov, D. A. 1988. *Atomnaya Energia* 65 (1): 12–17.) P_{in} = 25 MPa; $P_{in.h}$ = 10 MPa; T_{in} = 375°C; t_{in} = 100°C; G_0 = 2500 kg/h. Dark circle: experimental data; solid line: averaging curve.

coils on the unstable section of the thermal-hydraulic characteristic curve of the steam generator module at constant flow rates through all the coils. The reason for this is that, though parallel coils are identical, their thermal-hydraulic characteristic differs from that of the module (Figure 5.19) composed of these coils. Indeed, the determination of the module thermal-hydraulic characteristic shows the distribution of the heating medium temperatures to vary according to the heat balance. For the fixed heated medium flow rate through the module that corresponds to the ambiguity region of the thermal-hydraulic characteristic curve, a change of the flow rate through one of the parallel coils keeps the flow rate through the module constant. With good mixing of the heating medium in the module, the heating medium temperature variation along one coil with the varying heated medium flow rate remains unchanged. As has been shown before, such a method of heat input leads to an increased channel flow static stability and may cancel the ambiguity region (see Figure 5.14, curve 2).

The second series of experiments involved two identical parallel modules. Each module had two operating parallel coils. Figure 5.20 shows the thermal-hydraulic characteristic of two steam generator modules [87], as well as the distribution of flow rates and pressure drops for each module at the heated medium total flow rate variation. One can also see that with the total flow rate through two modules of ρW > 1500 kg/m²s, corresponding to the right unambiguous branch of the thermal-hydraulic characteristic curve, the modules operate stably and have almost the same flow rates. When the total flow rate decreases below the level corresponding to the minimum of the thermal-hydraulic characteristic curve (ρW < 1500 kg/m²s), a considerable module flow maldistribution is noted. A reduction in the total flow rate results in a considerably decreased flow rate through one of the parallel modules, while the flow rate through the other parallel module increases and stays on the right stable branch of the thermal-hydraulic characteristic

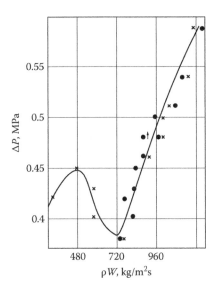

FIGURE 5.20
Thermal-hydraulic characteristics curve of a steam generator composed of two parallel modules. (From Belyakov, I. I., Breus, V. I., and Loginov, D. A. 1988. *Atomnaya Energia* 65 (1): 12–17.) P_{in} = 25 MPa; $P_{in.h}$ = 8.5 MPa; T_{in} = 375°C; t_{in} = 95°C. Dark circle: x are experimental data on flow rate and pressure drop, in first and second modules, respectively.

curve. At the reduction of flow rate through the module with maldistribution, the flow rate at first stays on the unstable branch of the thermal-hydraulic characteristic curve, keeping operation stable. With the further reduction of the total flow rate through the module with maldistribution, it goes to the left stable branch of the thermal-hydraulic characteristic curve. This caused oscillatory instability in the module and parallel coils. In this case, the flow rate fluctuation amplitude for the coils was much greater as compared to that for the module. The onset of large-amplitude fluctuations prevented further total flow rate reduction.

Stable operation of the module with maldistribution on the unstable branch of the thermal-hydraulic characteristic curve for the system investigated in Belyakov et al. [87] was caused by the influence of the parallel module operating on the right stable branch of the thermal-hydraulic characteristic curve. Indeed, it can be seen from Figure 5.20 that the magnitude of the $\partial \Delta P_m / \partial G$ derivative modulus on the right branch of the thermal-hydraulic characteristic curve is larger than that of $\partial \Delta P_m / \partial G$ on the unstable branch of the thermal-hydraulic characteristic curve. According to the condition of (5.12), at K = 1, flow stability is ensured on the unstable branch of the thermal-hydraulic characteristic curve. If the number of parallel modules in the system is larger than two, then the coefficient K, in the condition of static stability of (5.12), becomes smaller than unity in the right part of this inequality. Under certain conditions, static instability appears and leads to a situation when it is impossible to keep the flow rate on the unstable branch of the thermal-hydraulic characteristic curve. With a sufficiently large number of parallel modules, the coefficient in the right part of (5.12) tends to zero, and the condition of static stability for such a system becomes that of (5.10); that is, static instability appears when the minimum of the thermal-hydraulic characteristic is attained.

In contrast to the oscillatory instability, experimental investigation of static instability in a real system with a large number of parallel components by simulating it on a thermophysical test facility with two parallel components is incorrect.

Thus, in steam generators with convective heating, flow static instability in parallel channels and parallel steam generators is stabilized due to (1) heating medium mixing in the steam generator intertube space, (2) heating medium thermal inertia, and (3) hydrodynamic interaction of parallel modules.

It has been shown before that the appearance of the ambiguity region on the thermal-hydraulic characteristic when a partial power is used for convective heating may be caused by the decreasing flow rate of the heating medium. This is an indication of the fact that the method of parameter control, depending on the power level of a power plant with the convectively heated steam generators, may influence static flow stability of the system. As a rule, the basic steam generator controlled parameters, apart from the heated medium flow rate, are the heating medium flow rate and inlet temperature. Let us consider the effect of these parameters on static flow stability in the convectively heated steam generators.

When power of a power plant decreases, the flow rate of the heating medium usually decreases in proportion to the decreased flow rate of the heated medium. Simultaneously, the heating medium inlet temperature may decrease according to the given law, and the heated medium steam pressure at the steam generator exit may vary. The decreased heating medium flow rate may lead, as shown previously, to a stronger influence of the heated medium on the unfavorable distribution of heat input along the steam generator length and contribute to the initiation of flow static instability. Therefore, one of the ways of flow stabilization at partial power levels is to increase flow rate by decreasing the coefficient in the linear dependence of the heating medium flow rate on power or by presetting nonlinearity of flow variation.

A change of the heating medium inlet temperature at a preset flow rate also significantly influences the static stability boundary. It follows from the physical mechanism of static instability initiation in presence of the superheating section that the ambiguity region of the thermal-hydraulic characteristic corresponds to elimination of excessive steam generator surface when the increased heated medium flow rate starts sharply reducing the superheated steam temperature from that of the heating medium inlet temperature down to the heated medium saturation temperature. Therefore, when the steam generator's excessive surface is minimal at partial power levels and the working point is near the static stability boundary, the flow can be stabilized and the working point moved off from the stability boundary region by increasing the heating medium inlet temperature. In this case, the increased temperature difference between the heating and heated media increases the excessive superheating surface, thereby ensuring static stability of the system.

A similar effect on the static stability boundary is exerted by variation of the heated medium pressure. Indeed, the decreasing pressure of the heated medium reduces the saturation temperature, increases the temperature difference between the heating and heated media, and increases flow static stability—as the increasing inlet temperature of the heating medium does.

Also, let us briefly consider the effect of the variation of heat flux distribution in boiler unit furnaces with the changing load on flow static stability in components. Heat absorption may vary along the steam-generating tubes due to the changed furnace chamber aerodynamics at partial power levels. In this case, if the heat absorption redistributes at the heat load decrease toward a higher heat load in the evaporating section and a lower one in the economizer section, then, according to the condition of (5.28), the probability of static instability appearance increases. This circumstance should be taken into account in design work and, if necessary, different heating surface layouts or burner characteristics should be chosen.

5.3.4 The Ballast Zone

Belyakov, Kvetnyi, and Loginov [86,90] first discovered that, at a definite combination of parameters, steam generator convective heating may lead to the heated medium flow instability. The mechanism of its initiation is peculiar in some respects.

When the heated medium flow rate is reduced, or the heating medium inlet temperature is increased in a steam generator with the superheating section, a part of the superheating surface may get excluded from heat exchange due to a very small temperature difference between the heating and heated media in this section. This part of the surface located in the steam generator exit section is called the ballast zone. Such a position of the ballast zone is typical of steam generators, especially those with convective heating and countercurrent flow of media. However, the ballast zone may be located differently in steam generators—that is, in the evaporating–economizer section. It may happen because, with the countercurrent media flows, the heating medium temperature is becoming lower along the steam generator evaporating surface and approaches the saturation temperature of the heated medium at a certain distance from the boundary between the evaporating and economizer sections. Thus, the temperature difference between the heating and heated media is reduced to the minimum in the remaining part of the evaporating and a part of the economizer sections, with a resultant increase in the heat exchange surface in these sections of the steam generator.

Evidently, the possibility of the ballast zone formation in the evaporating–economizer section (at a preset geometry, inlet subcooling, and the heated medium flow rate) is contributed by (1) the reduced heating medium inlet temperature, (2) the increased pressure and hence the heated medium saturation temperature, and (3) the heating medium decreased flow rate, which leads to a sharper drop of the heating medium temperature along the steam generator surface.

Figure 5.21 [86] illustrates the distribution of temperatures and heat exchange areas during the low load operation in cases of two limiting options of the ballast zone position in the superheating and evaporating–economizer sections.

Following the analysis presented [86,90], let us show (a) the conditions at which the preceding options of the ballast zone location are realized at a fixed value of the heating medium inlet temperature, and (b) the mutual effect of these zones on the steam generator stable operation.

In accordance with Belyakov et al. [86], let us show the relationship between parameters that determines the condition of transfer from the ballast zone in

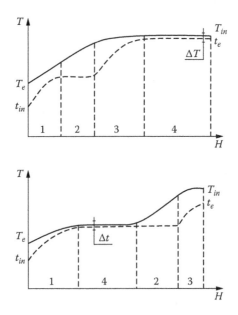

FIGURE 5.21
Ballast zones' location in a steam generator [86]. (a) Superheating section ballast zone; (b) economizer-evaporating section ballast zone; ——— heating medium temperature; ------- heated medium temperature; 1, 2, 3, 4: economizer, evaporating, superheating, and ballast sections, respectively.

the superheating section to that in the evaporating–economizer section and vice versa.

The maximum heat input from the heating medium in the superheating–evaporating section is

$$Q = G \cdot (I_{in} - I_0), \tag{5.29}$$

where G is the heating medium flow rate, I_{in} is the heating medium inlet enthalpy, $I_0 = I(t_s + \Delta t)$ is the minimal enthalpy of the heating medium in the ballast zone, t_s is the heated medium temperature at the saturation line, and Δt is the minimum temperature difference between the heating and heated media in the evaporating–economizer section ballast zone.

The maximum heat that may be obtained by the heated medium in the evaporating–superheating section is

$$Q = D(i_e - i_s), \tag{5.30}$$

where D is the heated medium flow rate, $i_e = i(T_{in} - \Delta T)$ is the steam enthalpy at the steam generator exit, i_s is the heated medium saturation enthalpy, and ΔT is the minimum temperature difference between the heating and heated media in the superheating section ballast zone.

By equating (5.29) and (5.30) and assuming Δt and T close to zero, it is possible to derive the condition of unstable relationship between parameters, a departure from which either side may lead to the ballast zone appearance in the superheating or in the evaporating–economizer sections.

The relationship has the following form:

$$\left(\frac{G}{D}\right)_{cz} = \frac{i_e(T_{in}) - i_s}{I_{in} - I(t_s)}. \tag{5.31}$$

When flow rate ratio is above critical,

$$\left(\frac{G}{D}\right) > \left[i_e(I_{in}) - i_s\right]\bigg/\left[I_{in} - I(t_s)\right]'$$

the ballast zone is located in the superheating section, and in the evaporating-economizer section when it is below critical.

According to Belyakov et al. [86,90], the ratio between the heating and heated medium flow rates in the once-through steam generators (with water as the heating medium) under operating conditions is generally much above critical, and the conditions of the ballast zone appearance in the evaporating–economizer section are absent, as a rule. In sodium-heated once-through steam generators in startup and some emergency conditions, as well as under partial-load operation, the flow ratio may approach the critical. In such a case, random or short-term deviations of such parameters as inlet temperature and the heating medium flow rate, pressure, and the heated medium flow rate may cause a shift of the ballast zone from the superheating to the evaporating–economizer section, and vice versa.

Let us consider the interrelation between the formation and shift of ballast zones and the heated medium static flow instability. The convectively heated steam generators usually use a compact heat exchange surface in the form of sharply bent coiled tubes with high hydraulic friction in the evaporating and especially in the superheating heat exchange sections. When the ballast zone is located in the superheating section, the pressure losses there determine the pressure drop across the steam generator. If a random or short-term deviation of parameters results in a flow rate ratio below the critical (G/D < $(G/D)_{cz}$), then the ballast zone starts shifting. In this case, the evaporating and economizer sections increase substantially at the expense of the ballast zone formation, while the superheating section decreases considerably.

The superheating section decrease lowers pressure losses across the steam generator and increases the heating medium flow rate at a preset pressure drop (i.e., an ambiguity region appears on the hydraulic characteristic curve with a sharp negative slope of the unstable branch). As an illustration, Figure 5.22 shows a thermal-hydraulic characteristic of a liquid sodium-heated once-through steam generator [86] calculated on the basis of the heated and heating media flow rates' ratio close to critical. One can see that the unstable branch of the thermal-hydraulic characteristic has a large negative gradient. A small variation of one of the determining parameters (a 5% increase in the heating medium flow rate) shifts the ambiguity region of the thermal-hydraulic characteristic toward higher flow rates of the heated medium.

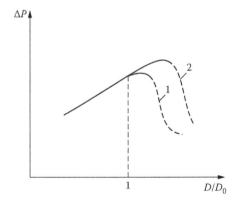

FIGURE 5.22
Thermal-hydraulic characteristic of a counterflow sodium-water steam generator with a developed ballast zone [86]. 1: $G = G_0$; 2: $G = 1.05\,G_0$.

As is seen from the condition of (5.31), when the ballast zone shifts to the evaporating–economizer section due to the superheating section reduction, the heated medium flow rate increases in a nonlinear fashion and steam enthalpy at the steam generator exit decreases. When thermal inertia is high, the G/D flow rates' ratio may be above critical in the dynamic mode, and a reverse process of the ballast zone decrease in the evaporating–economizer section and its formation in the superheating section starts. The zone may shift periodically; the rate of ballast zone shifting depends on thermal inertia of the heating and heated media, heat transfer wall characteristics, and the kind of deviating parameter.

The periodic shifting of the ballast zone across the heating surface may cause temperature fluctuations on the heating surface with a peak-to-peak amplitude between the heating medium inlet temperature and the heated medium saturation temperature, as well as a possible ejection of the steam–water mixture into the exit common header. Apart from instability in steam generator operation under the described conditions and associated deterioration of thermal reliability, the ballast zone appearance in the evaporating–economizer section essentially reduces thermal efficiency of a steam generator.

In these conditions, a large negative slope of the unstable branch of the thermal-hydraulic characteristic curve impairs the efficiency of inlet throttling in flow stabilization in the steam generator, as, according to Belyakov et al. [86], the local resistance coefficient required for stabilizing the flow may be so large that the steam generator pressure drop in the nominal operation mode may be impermissibly great.

Evidently, this kind of instability may be stabilized most efficiently by changing the method of controlling parameters, which extend the working region with the unambiguous thermal-hydraulic characteristic in order to prevent the operating parameters from approaching the dangerous $(G/D)_{cz}$ value too closely.

Belyakov et al. [86] recommend ensuring flow stability by maintaining the relationship

$$G/D > 1.1(G/D)_{cz} \qquad (5.32)$$

for the operating region parameters.

For an approximate evaluation of $(G/D)_{cz}$ in sodium–water steam generators, it is possible to use the diagram numerically obtained in Belyakov et al. [90] and presented in Figure 5.23. In it, the $(G/D)_{cz}$ value dependence is shown in dependence on the heating medium (Na) inlet temperature at different pressures of heated water coolant.

In conclusion, it should be mentioned that the preceding analysis of operation of steam generators with ballast zones was made in Belyakov et al. [86,90] on the basis of numerical investigations and has been validated

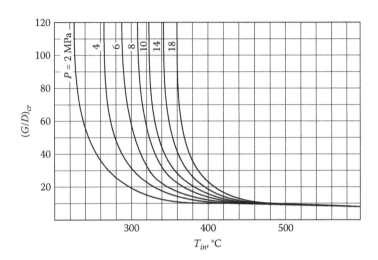

FIGURE 5.23
Flow rates' critical ratio for the counterflow sodium-water steam generator with a developed ballast zone. (From Belyakov, I. I., Kvetny, M. A., and Loginov, D. A. 1985. *Atomnaya Energia* 58 (3): 155–159.)

experimentally only indirectly. In the future, it will be necessary to conduct special experimental investigations of operation modes with the unstable ballast zone.

5.4 Hydraulic Characteristic Ambiguity in Cases of Coolant Downflow and Upflow–Downflow

Heat exchangers and steam generators with two-phase and single-phase coolant downflow movement have found wide application in some areas of engineering. In particular, such systems include boiler heating surfaces with N, U, and ∩-shaped panel arrangement; economizer downflow sections in some types of steam generators; and heat exchangers of NPP auxiliary systems and safety systems, as well as the two-phase coolant systems in chemical, space, and other branches of engineering.

At low and moderate coolant velocities when variation of the pressure drop gravity component becomes commensurate with (or even higher than) variation of friction pressure losses at the flow rate change in the heated channel with the coolant downflow (or upflow–downflow), an ambiguity region may appear on the hydraulic characteristic, causing static instability.

Let us consider the case in question in more detail concerning the distribution of pressure drop components in the channel with the variation of coolant flow rate. The pressure drop variation in the vertical channel from the upper chamber down to the lower one, if the coolant downflow is assumed as positive flow and the acceleration losses are neglected, may be written as

$$\Delta P_c = -\int_0^H \rho g\, dz + \int_0^H \xi \frac{G_c|G_c|}{2\rho d_h A^2} dz + \xi_{in.t} \frac{G_c|G_c|}{2\rho_L A^2}. \tag{5.33}$$

Qualitative behavior of pressure drop components in (5.33) is shown in Figure 5.24. As one can see from (5.33) and Figure 5.24, the friction and gravity pressure drop components have different signs and different nonlinear manners of variation with the changing coolant flow rate.

At high coolant flow rates, the heated channel is first filled with water and the friction pressure loss exceeds that due to gravity. With the decreasing flow rate, the friction pressure drop component decreases in accordance with the quadratic law, while the gravity component increases but insignificantly until coolant is in the single-phase state. As the flow rate keeps reducing and coolant starts boiling, the gravity component tends to increase more intensively. When the increasing pressure drop gravity component absolute value exceeds that of the decreasing pressure drop friction component, while the flow rate keeps reducing, a section with the negative slope appears on the hydraulic characteristic curve. The section may pass through the flow rate zero line and disappear with the negative coolant flow rate (see Figure 5.24).

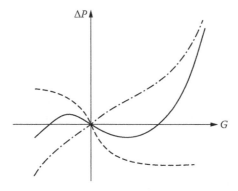

FIGURE 5.24
Qualitative change of pressure drop components in a channel with coolant downflow: ——— variation of the total channel pressure drop (ΔP_C); _._._ variation of the friction pressure drop component (ΔP_f); ------- variation of the gravity pressure drop component (ΔP_g).

The results of, for example, Proshutinsky and Lobachev [91] may serve as experimental proof of the ambiguous region appearance on the hydraulic characteristic curve of the channel with the coolant downflow when the boiling region appears at the channel exit entirely due to the gravity pressure drop component behavior. Their work describes the experimentally obtained hydraulic characteristics of identical single channels with the coolant upflow and downflow. The experimental hydraulic characteristic of a channel with the coolant upflow was represented by a monotonous curve that was rising with the increasing flow rates of single- and two-phase coolants. The hydraulic characteristic of the identical channel with the coolant downflow was ambiguous. In this case, the characteristic curve minimum corresponded to the beginning of coolant boiling at the channel exit. This fact proves that the basic mechanism of the ambiguity region formation on the hydraulic characteristic curve in this case is the regularity in the gravity pressure drop change—instead of the previously considered (see Section 5.3) mechanism of the ambiguity region formation on the hydraulic characteristic curve at the expense of the increased friction due to the appearance of the boiling region at the channel exit.

In the general case, it should be noted that the formation of the ambiguity region on the hydraulic characteristic can be contributed by all three mechanisms described before: (1) formation of the boiling region at the channel exit, (2) nonlinear variation of section lengths in the channel with the heated medium superheating, and (3) variation of the gravity pressure drop component with the coolant downflow. In some cases, this may even lead to formation of several ambiguity regions on the hydraulic characteristic curve.

The preceding mechanisms of ambiguity regions' formation have been considered separately because each of them may be dominant, and only one of them may determine the onset of static instability in a particular situation.

It has been shown [91–93,193,194] that at small heat flux densities and low loop circulation flow rates, static instability with the reversal of coolant circulation in the parallel-channel system with coolant downflow is possible even with the single-phase coolant and with no boiling. Since the development of static instability of the single and two-phase downflow coolant has some specific features as regards thermal-hydraulics, let us consider the mechanism of instability onset under these conditions in separate sections.

5.4.1 Two-Phase Coolant Flow Static Instability

As a rule, power plants and other heat exchange systems have no components with two-phase downflow at nominal power levels. However, it may be found at partial-load operation, in startup, and in emergency conditions, with some heat exchange sections disconnected or with partially operating equipment.

For example, the coolant downflow may be observed in some or in all channels of the NPP core or in pool-type research reactors with the downflow circulation during accidents due to the primary circuit depressurization or complete loss of power supply. Such accidents lead to the cessation of loop circulation or to the change in its direction following the safety systems actuation. Static instability may occur in channels with coolant downflow under the described conditions in the form of circulation reversal in some channels, or interchannel circulation appearance in the absence of loop circulation, or when loop flow rates are low.

It has been stated before that the prevailing mechanism of the considered type of static instability in the case of coolant downflow is the dominating variation of the gravity pressure drop component at the change of flow rate. Therefore, the boundary of the ambiguity region of the hydraulic characteristic curve, corresponding to its minimum, is shown in experimental and numerical investigations [46,85,91,92,94,95,195] to appear when the boiling section forms at the downflow channel exit (i.e., at the moment of the gravity pressure drop component's sharp rise with the decreasing flow rate). These peculiarities of this mechanism of the ambiguity region formation on the channel hydraulic characteristic curve make it possible to perform a sufficiently simple qualitative analysis of the influence of various flow and design parameters on the conditions of static instability appearance.

Let us first determine the heat input corresponding to the start of steam generation at the channel exit (Nigs) which, as has been stated previously, shall determine the relation of parameters at the point marking the minimum of the hydraulic characteristic curve; that is,

$$\text{Nigs} = N_{c.Bn},$$

where $N_{c.Bn}$ is the channel boundary power corresponding to the minimum of the hydraulic characteristic curve.

This power may be easily derived from the balance relationship

$$\text{Nigs} = \rho W \, A \, (i_{IGS} - i_{in}) = \rho W \, A \, \Delta i_{in} - \rho W \cdot A \, \Delta i_{IGS}, \tag{5.34}$$

where ρW is the coolant mass velocity, kg/m^2s; A is the channel flow area, m^2; i_{in} and i_{IGS} are coolant enthalpies at the channel inlet and at the cross section where steam generation starts, respectively, J/kg; and Δi_{in} and Δi_{IGS} are enthalpies of coolant subcooling below the saturation temperature at the channel inlet and at the cross section where steam generation starts, J/kg.

It has been shown [196] that, for the preset channel geometry, we may approximately write

$$\Delta i_{IGS} = Cq/\rho W, \tag{5.35}$$

where C is the constant dependent on channel geometry and found within C ≈ 150–300 and q is the surface heat flux density at the cross section where steam generation starts, W/m².

For the heat flux with a density constant along the height and taking (5.35) into consideration, (5.34) may be transformed into

$$N_{IGS} = N_{C.Bn} = \frac{1}{4}\pi\left(1+\frac{C\,d_h}{4\,H}\right)^{-1}\cdot\rho W\cdot d_h^2\Delta i_{in} \qquad (5.36)$$

or

$$\bar{q} = \frac{1}{4}\left(1+\frac{C\,d_h}{4\,H}\right)^{-1}\cdot\rho W\cdot\frac{d_h}{H}\cdot\Delta i_{in}, \qquad (5.37)$$

where d_h is the channel equivalent diameter, m, and H is the channel heated height, m.

Expressions (5.36) and (5.37) represent the relationships of design and flow parameters that determine the minimum of the hydraulic characteristic in presence of the ambiguity region and correspond to the right stability region boundary in the parallel-channel system.

For analyzing the influence of parameters on the right boundary of the static stability region, let us rewrite (5.36) and (5.37) as

$$(\rho W)_{Bn} = 4\left(1+\frac{C\,d_h}{4\,H}\right)\cdot N_{C.Bn}\left(\pi d_h^2\Delta i_{in}\right)^{-1} \qquad (5.38)$$

or

$$(\rho W)_{Bn} = 4\left(1+\frac{C\,d_h}{4\,H}\right)\frac{H}{d_h}\cdot\left(\bar{q}\,/\,\Delta i_{in}\right), \qquad (5.39)$$

where $(\rho W)_{Bn}$ is the coolant mass velocity corresponding to the minimum of the hydraulic characteristic curve in presence of the ambiguity region.

Now, let us consider the effect of design and flow parameters on flow static instability in the channel with coolant downflow.

5.4.1.1 Effect of the Channel Heated Height

Let us assume that the preset parameters are the constant power (N_c), equivalent diameter (d_h), inlet subcooling (Δi_{in}), and pressure (P). Then, as is seen from (5.38), an increase of H in the presence of the ambiguity region on the hydraulic characteristic decreases the limiting coolant flow rate corresponding to the right boundary of the static stability region. On the one hand, it somewhat stabilizes the system as regards the static instability, since the ambiguity region of the hydraulic characteristic curve becomes decreased. However, on the other hand, the relative contribution of the friction pressure drop component reduces when the right stability boundary shifts toward lower flow rates, while the increased channel height increases the relative contribution of the gravity pressure drop component. Both factors increase steepness of the negative slope of the hydraulic characteristic curve toward the ambiguity region, thereby destabilizing the system. Thus, the increased channel heated height may somewhat decrease the region of flow static instability in the channel by means of the reduced right boundary flow and the simultaneously increased steepness of the slope of the unstable branch of the hydraulic characteristic curve.

However, as it follows from (5.38), such an effect of the channel heated height variation on static instability occurs at relatively small H/d_h values ($H/d_h < 150$). At $H/d_h > 150$, the influence of the increased channel heated height on $(\rho W)_{Bn}$ starts to decay (see Equation 5.38). At $\dfrac{C}{4} \dfrac{d_h}{H} \ll 1$, the increasing H has almost no effect on the stability boundary, and only the channel height influences the increasing negative slope of the ambiguity branch of the hydraulic characteristic curve (i.e., the increasing channel height definitely destabilizes the system).

Now, let us consider a case when constant values have been preset for $(\bar{q})$, d_h, Δi_{in}, and P. As is seen from (5.39), a higher H in this case may cause an increase in $(\rho W)_{Bn}$ (i.e., an increase of the region of static instability without a decrease or even with an increase in steepness of the hydraulic characteristic curve in the ambiguity region). The latter is due to the fact that when $(\rho W)_{Bn}$ is increasing and the relative contribution of friction pressure drop grows accordingly, the relative contribution of the gravity pressure drop component also increases with the growth of H. This effect of the channel heated height on static instability at constant q manifests itself almost linearly at large H/d_h values and decays at small values (see Equation 5.38).

5.4.1.2 Effect of Channel Equivalent Diameter

This effect on static instability is clearly seen from (5.38) and (5.39). At given constant values of N_c (or $\bar{q}$), H, Δi_{in}, and P, an increased d_h leads, on the one hand, to a reduced $(\rho W)_{Bn}$ (i.e., narrows the range of static instability and stabilizes the system) and, on the other hand, increases the negative slope of the hydraulic characteristic curve in the ambiguity region due to the decreased relative share of the friction pressure drop component in the total channel pressure drop. The friction pressure drop component is reduced due to two reasons: (1) the ambiguity region of the hydraulic characteristic shifts into the lower flow rates area, and (2) the reduced friction pressure drop absolute value because of the increased channel equivalent diameter. The described influence of d_h on static instability is manifested in full at large H/d_h values. At small values of this ratio (especially at q = const.), this effect is not so strong (see Equations 5.38 and 5.39).

5.4.1.3 Effect of Pressure

With other parameters (N_c, H, d_h, Δi_{in}) unchanged, the decreased pressure destabilizes the channel flow as regards the static instability. It does not substantially affect the boundary mass velocity that determines the minimum of the hydraulic characteristic curve. However, the decreased pressure increases the negative slope of the hydraulic characteristic in the ambiguity region or causes the ambiguity region appearance on the hydraulic characteristic curve where it was absent prior to the pressure drop. Such an effect of pressure is determined by a considerable decrease in the two-phase flow density at the decrease of pressure, which results in a larger variation of the gravity pressure drop component at the beginning of steam generation at the channel exit. The effect of pressure is most pronounced at low pressure values, when steam density is more strongly dependent on pressure. The preceding qualitative analysis proves experimental results of Proshutinsky and Lobachev [91] and Proshutinsky and Timofeeva [92].

5.4.1.4 Effect of Heat Flux Surface Density

The increased heat flux density, on the one hand, shifts the static stability boundary toward the higher flow rate area (see Equation 5.39), thereby extending the ambiguity region and destabilizing the system. On the other hand, it can reduce steepness of the slope of the hydraulic characteristic curve ambiguity branch or eliminate the ambiguity region at all (i.e., stabilize the system). This is due to the fact that an intensive variation of the gravity pressure drop component may occur in the high flow rate area, where the friction pressure drop has a higher absolute value and a steep dependence on the flow rate. Under conditions with the decreasing flow rate, the increase of the gravity pressure drop component in the area of its abrupt change may be smaller modulo than the friction pressure drop reduction.

An example of degeneration of the ambiguity region on the hydraulic characteristic curve at the increase of heat flux density, Figure 5.25 presents the experimental results of Proshutinsky and Lobachev [91] and Proshutinsky and Timofeeva [92]. As one can see, the increasing channel input power (heat load) is accompanied by the mass velocity increase, which corresponds to the minimum of the hydraulic characteristic curve, and at a certain power value and corresponding mass velocity, static instability is absent and does not appear as power increases further.

In Proshutinsky and Lobachev [91] and Proshutinsky and Timofeeva [92], the full-scale model of superheating channels at the Beloyarskaya NPP was used to obtain mass velocities at which no circulation reversal occurred at the beginning of steam generation at the channel exit. As an example, Figure 5.26 shows the experimentally obtained dependencies (from these authors) of mass velocity corresponding to the boundary of circulation reversal at different pressure levels.

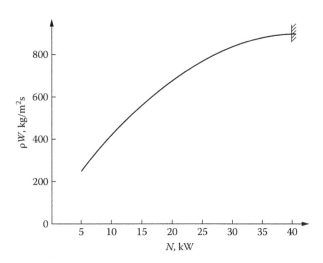

FIGURE 5.25
Dependence of the boundary mass velocity in the downflow channel on the input power [92].
$P = 12$ MPa; $\Delta T_{in} = 25°C$.

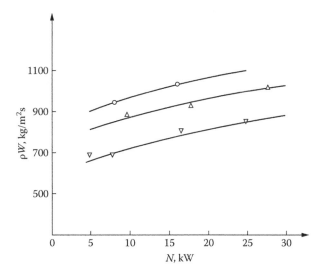

FIGURE 5.26
Dependence of mass velocity, corresponding to the boundary of circulation reversal absence
in the channel, on heat power [92]. Circle: experimental data at $P = 7$ MPa; ordinary triangle:
experimental data at $P = 10$ MPa; reversed triangle: experimental data at $P = 12$ MPa; solid line:
averaging curve.

However, degeneration of the ambiguity region of the hydraulic characteristic curve at the increase of the heat flux surface density is far from occurring at all conditions. For this to happen, coolant pressure should be sufficiently high in the parallel channels system; the channels should have a considerably large heated height, small equivalent diameter, and low inlet coolant subcooling. Then, steam generation at the channel exit starts at high mass velocities, weak effect of gravity due to the increased pressure, and small heat flux surface density. For example, in experimental investigations [91,92], the heated height of the channel (H) was 6 m, H/d_h was about 400, and coolant pressure (P) > 7 MPa. Under these conditions, the mass velocity at which degeneration of the ambiguity section on the hydraulic characteristic curve occurred $(\rho W)^*$ was above 700 kg/m²s, while the corresponding surface heat flux density (q^*) was less than 150 kW/m².

At low coolant pressures (which are typical, for example, of the conditions of pressurized water reactors' cooldown during accidents involving the primary circuit depressurization or of the pool-type reactors), as well as in cases with relatively small channel heated height, there may be no degeneration of the ambiguity region on the hydraulic characteristic curve. As it will be shown later for such conditions with the increasing flow rate, an increase of the heat flux density with the resultant steam generation at the channel exit is accompanied by the flow rate reversal. Starting from certain values of $\bar{q}$ and corresponding $(\rho W)_{Bn}$, the onset of static instability at the beginning of steam generation at the channel exit is accompanied by channel burnout.

5.4.1.5 Effect of Heat Flux Nonuniformity along the Height

The relation (5.39) was obtained assuming uniform surface density of the heat flux in the channel. If the heat flux density distribution along the height is nonuniform, (5.39) may be transformed into

$$(\rho W)_{Bn} = 4\left(1 + \frac{C\,d_h}{4\,H}\frac{q_e}{\bar{q}}\right)\bar{q}\,\frac{H}{d_h}\Big/\Delta i_{in}, \qquad (5.40)$$

where q_e is the heat flux surface density at the channel exit section, W/m².

It follows from (5.40) that the heat flux density increasing toward the channel exit section may destabilize the system as regards the static stability, since it increases $(\rho W)_{Bn}$, thereby extending the ambiguity region of hydraulic characteristic curve. This is due to the fact that, when the heat load is increased at the exit section while the flow exit enthalpy is maintained constant, a higher mass velocity in the channel is required to ensure the beginning of steam generation in accordance with (5.35).

However, height nonuniformity of q may reveal itself at limited H/d_h values $(H/d_h < 300)$. At $H/d_h > 300$, as was the case in the experiments of Akyuzlu et al.

[29] and Proshutinsky and Lobachev [91], the influence of height nonuniformity decayed and steam generation started at x_e approaching zero.

5.4.1.6 Effect of Heating Surfaces' Arrangement in Channels with Coolant Upflow–Downflow

Usually, such designs of heating surfaces are used in large boiler units. The results of the investigations of flow static instability and the effect of design and flow parameters on the stability region boundary in such systems have been published [46,47,94,95].

Location of ambiguity regions on the hydraulic characteristic curve and the value of the negative slope of unstable branch in cases of the U-, ∩-, and N-shaped arrangement of heating surfaces are determined by the height of the panels, ratios between the height and the heated length in the multipass designs, mutual arrangement of the distributing and collecting headers, heat flux density and distribution, and other factors.

As an example, Figures 5.27 and 5.28 show hydraulic characteristic curves and pressure drop components for the panels with the U- and ∩-shaped arrangement of heating surfaces [46].

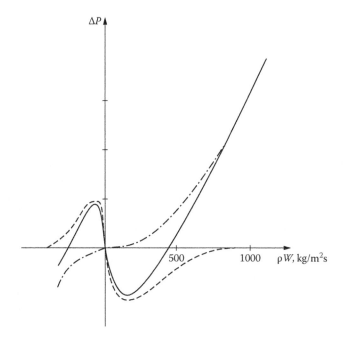

FIGURE 5.27
Hydraulic characteristic curve and variation of pressure drop components in a channel with the U-shaped heating surface arrangement [46]. ——— total pressure drop variation (ΔP_C); _._._ friction pressure drop component variation (ΔP_f); ------ gravity pressure drop component variation (ΔP_g).

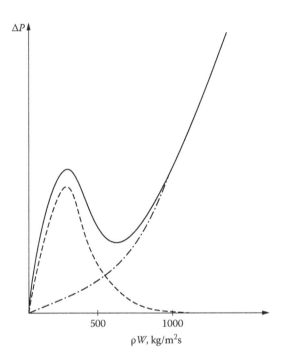

FIGURE 5.28
Hydraulic characteristic curve and variation of pressure drop components in a channel with the Π-shaped heating surface arrangement [46]. ——— total pressure drop variation (ΔPc); _._._ friction pressure drop component variation (ΔPf); ------- gravity pressure drop component variation (ΔPg).

It is clear that with the preset parameters, hydraulic characteristic curves have ambiguity regions. In the case of the U-shaped heating surfaces' arrangement with the given parameters and preset pressure drop, the onset of static instability causes channel circulation reversal. In the case of the ∩-shaped arrangement, the location of the ambiguity region in the first quadrant of the plane with [ΔP, (ρW)] coordinates leads to the decreased flow rate with the coolant flow direction unchanged.

The cause of such a location of ambiguity regions is clearly seen from the nature of pressure drop components' variation (Figures 5.27 and 5.28). In the general case, a comparison of the friction and gravity pressure drop components makes it possible to perform a sufficiently simple qualitative analysis of the ambiguity region location on the hydraulic characteristic curve for any arrangement of heating surfaces.

The effect of design and flow parameters on flow static instability considered before for the vertical channels with coolant downflow may be qualitatively similar for downflow sections of heating surfaces with coolant upflow–downflow. Therefore, only some peculiarities of static instability related to heating surfaces with coolant upflow–downflow will be noted.

First, when the number of passes in panels with coolant upflow–downflow increases—that is, when L/H grows (where L is the heated length, L = nH, n is the number of passes) due to the increased contribution of the friction pressure drop component, the flow becomes statically stable due to coolant downflow.

Second, in the case of multipass panels with coolant upflow–downflow, several ambiguity regions may appear on the hydraulic characteristic curve; they will correspond to the beginning of steam generation at the exit from separate downflow sections of the multipass heating panel.

Third, depending on the mutual arrangement of the distribution and collecting headers (both at the bottom, or both at the top, or the distributing header at the bottom and the collecting one at the top, or vice versa), the onset of static instability may cause either circulation reversal or flow rate reduction in the unstable channel, while the prescribed coolant flow direction in the parallel-channel system remains unchanged.

It has been shown [46,47,85,94,95] that the onset of static instability in such systems, especially in the channels with reversed circulation, is accompanied by the establishment of a low flow rate and heavy-duty temperature conditions on the heating surface. Therefore, static instability in these systems is extremely undesirable. In addition, when determining the temperature regime of the channel, the hydraulic characteristic curve in the region of negative flows should be plotted, taking into account that the coolant inlet temperature at the reversal of circulation is equal to that of the collecting header. With convective heating, the value and axial distribution of heat flux density change because of the changed temperature difference.

5.4.1.7 Effect of Throttling

The increased inlet throttling facilitates flow static stability in a system. However, since static instability caused by variation of the gravity pressure drop component with coolant downflow generally occurs at low coolant mass velocities ($\rho W < 1000$ kg/m^{2}s), the efficiency of influence of throttling on the magnitude of friction pressure drop change reduces as the flow rate decreases. Additionally, the ambiguity region cannot be completely eliminated under these conditions. The increasing throttling shifts the ambiguity region on the hydraulic characteristic curve toward the area of lower flow rates.

Thus, the use of inlet throttling to stabilize the flow in the described conditions is not justified because

1. Low throttling efficiency demands a large friction coefficient of the throttling orifice to increase static stability, but it creates a high pressure drop during the nominal mode and reduces power plant efficiency.

2. Throttling fails to eliminate the ambiguity region on the hydraulic characteristic curve completely and only shifts it toward lower flow rates.

3. Each of the parallel channels requires individual throttling, thus causing design problems.

Therefore, the most optimal method of avoiding static instability is either the arrangement of heating surfaces or selection of operation modes and means of power plant control that prevent parameters drifting toward the region close to static instability.

Like inlet throttling, exit throttling facilitates flow static stability. However, apart from the preceding constraints associated with both inlet and exit throttling, the increasing inlet throttling leads to additional system destabilization as regards the oscillatory instability. Therefore, this method of static instability stabilization is not applied.

5.4.1.8 Effect of Steam Slip

Steam slip in the vertical channel with coolant downflow causes flow destabilization as regards the static instability. It does not strongly affect the boundary mass velocity corresponding to the minimum of the hydraulic characteristic curve; however, the steam slip increases slope steepness of the unstable branch of the hydraulic characteristic curve at the expense of the increasing steam quality in the channel with coolant downflow and flow rate below $(\rho W)_{Bn}$ (i.e., at the expense of a sharper increase of the gravity pressure drop component following the flow rate decrease after the beginning of steam generation at the channel exit).

The destabilizing effect of steam slip is most pronounced at downflow coolant low mass velocities ($\rho W < 500$ kg/m²s) and at low pressures, when the slip factor has a noticeable value.

It should be noted that calculations of the channel downflow branch hydraulic characteristic in the two-phase area are associated with certain difficulties due to the lack of reliable empirical correlations for determining the steam void fraction α and friction pressure losses.

5.4.1.9 Effect of Channel Inlet Coolant Subcooling

In the case of inlet enthalpy that does not exceed the enthalpy of the beginning of steam generation, an increase in Δi_{in} at constant design parameters and values of P and $\bar{q}$, as is seen from (5.39), decreases $(\rho W)_{bn}$ (i.e., shifts the minimum of the hydraulic characteristic curve toward lower flow rates), thereby decreasing the ambiguity region and stabilizing the system as regards the static instability. However, when inlet subcooling is small and is close to that corresponding to the onset of steam generation, an opposite effect of inlet subcooling is observed at preset design and flow parameters (i.e., the increasing inlet subcooling stabilizes the flow as regards the static instability). To illustrate this effect of inlet subcooling, Figure 5.29

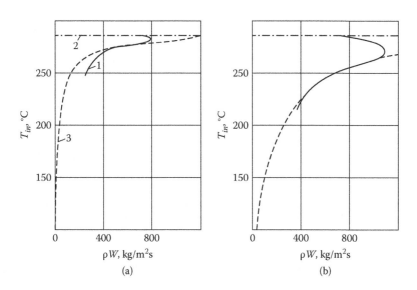

FIGURE 5.29
Static stability boundary in the channel with coolant downflow [92]. P = 7 MPa; (a) heat power, N = 6 kW; (b) heat power, N = 17.8 kW; 1: stability boundary; 2: X_{in} line = 0; 3: X_e line = 0.

shows the results of experimental-numerical investigation [92]. One can see the ambiguous effect of inlet subcooling on static flow instability in the case of coolant downflow.

To conclude this section, let us consider the effect of static flow instability on coolant flow regimes in the system of parallel heated channels with no loop circulation or at low velocities of circulation through this system. The following analysis is based on the results of experimental investigations presented in detail in Khabensky, Migrov, and Efimov [195].

The greatest problems in a study and quantification of thermal-hydraulics and critical situations in the system of parallel heated channels (e.g., in the reactor core, with no loop circulation or at low velocities) are associated with the necessity of taking into account hydrodynamic interaction of parallel operating channels. It has been shown [91–93,193–195,197,202] that the behavior of thermal-hydraulic parameters in the system and in separate channels under quasisteady-state and unsteady conditions may qualitatively differ from similar processes at high circulation velocities, or in a single channel with a strictly preset flow rate. The onset of static instability under such conditions in the reactor core may cause the interchannel circulation with an intensity depending on the energy release and its nonuniformity, flow rate direction and magnitude, coolant temperature in the upper and lower plenums, and other factors. In this case, some channels (with a lower heat input) exhibit single-phase downflow and in some channels the coolant flows upward together with steam generation, while, in some channels, with the intermediate heat input, multiple flow reversals may be observed.

Experimental investigations [195] were conducted to study flow regimes and critical situations in a multichannel reactor core with no loop circulation, or at low circulation velocities with coolant downflow, taking simulation of hydrodynamic interaction of parallel channels into account.

There were several reasons for taking the hydrodynamic interaction into account in Khabensky et al. [195]. Under steady-state and unsteady conditions, thermal-hydraulic and critical phenomena in each separate heated channel are determined by the pressure drop in the core (or other system of parallel heated channels) and by temperature distribution in the upper and lower plenums, which depends on the interchannel interaction. However, variation of flow regimes and critical phenomena originates first in a small group of separate channels (i.e., these processes in channels with different thermal-hydraulic characteristics may be considered under conditions of the preset pressure drop).

Therefore, the experimental simulation of thermal-hydraulics in separate channels of a multichannel reactor core using a thermophysical installation [195] employed a system consisting of the channel simulator with the nonheated bypass, which ensured channel simulator investigations with variation of preset constant values of the channel pressure drop and of coolant temperature in the upper plenum. The simulator pressure drop was preset by the downflow coolant flow rate via two bypass tubes of 70 mm in diameter equipped with control valves. The ratio between the bypass and channel simulator flow rates (G_{By}/G_c) was 10–80 and was found to have no influence on the results in this range. With the pressure drop and temperature preset in the upper plenum, the initial simulator power was zero and then increased in small steps. In Khabensky et al. [195], the simulator was made as a rectangular section channel 3.5×70 mm^2 with the heated height (H) of 1000 mm, and distilled water at atmospheric pressure was used as coolant.

Figure 5.30 presents the experimentally obtained [195] typical diagram of the coolant flow regimes in a channel simulator at a fixed coolant temperature in the upper plenum ($T_{in} = 35°C$) in coordinates: heat flux surface density q and pressure drop ΔP measured by the differential pressure gauge. In this case, $\Delta P = \Delta P_c + \rho_{in}g\,H_c$, where ΔP_c is the total channel pressure drop and H_c is the height. For the sake of clarity, the X-axis provides a scale for scoring downflow velocity in the channel simulator (W_{Beg}), which the coolant had at zero power. In this case, the zero value of W_{Beg} corresponds to no loop circulation and is simulated in the experiments [195] by closing the valve in the installation circulation loop.

As is seen from Figure 5.30, the values of q and ΔP and the direction of their change determine different coolant flow regimes in a channel when its

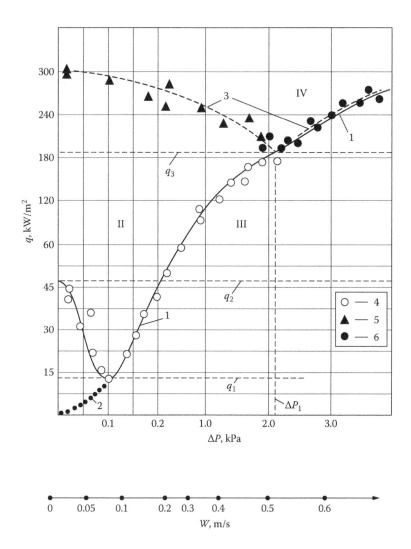

FIGURE 5.30
Coolant flow pattern map in the slit-shaped channel, its parallel operation simulated [195]. P = 0.1 MPa; T_{in} = 30°C–35°C. Boundaries: 1: incipient boiling; 2: single-phase stable downflow; 3: burnout; regions I, II: single- and two-phase upflow, respectively; III: single-phase downflow; IV: burnout.

parallel operation is simulated. The regions with different flow regimes and heat transfer are separated by typical curves having clear physical sense:

Curve 1 corresponds to the beginning of steam generation at the channel exit and separates the single-phase flow region (below curve 1) and the two-phase flow region (above curve 1). In the case of single-phase coolant downflow, a section of curve 1 (on the right of the minimum point) is the static instability boundary.

Curve 2, being short and connecting the coordinate origin with the minimum point on the curve at the beginning of steam generation, is the single-phase coolant state static instability boundary that separates the upflow (on the left of curve 2) and downflow.

Curve 3 corresponds to the onset of burnout in the slit-shaped channel when its parallel operation is simulated. On curve 3 to the right of the minimum point, the onset of burnout coincides with the boundary of the beginning of steam generation and is caused by static instability onset.

Let us consider specifics of thermal-hydraulics in these regions in more detail. When heat flux values are in the $0 < q < q_1$ range (Figure 5.30) (in conditions of Khabensky et al. [195] with $q_1 = 14$ kW/m^2), the coolant is single phase within the entire range of pressure drop variation in the test section. When the pressure drop decreases at constant q (as one can see from Figure 5.30), the channel coolant flow may change from the initial downflow to upflow in conditions corresponding to the flow static instability boundary (i.e., at an intersection with curve 2). In this case, the coolant upflow without boiling (region 1 in Figure 5.30) becomes established in the heated channel along with the bypass downflow. The single-phase coolant static instability and peculiarities of the associated thermal-hydraulic processes are described in more detail in Section 5.4.2.

When heat flux values are in the $q_1 < q < q_2$ range (in conditions of Khabensky et al. [195] with $q_2 = 46$ kW/m^2) at constant q and decreasing ΔP corresponding to the decreasing loop circulation, the single-phase coolant in the downflow channel starts to boil and the coolant flow changes its direction at the crossing of the right branch of curve 1 (Figure 5.30). In this case, the two-phase upflow with low exit quality is established in the channel. The further decrease of pressure drop in the channel results in the increased coolant upflow velocity and boiling termination at the crossing of the left branch of curve 1 (Figure 5.30), while the single-phase upflow is maintained. Such behavior of parameters in the considered range may be simply explained when analyzing the channel hydraulic characteristic curves and coolant exit quality variation given in Figure 5.31 (curve 1).

When heat flux values are in the $q_2 < q < q_3$ range (in conditions of Khabensky et al. [195] with $q_3 = 185$ kW/m^2) at constant q and decreasing ΔP, the single-phase coolant downflow in the channel is found to boil and change its direction for upflow when boundary 1 is reached (Figure 5.30). However, boiling is this region does not terminate as ΔP keeps reducing until the complete cessation of coolant loop circulation. This fact is proved by the results of the channel static thermal-hydraulic calculation (see curve 2 in Figure 5.31). In this case, the two-phase upflow (along with the bypass single-phase downflow) is accompanied by the "geysering" flow instability with coolant ejection through the channel upper and lower

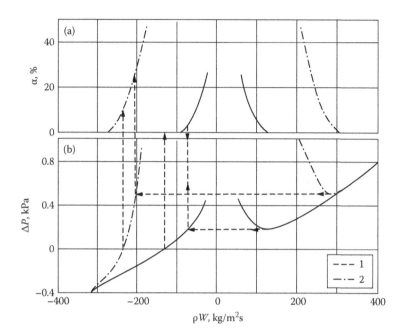

FIGURE 5.31
Calculated dependencies of true exit flow quality (a) and channel pressure drop (b) on the coolant mass flow rate at different heat flux densities. 1: $q = 30$ kW/m^2; 2: $q = 80$ kW/m^2.

sections. Steam slugs entrained to the upper plenum by the subcooled coolant start to condense intensively, thus leading in Khabensky et al. [195] to hydraulic hammers with strong mechanical vibrations in the test section and test facility as a whole. Figure 5.32 shows variation of some parameters in one of these regimes.

Finally, when heat fluxes are in the $q > q_3$ range and $\Delta P > \Delta P_1$ (for the given example from Khabensky et al. [195], the value of ΔP_1 corresponds to $W_{Beg} = 0.5$ m/s), coolant channel boiling (curve 1 in Figure 5.30) at the decrease of ΔP causes deterioration of downflow stability with simultaneous onset of burnout (i.e., the coolant incipient boiling boundary is at the same time the burnout boundary), while curves 1 and 3 coincide. When $\Delta P < \Delta P_1$ and $q > q_3$ is increasing, the parameter fluctuation frequency increases in the case of the geysering instability, the exit quality increases along with the growing upflow flow rate, and the burnout occurs at a certain value of $q = q_{cz}$.

The experimental results presented in Figure 5.30 show that the q_{cz} value has a clear minimum at a certain pressure drop value $\Delta P = \Delta P_1$ and the corresponding finite value of coolant downflow velocity W_{Beg}^{Bn}. This is due to the oppositely acting gravity and inertia forces.

Indeed, when $\Delta P = 0$ ($W_{Beg} = 0$) and thermal load is present, natural circulation starts between the bypass and the heated channel (or interchannel

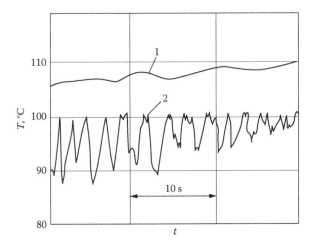

FIGURE 5.32
Experimental time variation of the test channel surface temperature (1) and coolant temperature (2) in the central channel section during "geysering" instability. q = 95 kW/m²; ΔP = 0.7 kPa.

circulation in the system of heated channels), thereby ensuring favorable conditions for steam evacuation from the channel. The further startup of the pump and the forced downflow circulation via the test section lead to the decreased effect of natural circulation due to the oppositely directed lifting and inertial forces. At a certain W_{Beg}^{Bn} value, the flow hydrodynamic blockage conditions are established: The upflow channel feeding due to natural circulation substantially decreases and becomes commensurate with the rate of coolant flow from the upper plenum to the location of steam exit from the channel. As the downflow velocity ($W_{Beg} > W_{Beg}^{Bn}$) keeps increasing, natural circulation loses its influence on heat removal from the channel, and the value of q_{cz} starts increasing because of the influence of forced downflow on coolant boiling.

The qualitative nature of the diagram in Figure 5.30 is maintained when coolant temperature in the upper plenum changes. However, as one can see from Figure 5.33, the curves corresponding to the boundaries of downflow stability, of coolant incipient boiling, and of the burnout become deformed. In this case, as the degree of coolant subcooling in the upper plenum decreases, the minimum value of q_{cz}^{min} reduces and shifts toward higher downflow flow rates. In the limit, when the coolant temperature in the upper plenum approaches that of saturation, experimental investigations of Khabensky et al. [195] have shown the minimum value of q_{cz}^{min} to become commensurate with the critical heat flux value in the blind (i.e., plugged from the bottom) channel (q_{cz}^b). Naturally, the value of q_{cz}^{min} depends on design features of the channels and flow thermal-hydraulic parameters, which are to be modeled for the experimental determination of q_{cz}^{min}.

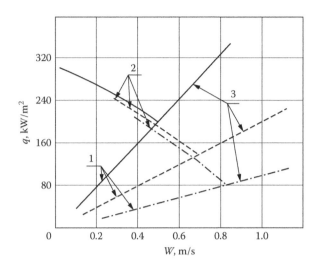

FIGURE 5.33
Effect of coolant subcooling in the upper plenum on the boundaries of flow pattern regions. ——— $-\Delta T_{in} = T_s - T_{in} = 70°C$; ------- $-40°C$; _._._ $-10°C$; 1 – static stability boundary; 2: upflow burnout boundary; 3: downflow burnout boundary.

The performed analysis shows the following:

1. Depending on design and thermal-hydraulic parameters of the parallel heated channels, different flow regimes are determined in separate channels by flow static instability at low loop circulation velocities.

2. The worst situation from the point of view of channels' cooling may realize not so much in the absence of loop circulation via the reactor core but rather in the presence of some forced downflow leading to hydrodynamic blockage of hot channels and burnout.

3. The critical heat flux value corresponding to the burnout in the blind channel, (q_{cz}^B), may be used as the limiting lower boundary for assessing heat engineering reliability of pressurized water reactor (PWR) cores' cooling at the cessation of forced circulation.

5.4.2 Single-Phase Coolant Flow Static Instability

The experimental and numerical investigations described in references 91–93, 193–195, and 197–203 show that the onset of static instability is possible under certain conditions at low forced downflow velocities in parallel heated channels, even with single-phase coolant.

Static instability in heated parallel channels may be observed, for instance, in cores of liquid-metal and water coolant reactors in emergency situations at cessation of forced circulation, during the heating-up regime under natural circulation, and in primary circuit depressurization accidents, as well as in heat exchangers with forced coolant downflow at low flow velocities.

The onset of static instability of the single-phase downflow coolant at the forced flow rate variation is determined by the change of the gravity pressure drop component that dominates over the friction pressure drop. Therefore, the influence of some parameters on the stability boundary may be qualitatively assessed in a simple way.

Indeed, the ambiguity region that appears on the hydraulic characteristic curve and determines the onset of static instability facilitates an increase in the change of the gravity pressure drop component along with varying flow rate.

Therefore, an increase of the channel heated height and of the heat load destabilize the single-phase coolant flow, while variation of inlet subcooling and coolant pressure have an insignificant effect on the static stability boundary. The latter is due to weak nonlinear dependence of the single-phase coolant density on temperature and pressure. The experimental confirmation of such influence of parameters on the static stability boundary is given in references 91 and 92. As an example, Figure 5.34 presents the experimental results from the latter.

However, the static stability boundary of the single-phase coolant flow and variation of thermal-hydraulic parameters in the unstable region may be evaluated numerically using the one-dimensional mathematical model of forced flow only at forced flow velocity corresponding to the Reynolds number $(\text{Re}_f) > 1,000$. At lower velocities, complex interaction between gravity and inertia-viscous forces give rise to multidimensional effects. These effects may lead to the appearance of thermal density stratification in the upper

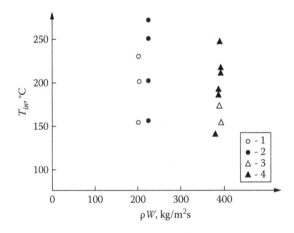

FIGURE 5.34
Experimental values of boundary mass velocity corresponding to the static instability boundary in a channel with single-phase coolant downflow [92]. 1, 2, 3, 4: experimental values. 1: P = 7 MPa, N = 6 kW; 2: P = 12 MPa, N = 6 kW; 3: P = 7 MPa, N = 17.9 kW; 4: P = 12 MPa, N = 17.9 kW.

and lower plenums [190], violation of the laminar flow stability in separate heated channels [192,204,205], and heat and mass transfer between the heated channel and the upper plenum [191,193].

In addition to the quantitative influence on the static instability boundary in a channel with the single-phase coolant, such effects can change the manner of unstable processes' development in the channel. Therefore, let us consider in more detail peculiarities of the single-phase flow static instability in the system of parallel heated channels with low velocities of the forced coolant flow ($Re_f < 1,000$) when the influence of multidimensional effects is manifested. References 93, 192–194, and 197 present the technique of static stability boundary numerical assessment and another one of the channel thermal-hydraulic parameters' calculation in the unstable region, with approximate accounting for multidimensional effects under unsteady and quasisteady-state conditions.

The cited works also contain the results of experimental investigations on the onset and development of the single-phase flow static instability in the region where multidimensional effects occur, and they prove sufficient accuracy of predictive methods in describing the experimental results obtained in unsteady and steady-state conditions. Here, we omit the detailed description of predictive techniques, as they are available in the mentioned sources, and concentrate on peculiarities of the physical mechanism and qualitative analysis of multidimensional effects on the single-phase flow static instability in the heated channel.

It has been stated previously that three basic processes, induced by the combined convection and usually disregarded in the widely used one-dimensional mathematical models of the heated channel thermal-hydraulics, produce a considerable effect on the static stability boundaries in the single-phase flow channel and on the manner of thermal-hydraulic parameters' variation in the unstable region at low coolant velocities ($Re_f < 1,000$). These are as follows:

1. Violation of the channel laminar flow stability at low coolant down-flow velocities
2. Heat and mass transfer between the channel and the upper plenum
3. Coolant temperature distribution in the upper and lower plenums with due account for stratification effects

These processes influence static instability by means of the coolant thermal state variation in the channel; hence, they influence the regularity of variation of the gravity pressure drop component and the total channel pressure drop with the varying channel flow rate. Let us consider the qualitative effect of each of these processes separately.

It has been shown [204,205] that in the heated channel with forced coolant downflow at $Re_f < Re_{f.cz}$ and $Gr_q/Re_f \geq C$, the channel section features the loss of laminar flow stability and the combined convection appears. In this case,

$$\mathrm{Re} = \frac{W_f \cdot d_h}{\nu}, \quad Gr_q = \frac{g\beta q d_h^4}{\lambda \nu^2}$$

are the Reynolds and Grashof numbers, respectively; β is the coefficient of fluid volumetric expansion; q is the heat flux density; W_f is the average forced fluid flow velocity; ν and λ are the kinematic viscosity and fluid thermal conductivity, respectively; d_h is the equivalent channel diameter; and C is the constant dependent on the channel shape. In accordance with recommendations of reference 192 for slit and annular channels, $\mathrm{Re}_{f.cz} = 600$ and C = 240, while for tubes (from reference 205), $\mathrm{Re}_{f.cz} = 250$ and C = 270.

In the case of coolant downflow and uniform heat release along the channel height, the loss of flow laminar stability is first observed in the channel lower exit section, where the coolant temperature and hence the Gr_q are the largest. When the forced flow decreases in the channel or heat flux density increases, the laminar flow stability boundary shifts toward the channel upper section. In the channel section below the stability boundary, due to combined convection, intensive macrovortex recurrent currents occur that sharply change the coolant temperature distribution along the channel height, which is close to uniform if averaged over the cross section [192]. Figure 5.35 shows examples of height distribution of the cross section-averaged coolant temperatures at different heat flux densities obtained experimentally [192]. The coolant downflow is adopted in Figure 5.35 as the positive direction of the axial coordinate. As one can see from the figure, the cross section-averaged

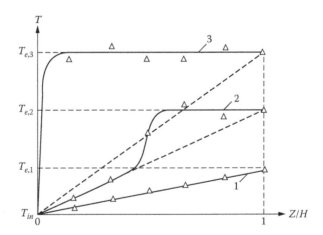

FIGURE 5.35
Distribution of the cross section-averaged coolant temperature along the channel height. 1: stable flow; 2: unstable flow in the channel lower half; 3: unstable flow in the entire channel; triangle: averaged thermocouple readings; solid line: averaging curve; $q_1 < q_2 < q_3$.

coolant temperature is practically constant along the height in the region of convective instability and is close to the channel exit coolant temperature.

Thus, the disturbance of the laminar flow stability and initiation of combined convection in the channel increase the integral coolant temperature in the channel, thereby increasing the gravity pressure drop component and destabilizing the flow with regard to static instability. As an example, Figure 5.36 presents a typical hydraulic characteristic curve of a channel with single-phase coolant experimentally obtained [93]. For a comparison, the figure shows a hydraulic characteristic curve obtained using a one-dimensional model for the forced flow disregarding the combined convection. The coolant downflow is adopted in Figure 5.36 as the positive direction, while the pressure drop on the ordinate axis was determined, for the sake of clarity, by $\Delta P = \Delta P_f + gH \cdot (\rho_{in} - \bar{\rho})$, where ΔP_f is the friction pressure drop; ρ_{in} and $\bar{\rho}$ are the coolant density at the channel inlet (in the upper plenum) and the average integral coolant density in the channel, respectively. It is obvious from Figure 5.36 that a reduction of the coolant column weight at the expense of convective instability shifts the minimum point of the hydraulic characteristic curve toward larger mass velocities. In this case, the boundary mass velocity that determines the onset of static instability may be, with due account for convective instability, almost twice as high as the boundary mass velocity determined numerically applying a one-dimensional mathematical model for the coolant forced flow.

Another example of the multidimensional effects' influence on the single-phase coolant flow static instability may be found in the results of unsteady experiments [194], in which the experimental channel heat load was increased in a stepwise manner by different values, while the constant pressure drop in the channel was set by the bypass downflow flow rate. In this case, unsteady channel flow variation was investigated starting from the initial downflow,

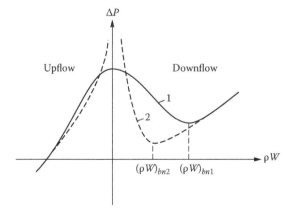

FIGURE 5.36
Hydraulic characteristic curve of a channel with single-phase coolant at the constant heat flux and low coolant velocities. 1: Free convection taken into account; 2: free convection disregarded.

corresponding to zero thermal load, up to the value corresponding to the preset load increment.

Typical experimental results are presented in Figure 5.37, which also shows for comparison the numerical results obtained using a one-dimensional model, taking into account heat removal due to forced convection only. The coolant downflow is adopted in Figure 5.37 as the positive direction. One can see from this figure that a power increment (N) of 1 kW caused no circulation reversal in the channel and 100–150 s after the heat load increase, parameters were found to stabilize with coolant downflow in the experimental channel. Calculations employing a one-dimensional model describe the experiment quite well in this case. However, when N was equal to 1.3 kW, the unsteady channel heating continued up to 850 s and, finally, the channel coolant flow was reversed from down- to upflow. The stepwise power increments above 1.3 kW did not change the process qualitatively, though the rate of coolant heating up in the channel kept increasing and the time prior to circulation reversal reducing.

As is seen from Figure 5.37, the one-dimensional model predicted no circulation reversal with N = 1.3 and 1.7 kW, but on the contrary, it showed the flow rate to stabilize with downflow circulation. The onset of static instability and flow reversal in the channel occurred, according to the one-dimensional model predictions, at power increments of up to 1.8 kW (Figure 5.37)—that is, almost 1.5 times higher than the experimental power at which experimental circulation reversal started in the channel.

The difference between the experimental and predicted (using the one-dimensional mathematical model of forced flow) hydraulic characteristic

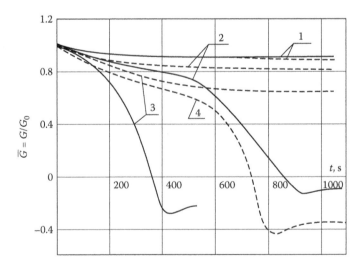

FIGURE 5.37
Time variation of the relative coolant flow rate in channels with different and stepwise heat load disturbances. Solid line: experimental data; dashed line: calculated employing the forced convection model; 1: N = 1 kW; 2: N = 1.3 kW; 3: N = 1.7 kW; 4: N = 1.8 kW.

curves in the region of forced flow low and zero mass velocities (Figure 5.36) is determined by one more multidimensional process caused by the onset of combined convection between the heated channel and upper plenum [93,191–194].

In the case of unstable density stratification between the heated channel and the upper plenum, this process leads to heat and mass transfer between the channel and upper plenum, and a natural convective upflow–downflow coolant countercurrent occurs against the background of *low* velocities of the coolant-forced down- or upflow. That is, the coolant from the upper plenum flows down through one part of the channel cross section, while the heated coolant flows up through the other part of the channel and enters the upper plenum.

Thorough experimental investigations of the process [93,191–194] show the following:

1. In addition to forced convection, free-convective mass transfer ensures additional heat removal from the channel to the upper plenum.

2. The maximum amount of free-convective mass transfer between the channel and upper plenum occurs with no forced flow in the channel. In this case, the channel heat and mass exchange with the upper plenum is the sole mechanism of heat removal from the channel and may ensure single-phase coolant state in the channel with no forced flow through the channel. As soon as the forced upflow or downflow appears, the magnitude of the free-convective countercurrent flow decreases and terminates when the forced coolant velocity in the channel exceeds that of the coolant exchange flow between the channel and the upper plenum determined in the absence of the forced flow.

According to recommendations of Malkin et al. [191] and Khabensky et al. [194], the countercurrent flow velocity with no forced coolant flow through the channel is derived from

$$\mathrm{Re}_{cou}^0 = 7 \cdot 10^{-3} \frac{H}{d_h} \mathrm{Pr}_L^{-1} Ra^{0.46}, \tag{5.41}$$

where

$\mathrm{Re}_{cou}^0 = \dfrac{W_{cou}^0 d_h}{\nu_L}$ is the Reynolds number

Pr_L is Prandtl number for the channel coolant

$Ra = Gr_q \cdot \mathrm{Pr}_L = \dfrac{g\beta C_p \rho_L d_h^4 q}{\lambda_L^2 \nu_L}$ is the Rayleigh number for the channel coolant

W_{cou} is the countercurrent flow velocity in the absence of forced flow in the channel

d_h and H are the channel equivalent diameter and heated height, respectively

v_L and λ_L are kinematic viscosity and thermal conductivity of the channel coolant, respectively

Parameters in (5.41) are determined at the coolant average temperature in the channel region where there exists free-convective mass transfer between the channel and the upper plenum.

In the presence of forced flow, the velocity of the countercurrent flow is determined according to recommendations of Malkin et al. [191] and Khabensky et al. [194] from

$$
W_{cou} = \begin{cases} W_{cou}^0 - |W_f|, & W_{cou}^0 > |W_f|, \\ 0 & W_{cou}^0 \leq |W_f|, \end{cases}
$$

(5.42)

where W_f is the coolant forced flow velocity in the channel, while W_{cou}^0 is derived from (5.41).

The amount of heat removed from the channel at the expense of free convective mass exchange with the upper plenum was determined [194] by

$$
Q_{cou} = \frac{1}{2} A \rho W_{cou} C_P \left(\overline{T}_c - T_{up} \right),
$$

(5.43)

where A is the channel flow passage, $\overline{T}_c$ is the average temperature of the channel section with free-convective mass transfer, and T_{up} is the temperature of coolant flowing from the upper plenum into the channel.

To determine the onset of free-convective countercurrent coolant flow between the channel and upper plenum, a relationship recommended in Jannello and Todreas [193] may be simplified into

$$
\frac{Gr_{\Delta t}}{Re_f^2} = 0.5, \quad 175 < Re_f < 590;
$$

here,

$Re_f = \dfrac{W d_h}{v_L}$ is the Reynolds number for the coolant forced flow in the channel

$Gr_{\Delta t} = \dfrac{g \beta \left(\overline{T}_c - T_{up} \right) d_h^3}{v_L^2}$ is the Grashof number

3. Under the conditions of mass exchange between the channel and upper plenum, an upflow–downflow countercurrent occurs in the channel against the background of the forced flow, when the coolant

from the upper plenum flows down through one part of the channel cross section, while the heated coolant flows up through the other part of the channel and enters the upper plenum. In this case, the countercurrent flow is unstable and wavy, which, according to numerous measurements [93,191,194] ensures a practically height-constant, cross section-averaged coolant temperature within the region of free-convective countercurrent flow.

In the absence of forced flow, the length of this region covers the entire channel heated height, while in the presence of forced upflow it depends on the upflow velocity and the rate of countercurrent mass transfer. As an example, Figure 5.38 shows the experimental heated wall temperature distribution along the channel height, which is typical for these conditions. The data were obtained [194] for the steady-state upflow after the reversal of the initial downflow circulation. The coolant downflow is adopted as the axis positive direction.

Figure 5.38 shows that the channel features two regions differing in the heat transfer mechanism. The regime of stable upflow becomes established in the lower part of the channel and energy is transferred by forced convection only. In the upper channel part, which in the given experiment amounts to over half of the channel, heat may be transferred both by the forced convection and by the free-convective countercurrent mass transfer between the channel and upper plenum, which originates because of the unstable fluid density stratification and "cuts off" the linear rise of coolant temperature.

The mass transfer between the channel and upper plenum decreases the average integral coolant temperature in the channel. Thus, as is obvious from

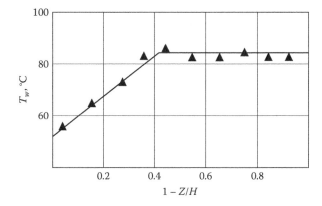

FIGURE 5.38
Distribution of the heated wall temperature along the channel height. Dark triangle: experimental data; solid line: calculated according to reference 194.

Figure 5.36, it decreases the angle of inclination of the hydraulic characteristic curve into the ambiguity region.

Also, the coolant temperature distribution in the upper and lower plenums at low coolant velocities may have a marked effect on the single-phase coolant temperature condition in the channel. It was shown, for example [194,197,206], that the temperature distribution in the plenums may substantially differ from uniform distribution and change the temperature boundary conditions at the channel inlet in the case of low coolant velocities and in the absence of both hydrodynamic mixing and a possibility of stable and unstable density stratification. For instance, PWRs with no loop circulation and the developing interchannel circulation demonstrate stable coolant density stratification in the lower plenum. In this case, only the upper part of the lower plenum adjacent to the channels may contribute to heat transfer occurring by means of the interchannel circulation; its remaining part is heated for a long time due to thermal conductivity only. This spatial process is also neglected in one-dimensional mathematical models of forced flow.

The mentioned multidimensional processes in single-phase coolant at low velocities of forced flow may not only quantitatively change boundaries of the static stability region as compared to the one-dimensional description of the process, but also qualitatively change behavior of thermal-hydraulic parameters. For example, the mere manifestation of spatial effects makes the single-phase coolant state possible in the absence of forced flow in the channel [93,191–195,206]. Additionally, spatial effects explain such behavior of parameters (experimentally observed in references 93 and 193) as the restoration of single-phase coolant downflow in one of the parallel channels at the quasistatic reduction of heat load [93] or of coolant flow rate through the system of parallel channels [193]. The cause of such behavior of thermal-hydraulic parameters is the strengthening of countercurrent mass transfer between the channel and upper plenum, which increases heat removal from the channel, thereby decreasing the gravity pressure drop component and leading to the onset of static instability with transition through the left boundary of the stability region (see Figure 5.36).

Thus, in the absence of loop flow or at low flow rates in a system of parallel heated channels with single-phase coolant, intensive interchannel circulation develops as a result of static instability. Depending on heat release distribution, hydraulic parameter values, and the kind of transient, channels with upflow and downflow circulation, as well as channels with coolant flow rate close to zero, may be found in the system of parallel channels.

A sufficiently simple dynamic mathematical model that takes multidimensional effects for single-phase coolant into account approximately and describes experimental results for unsteady and steady-state conditions with good accuracy was suggested [194,195].

Let us consider another peculiarity of multichannel systems with single-phase coolant. Flow reversal in unstable channels and the onset of interchannel

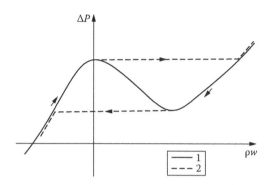

FIGURE 5.39
Hydraulic characteristic curves of the channel (1) and parallel channel system (2).

circulation in the system of parallel heated channels substantially influence not only thermal-hydraulic characteristics of separate channels, but also the total pressure drop across the plenums, thereby affecting the loop circulation parameters. Figure 5.39 offers a simple, clear example of static hydraulic characteristics of a separate channel and of the whole system made up by a large number of practically identical channels. As is seen from the figure, hydraulic characteristics of the system of identical channels and its separate channels differ greatly.

The physical nature of this difference is that, starting from a certain coolant flow rate in the system that corresponds to the hydraulic characteristic maximum or minimum depending on upflow or downflow, the further flow rate variation with the change of the loop circulation direction is achieved by reducing (with, for example, loop circulation reversal from downflow to upflow) the number of channels with coolant downflow and increasing the number of channels with coolant upflow. The loop downflow–upflow reversal occurs at a ΔP in the upflow and downflow channels that corresponds to the minimum of the hydraulic characteristic curve, while the loop upflow–downflow reversal takes place at a ΔP in the upflow and downflow channels that corresponds to the maximum of the hydraulic characteristic curve. Thus, the region of the channel multivalued hydraulic characteristic shows the hydraulic characteristic of the parallel channels' system to have a practically constant pressure drop across the two plenums, depending on the loop flow direction.

5.5 Pressure Drop Oscillations

In the presence of a compressible volume at the channel inlet, ambiguity of its hydraulic characteristic may lead under certain conditions to periodic steady-state oscillations of the coolant flow rate and pressure drop of a peculiar nature, which are referred to in the literature as "pressure drop oscillations" (PDOs).

Oscillations of the two-phase flow caused by a compressible volume at the channel inlet were originally discovered experimentally when investigating coolant flow stability on thermophysical test facilities (see, for example, references 207 and 208). A compressible volume at the inlet of the investigated heated channel on a thermophysical test facility is often used as the volume compensator or for smoothing pressure (flow rate) oscillations produced by the delivery pump. Sometimes, compressible volumes at the inlet and outlet of the working section serve to isolate the channel hydrodynamically.

Conditions provoking instabilities of the PDO-type can arise in various technical devices that comprise heat-transfer two-phase coolant systems. For instance, as has been suggested [209], in the nuclear power industry, the conditions provoking PDO-type instabilities can occur in steam generators of breeder reactors with a liquid-metal coolant circulating in the primary loop. Also, conditions for PDO occurrence can form in the emergency-cooling systems of the PWR NPPs, which comprise hydraulic reservoirs and complicated circulation loops for delivering coolant to the reactor and steam generators during an accident involving the primary loop depressurization. For instance, Godik et al. [210] suggest installing a fixed hydraulic reservoir at the steam generator inlet for cooling steam generators in emergency situations caused by the loss of the feed-water supply. The experimental investigations [210] showed that this reservoir can cause PDO oscillations even at normal operating conditions of the reactor.

The mechanism of the PDO instabilities' onset and development has been described in detail [15,209]. By referring to these sources, the onset of PDO can be illustrated by a simple diagram given in Figures 5.40 and 5.41. In the diagram, a pump (or a large-volume reservoir with a specified

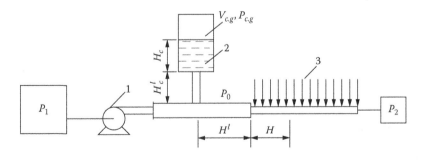

FIGURE 5.40
A hydraulic system with a compressible volume at the inlet. (1) Pump; (2) compressible-volume reservoir; (3) heated channel.

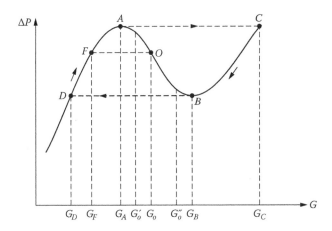

FIGURE 5.41
A PDO cycle.

pressure) provides a preset constant flow rate of the coolant through the system, while the outlet pressure P_2 is maintained constant for all regimes. Let the preset flow rate through the system decrease and reach the level G_0, which corresponds to the unstable branch of the hydraulic characteristic curve.

The compressible-volume reservoir at the inlet provides a practically constant inlet pressure during the initial time interval; therefore, in accordance with the hydraulic characteristic, the flow rate in the channel decreases to the G_F value (Figure 5.41). However, if the flow rate through the system is constant, the reduction of the flow rate through the channel is compensated for by the delivery of coolant to the compressible-volume reservoir. As time passes, the level and hence the pressure in compressible volume 2 increases, with the working point of the coolant flow rate following the FA line (Figure 5.41). Upon reaching point A, the flow rate through the channel grows abruptly along the AC line. At the constant flow rate through the system, this makes the coolant flow out of the compressible-volume reservoir into the channel. As a result, both the level and pressure in the compressible volume drop at a rate that depends on the system parameters, and the working point of the coolant flow rate moves along the CB line. When B is reached, an abrupt decrease in the flow rate of the coolant that flows along the BD line takes place, and then the closed DACBD cycle is repeated.

This diagram is ideal. In reality, the flow rate does not change instantaneously and the closed oscillatory cycle trajectory somewhat deforms because of the coolant inertia and dynamic characteristics of the compressible volume. This can be illustrated by experimental data [209] on oscillations of the flow rate and pressure drop (Figure 5.42) and on the limiting cycle of oscillations (Figure 5.43). Experimental studies [209]

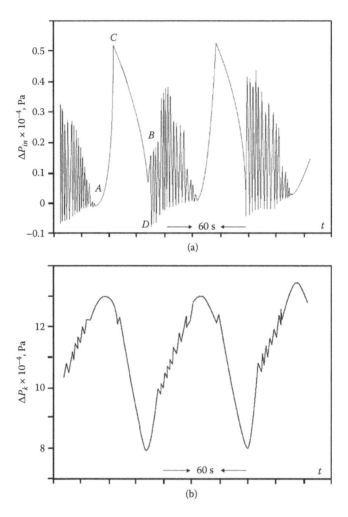

FIGURE 5.42
Experimental data on steady-state oscillations of (a) the velocity at the heated channel inlet and (b) the pressure drop.

were performed on a thermophysical facility whose basic diagram was similar to that in Figure 5.40. R-113 refrigerant was used as the working fluid. The flow rate at the channel inlet was measured by a Venturi tube. Therefore, the pressure drop values obtained by the Venturi tube and corresponding to the flow velocity at the channel inlet are plotted as ordinates in Figure 5.42.

Figure 5.43 contains the experimental static hydraulic characteristic of the channel and the dynamic limiting cycle of oscillations presented as the pressure drop across the channel versus the coolant flow rate through the channel. It is obvious from Figure 5.43 that the experimental limiting cycle

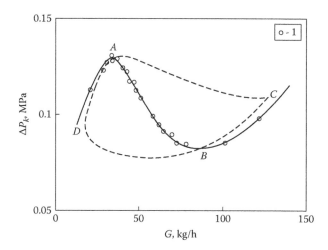

FIGURE 5.43
Experimental data on the static hydraulic characteristic and the oscillation cycle trajectory [209]. 1: Experimental data on the static pressure drop at different flow rates; dashed line: the oscillation cycle dynamic trajectory.

is somewhat deformed in comparison with the ideal one (Figure 5.41). The strongest deformation of the limiting cycle is observed when dynamic characteristics of the compressible volume give a moderate period of pressure oscillations, which is comparable with the coolant mechanical inertia. As a result, not only the limiting cycle is deformed, but also the PDO amplitude is reduced (Figure 5.44).

The oscillograms of experimental oscillations of the inlet flow velocity and of the pressure drop across the channel shown in Figure 5.42 are typical of PDO instabilities and have the following characteristic features:

1. An essentially nonlinear behavior of oscillations
2. Counterphase oscillations of the coolant velocity at the channel inlet and of the pressure drop across the channel
3. A long period of parameter oscillations sufficiently exceeding that of the density-wave flow instability
4. The higher frequency oscillations against the background of the pressure drop low-frequency oscillations

The high-frequency inlet-flow velocity oscillations (see the oscillogram in Figure 5.42) are caused by a thermohydraulic density-wave instability, which occurs when the coolant flow rate decreases to small values during the PDO cycle. The PDO instabilities are often accompanied in experimental studies by density-wave instabilities (see, for example, references 211 and 212). However, the latter instability does not necessarily appear

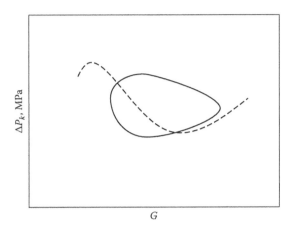

FIGURE 5.44
The limiting cycle trajectory in the case of the short-period steady-state PDO. Dashed line: the static hydraulic characteristic.

together with the PDO instability [29,30,213] and depends on the characteristics of the experimental setup and, primarily, on the values of the inlet and outlet flow throttling, on the system pressure level, and on the location of the area of parameters where the hydraulic characteristic section is negative.

The characteristics of the limiting PDO cycle can be easily analyzed upon applying a simple mathematical model developed in Ozava et al. [209], which allows a qualitative estimate of the influence of system parameters on the period, amplitude, and behavior of oscillations of the inlet velocity flow, and of the PDO across the channel. The equations for nonstationary thermohydrodynamics of a boiling channel with a compressible volume at the inlet are written in terms of lumped parameters [209]. When approximating the nonlinear multivalued static hydraulic characteristic of the channel by a third-degree algebraic equation and taking into account some assumptions and transformations described in detail in Ozava et al. [209], the authors reduce the nonstationary thermohydrodynamic equations for a heated channel with a compressible volume at its inlet to the Van der Pol differential equation. The obtained equation is shown [209] to be able to describe with sufficient accuracy the parameter oscillations under the PDO-type instabilities.

In Ozava et al. [209], it is written as

$$\frac{d^2y}{dt^{*2}} - \varepsilon(1 - 2\beta y - y^2)\frac{dy}{dt^*} + y = 0 \qquad (5.44)$$

$$
\left.
\begin{aligned}
& y = \frac{G^*}{\sqrt{G_A^* G_B^*}}, \quad t^* = \frac{t}{\sqrt{C(I + I_E)}}, \\[2mm]
& \varepsilon = \sqrt{\frac{C}{1 + I_E}} \alpha G_0^2 G_A^* G_B^*, \\[2mm]
& \beta = \frac{G_A^* - G_B^*}{2\sqrt{G_A^* G_B^*}}, \\[2mm]
& G^* = \frac{G - G_0}{G_0}, \quad G_A^* = \frac{G_0 - G_A}{G_0}, \quad G_B^* = \frac{G_B - G_0}{G_0},
\end{aligned}
\right\}
\tag{5.45}
$$

$$
I = \frac{H}{F} + \frac{H'}{F'}, \quad I_E = \frac{H_E}{F_E} + \frac{H_E'}{F_E'}, \quad C = \frac{\rho_l V_{c.g}}{\kappa P_{c.g}},
$$

$$
\alpha = \frac{1}{(G_0 - G_A)(G_0 - G_B)} \frac{d\Delta P_k}{dG}\bigg|_{G=G_0},
$$

where

G₀ is the preset coolant flow rate at the system inlet

G_A and G_B are the coolant flow rates corresponding to the maximum and minimum of the heated channel hydraulic characteristic curve (Figure 5.41)

α is the coefficient determining the gradient of the negative slope of the hydraulic characteristic curve

κ is the adiabatic exponent of gas in the compressible-volume reservoir

$V_{c.g}$ and $P_{c.g}$ are the volume and pressure of gas in the compressible-volume reservoir

H and F are the length and flow area of the heated channel economizer section (Figure 5.40)

H′ and F′ are the length and flow area of the pipeline from the compressible-volume reservoir to the heated channel inlet (Figure 5.40)

H_E and F_E are the length and flow area of the water volume in the compressible-volume reservoir (Figure 5.40)

H_E' and F_E' are the length and flow area of the pipeline connecting the system to the compressible-volume reservoir (Figure 5.40)

To facilitate the analysis, let us assume that the preset coolant flow rate at the system inlet is $G_0 \sim 0.5\,(G_A + G_B)$. Then, β = 0:

$$
\varepsilon = -\sqrt{\frac{C}{1 + I_E}} \frac{d\Delta P_k}{dG}\bigg|_{G=G_0},
\tag{5.46}
$$

and (5.44) changes to the Van der Pol equation:

$$\frac{d^2y}{dt^{*2}} - \varepsilon(1-y^2)\frac{dy}{dt^*} + y = 0,\qquad\qquad(5.47)$$

the behavior of solution of which has been thoroughly studied.

According to references 209 and 214, the form of the limiting cycle in (5.47)—that is, the nonlinearity of the cycle and the oscillation period—depends on the parameter ε. At small ε ($\varepsilon \ll 1$), the self-sustained oscillations do not significantly differ from the harmonic ones. As ε grows, the oscillation shape begins differing significantly from the sinusoidal, and the oscillation period increases. The evolution of the oscillation shape at the increase of ε in (5.47) is presented in Figure 5.45 [214]. At $\varepsilon = 0.1$ (Figure 5.45a), the oscillations are harmonic; at $\varepsilon = 1.0$ (Figure 5.45b), they are not sinusoidal, and at $\varepsilon = 10$ (Figure 5.45c), they are of the relaxation type (i.e., they include sections of both fast and slow changes in the flow rate). For example, in the case of PDO-type instability oscillations experimentally obtained in Ozava et al. [209] and shown in Figure 5.42, the value of the parameter ε is 34.2.

Knowing the relationship between ε and the design and flow parameters of the system (see Equations 5.45 and 5.46), their influence on the oscillations of the coolant flow rate through the channel and the pressure drop can be analyzed qualitatively. It is obvious from (5.45) and (5.46) that the parameter ε grows together with the gas volume V_{cg} in reservoir 2 (Figure 5.40) and with the increasing slope gradient of the negative branch of the hydraulic

characteristic curve $\left(\left.\dfrac{d\Delta P_k}{dG}\right|_{G=G_0}\right)$.

The parameter ε decreases when the gas pressure $P_{c.g}$ in the compressible-volume reservoir is growing and when the $(I + I_c)$ parameter is increasing. Thus, an increase in $V_{c.g}$ and in the slope gradient of the negative branch of the hydraulic characteristic curve, as well as a decrease in $P_{c.g}$, should augment nonlinearity of oscillations of the coolant flow rate and of the pressure drop, as well as cause an increase in the parameters' fluctuation period. This is illustrated in Figure 5.46 by an experimentally obtained dependence of the PDO period and coolant flow rate at the channel inlet on the volume of gas in the compressible-volume reservoir [209]. The period of oscillations varies linearly with the gas volume change. A decrease in the volume or increase of gas pressure in the reservoir results in a greater rigidity of the compressible volume, which, in turn, reduces the time lag for the pressure change in the compressible volume as the flow rate through the channel varies. Thus, the period of oscillations decreases and the system becomes stabilized with respect to the PDO instability.

Let us consider now the influence of the $(I + I_c)$ parameter. It follows from (5.45) and (5.46) that a change in its value significantly affects oscillations'

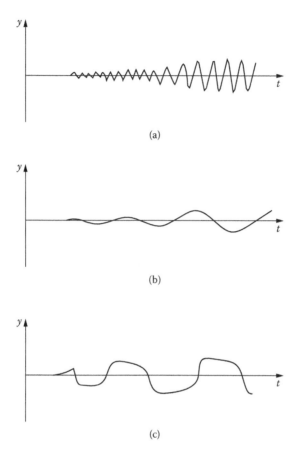

FIGURE 5.45
Oscillogram of self-sustained oscillations of the solution of the Van der Pol equation (5.47) at different values of ε [214]: (a) ε = 0.1; (b) ε = 1.0; (c) ε = 10.

nonlinearity, but not their period, as an increase in $(I + I_c)$, on the one hand, reduces the parameter ε and hence the period of oscillations τ^* and, on the other hand, increases the real-time scale $t = t^* \sqrt{C(I + I_c)}$.

Now we should return to Equation (5.44), where the parameter β is present, and find its influence on the oscillation cycle characteristics. If $G_0 \neq 0.5 (G_A + G_B)$, then $\beta \neq 0$. Based on (5.45) determining β, the value of $\beta < 0$ corresponds to the condition $G_A^* < G_B^*$ (i.e., when G_0' shifts toward the coolant flow rate corresponding to the maximum of the hydraulic characteristic curve, $G_0' - G_A < G_B - G_0'$) (Figure 5.41). At $\beta > 0$, we have $G_A^* > G_B^*$ $(G_0'' - G_A > G_B - G_0'')$; that is, the preset coolant flow rate in the system is closer to that corresponding to the minimum on the hydraulic characteristic curve (G_B). The effect of β on the limiting cycle of oscillations was found [209] on the basis of a numerical solution. If β is negative, the part of the oscillation period with $G < G_0'$ is longer than that

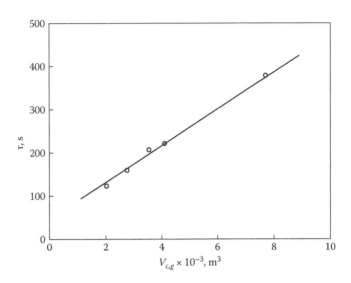

FIGURE 5.46
Dependence of the parameters' fluctuation period on the gas volume in the reservoir [209].

with $G > G'_0$. On the contrary, at $\beta > 0$, the part of the period with $G > G''_0$ is lon-ger, while that with $G < G''_0$ is shorter. Thus, at $\beta < 0$, the coolant flow rate at the channel inlet in considering the cycle of oscillations has a lower value during most of the period than the preset flow rate in the system. Possible overlapping with the higher frequency density wave oscillations takes place in this case over a longer time interval. When $\beta > 0$, the inlet flow rate is higher than the preset one in the system during most of the oscillation period. The described regularity is illustrated in Figure 5.47.

The period of oscillations also increases with the decreasing β because the fraction of the period with the reduced coolant flow rate becomes larger. As an example, Figure 5.48 presents the PDO period and amplitude experimentally obtained in Yuncu, Yildirim, and Kakac [212] as a function of the preset coolant flow rate in the system. It is evident that, all other factors being equal, a decrease in the preset flow rate (equivalent to a decrease in β) results in an increase in the period of oscillations.

According to the PDO instability mechanism, if the oscillation periods are much greater than the time constant of the coolant mechanical inertia, the PDO amplitude, all other factors being equal, is independent from the preset coolant flow rate in the system and is equal to the difference between pressure drops corresponding to the maximum and minimum of the hydraulic characteristic curve. If the oscillation period is comparable with the time constant of the coolant mechanical inertia, the PDO amplitude may decrease because of the deformation of the limiting cycle (see Figure 5.44).

The mechanism of the preceding effect of β is physically meaningful. At $\beta < 0$, the preset coolant flow rate G'_0 in the system (see Figure 5.41) is closer

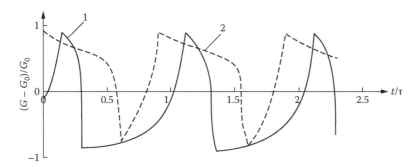

FIGURE 5.47
Flow rate oscillations' behavior at the channel inlet in the case of unstable PDO oscillations at different values of parameter β. (1) β < 0; (2) β > 0.

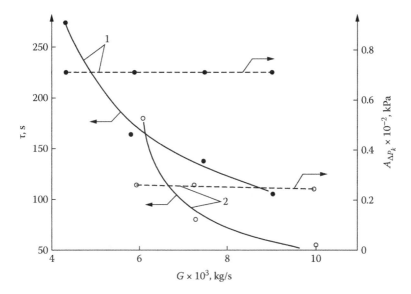

FIGURE 5.48
Dependence of the period and amplitude of the PDO oscillations on the preset coolant flow rate in the system [212] at Nk = 500 W. The outlet diameter of the orifice d_{or}^{out} is (1) 1.8 and (2) 1.4 mm. Dashed line: oscillations' amplitude; solid line: oscillations' period. ○, ●: Experimental values.

to the flow rate corresponding to the peak of the hydraulic characteristic curve (G_A). Therefore, when the coolant flow rate at the channel inlet decreases to the minimum value G_D in the oscillation cycle, the coolant flow rate into the compressible volume is the minimum possible and equals $G_c = G'_0 - G_D$. Because of this, the level and hence pressure in the compressible volume increase at the minimal possible rate, which decreases along the DA trajectory (Figure 5.41) and lengthens the fraction of the period with the low coolant flow rate at the channel inlet. Upon reaching point A and abruptly changing the flow rate at the channel inlet to the G_c value, the coolant flows out of the compressible volume at the maximal possible flow rate $G_E = G_c - G'_0$, which leads to a fast drop of the level and pressure in the compressible volume, thus reducing the fraction of the period when the coolant flow rate at the channel inlet is greater than G'_0. Similar reasoning shows that at $\beta > 0$, when G''_0 is closer to G_B, the fraction of the oscillation period with the coolant flow rate at the channel inlet $G > G''_0$ is greater than the fraction with $G < G''_0$.

The described regularities make it easy to analyze qualitatively the effect of regime parameters such as inlet subcooling, power supplied to the channel, outlet throttling, etc. on the limiting PDO cycle parameters. Experimental studies of the effect of these parameters on the PDO characteristics have been carried out [29,30,211–213]. Let us consider the effect of the main regime parameters.

The effect of the coolant subcooling temperature ΔT_{sub} at the channel inlet has been experimentally studied [29,211] and found mainly to influence the ambiguity region of the hydraulic characteristic curve. The working section in Mentes et al. [211] represented a vertical heated pipe, while the experimental facility diagram is given in Figure 5.40. Freon-11 was used as the working fluid. The experiments were carried out at a constant power input of approximately 413 W and at different inlet coolant subcooling in the –9.8°C through 38°C range. Figure 5.49 demonstrates a typical experimental distribution of the channel hydraulic characteristics and of the PDO instability region boundaries. The effect of the inlet coolant subcooling on the PDO instability characteristics obtained in Mentes et al. can be easily explained by the mechanism of instability and Figure 5.49. The main conclusions of Mentes et al. are presented next.

1. At high and moderate inlet subcooling, an increase of inlet subcooling, on the one hand, reduces the range of preset flow rates in the system associated with the occurrence of the PDO (i.e., it stabilizes the system); on the other hand, an increase of inlet subcooling increases the PDO amplitude and period in the channel. Such an influence of inlet subcooling is obvious from Figure 5.49 and conclusions made in previous sections: An increase of inlet subcooling, on one hand, causes the point of hydraulic characteristic minimum to shift to the region of lower coolant flow rates, resulting in a narrower range of flow rates, at which the negative slope of the hydraulic characteristic curve that defines the PDO instability region can occur. The other result is an increase in the slope of the negative branch $\left(\text{an increase in} \left| \dfrac{d\Delta P_k}{dG} \right| \right)$, as well as a decrease in the overall pressure level (given a preset constant pressure at the system outlet) and therefore a decrease in the pressure of gas in the compressible volume. Both factors are responsible for the increasing ε in (5.44)

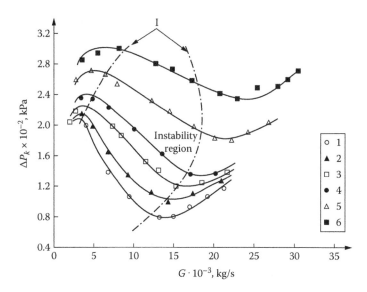

FIGURE 5.49
Effect of inlet coolant subcooling on the PDO instability region [211]. Coolant: Freon-11, Nk = 416 W; (I) the PDO stability boundaries. Inlet temperature,°C: (1) –7.4; (2) –0.9; (3) 7.3; (4) 15.6; (5) 26.0; (6) 37.6.

and, consequently, for increasing the PDO period and emphasizing the oscillation nonlinearity. Figure 5.49 shows that the difference in pressure drops corresponding to the maximum and minimum of the hydraulic characteristic curve grows together with inlet subcooling, resulting in the PDO amplitude increase.

2. At low inlet subcooling, decrease of its value reduces the PDO instability region, as can be seen from Figure 5.49, and stabilizes the system. This effect of the inlet subcooling derives from the fact that, at certain low values of inlet subcooling, the negative section of the hydraulic characteristic curve becomes gently sloping (i.e., the slope gradient decreases together with the parameter ε). In this case, the period of parameters' oscillations decreases to become comparable with the time constant of the coolant mechanical inertia. As a result, the system becomes stabilized.

As applied to the experimental setup in Mentes et al. [211], an additional stabilizing effect is achieved by a buildup of gas pressure in the compressible volume when the inlet subcooling decreases due to an increase in the channel pressure drop.

It is shown in Figure 5.49 that the PDO stability boundaries pass through the points where the gradient of the derivative in the negative section of the hydraulic characteristic curve is significant in magnitude, rather than through the maximum and minimum of the characteristic curves. The more flattened the hydraulic characteristic curve is in the vicinity of its maximum or minimum, the farther away from the maximum and minimum points the stability boundary will be. Thus, a certain value of inlet subcooling causes the instability region to expand, all other conditions being equal. An increase or decrease of inlet subcooling from this value results in a reduction of the PDO instability region.

The effect of the power input to a heated channel on the PDO stability boundary has been experimentally studied [212] and found to display itself in the deformation of the hydraulic characteristic curve at the power input variation. The typical hydraulic characteristic curves [212] for different input power levels (the other parameters being equal), together with the PDO instability region, are presented in Figure 5.50. As the power input grows, the system stability decreases, since, in this case, (1) the range of flow rates at which the negative slope section of the hydraulic characteristic curve can occur broadens, and (2) the slope gradient of the unstable branch of the hydraulic characteristic curve increases. Also, the previously mentioned effects enhance nonlinearity of the period and amplitude of the coolant flow rate oscillations at the channel inlet and of the pressure drop across the channel. It follows from Yuncu et al. [212] that the oscillation period grows together with the power input even if the gas pressure in the compressible volume somewhat increases. This happens because the influence of the increasing slope gradient of the unstable branch of the hydraulic characteristic curve, which leads to an increased oscillation period, is greater than the effect of pressure increase in the compressible volume.

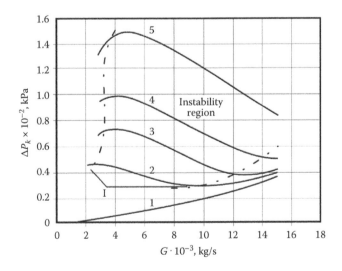

FIGURE 5.50
Effect of the input power on the PDO instability region [212] for $d_{or}^{out} = 1.6$ mm; Nk, W. (1) 0; (2) 321.75; (3) 421.8; (4) 498.15; (5) 607.6. Dotted dashed line: the stability boundary.

The effect of flow throttling at the heated channel outlet has been experimentally studied [212]. The typical hydraulic characteristic curves experimentally obtained for a channel with the fixed power input and different throttling orifice diameters (1.4, 1.6, and 1.8 mm) at the channel outlet are given in Figure 5.51, in which the PDO instability region boundaries are also plotted. Figure 5.51 shows that the increased throttling at the channel outlet increases the slope of the negative branch of the hydraulic characteristic curve. The simultaneous pressure drop growth causes the inlet subcooling to increase; the coolant temperature at the channel inlet remains unchanged. As the outlet throttling grows, an increase in the inlet subcooling shifts the minimum of the hydraulic characteristic curve to the lower flow rate region, thereby shifting the PDO instability boundaries likewise.

The previously discussed influence of outlet throttling on the deformation of the negative branch of the hydraulic characteristic curve, as well as the previously considered effect of the hydraulic characteristic parameters on the PDO mechanism, enable us to interpret the experimental results from Yuncu et al. [212] concerning the effect of outlet throttling on PDO characteristics. For instance, according to Yuncu et al., the PDO amplitudes increase with outlet throttling. This is caused by the increasing difference between the pressure drop maximum and minimum points on the hydraulic characteristic curve as the outlet throttling increases (Figure 5.51). The PDO period shows an increase with an increase in outlet throttling because of a greater slope gradient of the negative slope branch of the hydraulic characteristic curve caused by the increased outlet throttling and the displacement of stability boundaries to the lower flow rate region. The PDO growth along with the

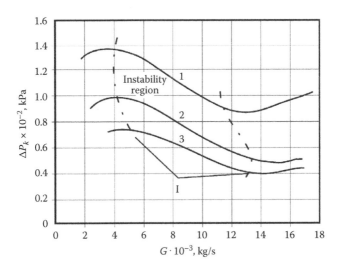

FIGURE 5.51
Effect of outlet throttling on the PDO instability region [212]; $Nk = 500$ W. d_{or}^{out}, mm. (1) 1.8; (2) 1.6; (3) 1.4. Dotted dashed line: the stability boundary.

increase of outlet throttling does not compensate for a certain increase in the gas pressure in the compressible volume, which reduces, for the preceding reasons, the PDO period. The nonmonotonic behavior of the right boundary of the stability region shown in Figure 5.51 can be explained by the fact that the decreased outlet throttling reduces the slope gradient of the negative branch of the hydraulic characteristic curve, which becomes flat in the region of its minimum. In this case, the stability boundary shifts to the left of the minimum point of the hydraulic characteristic curve, thereby reducing the instability region.

5.6 Some Other Mechanisms Inducing Static Instability

In the case of laminar gas flow in the heated channel, the pressure drop is largely determined by the shear stress and, to a lesser extent, by the gas column weight. For the shear stress, the relation

$$\tau \sim \mu G \tag{5.48}$$

holds true, where $\mu \sim T^n$ ($n > 1$ is the power factor).

When the coolant flow rate is increasing (at constant heat flux), on the one hand, the proportionate dependence of τ on the flow rate contributes to the increase of pressure drop across the channel. On the other hand, second multiplier μ in (5.48) decreases due to a drop in gas temperature, thus corresponding to a pressure drop decrease. In some region of parameters, the opposite effect of multipliers in (5.48) may lead to the appearance of the negative slope of the channel hydraulic characteristic curve and the onset of static instability. In the case of the gaseous coolant downflow, the gravity pressure drop component also exerts a destabilizing effect on static instability.

For the first time, the problem of static instability in the laminar gas flow (hydrogen) was considered [96,97] when seeking a solution to the problem of thermal reliability of nuclear rocket engines. Calculations of the static stability boundary for CO_2 as the gaseous coolant may be found in Antonyuk and Korolev [98]. The results show that the increasing heat flux density leads to the stability region narrowing.

Another mechanism causing static instability in the system of parallel channels is the change of flow parameters with the occurrence of thermal-acoustic instability. It is known that when the section of surface boiling is large and the heat flux density is high, thermal-acoustic oscillations may arise in the heated channel. The experiments show that the channel friction tends to increase at the onset of thermal-acoustic oscillations. When the pressure drop across the channel is constant, this effect induces a periodic

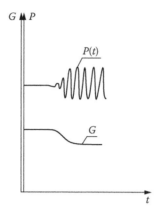

FIGURE 5.52
Variation of pressure and averaged coolant flow rate with respect to the oscillation cycle in the case of thermal-acoustic oscillations (TAOs).

reduction of the period-average coolant flow rate fluctuations. The reduced coolant flow rate may lead to burnout.

The oscillogram of varying parameters in the case of a static instability of this type is shown in Figure 5.52. Similar phenomena may also be observed at supercritical pressures.

6

Thermal-Acoustic Oscillations in Heated Channels

6.1 Thermal-Acoustic Oscillations at Subcritical Pressures

6.1.1 Oscillations' Development Pattern and Initiation Mechanism

The so-called thermal-acoustic oscillations (TAOs) may appear under certain conditions in heated channels with the boiling coolant subcooling below the saturation temperature [102,103]. Much work, mostly experimental, has been dedicated to these oscillations. In this chapter, we shall refer to the work in which a heated channel (generally, an electrically heated tube) was acoustically isolated (acoustically open) from other components of the test facility. This was mainly achieved by locating vessels at the channel inlet and outlet. With their volumes, the vessels ensured acoustic channel isolation due to coolant compressibility. That is, equal pressure was provided upstream and downstream from the channel in the case of spontaneous initiation of the flow parameters' high-frequency oscillations of an acoustic nature.

Acoustic isolation of the heated channel is not used in the majority of work dealing with TAO experimental investigations. Instead of studying the single tube TAOs, the initiation and characteristics of TAOs in a test facility are investigated. In the latter case, TAOs induced in the heated channel penetrate into the test facility lines, reflect from the components, and superimpose one another to create a rather complex phenomenon. It should be noted that characteristics of stability regions in the flow parameter space and oscillation amplitudes dependent on flow parameters of TAOs emerging in acoustically isolated and not isolated channels may differ qualitatively.

The pattern of TAO initiation is as follows. When the heat load is smoothly increasing from zero, the no-boiling regime establishes in the channel, and the pressure pickup located at a certain distance from the channel ends (constant pressure points) records only the noise caused by flow turbulence. Further, as the heat load increases, a surface boiling area is observed in the channel. The pressure pickup records the total noise generated by boiling and flow turbulence. At a still higher planned heat load and sufficiently significant coolant subcooling (herein, water is considered as the coolant) below the saturation temperature, there comes a moment at which the pickup starts recording the spontaneously developing high-frequency pressure oscillations. An example of pressure peak-to-peak oscillations' (2A) dependence on the heat flux is illustrated in Figure 6.1. Oscillation amplitudes (A) were found first to increase with the increasing heat flux and then to drop. Prior to burnout, regular high-frequency oscillations are usually absent and the pickup records only random low-frequency noises whose intensity increases before the heated tube reddening. The maximum pressure oscillation amplitudes may reach, in some cases (with relatively small heated section lengths and high subcooling), the time-averaged channel pressure.

Some investigators report [104,105] that TAOs cause destruction of tubular channel walls. The time of heated thin-walled tubes' operation under the conditions of high-amplitude fluctuations numbers several hours before their failure. TAOs are especially dangerous in nuclear reactor fuel assemblies, where a gas gap may form over time between fuel pellets and cladding. In such a case, TAOs induce high-frequency transverse oscillations of the thin-walled cladding and may lead to deterioration of its integrity with radioactive products release into the loop.

An analysis of the pressure oscillation spectrum [102,106,107] shows that oscillations are made up of a sum of standing waves of different harmonics with frequencies $\omega_n^{-1} = 2H/na$, n = 1, 2, 3, ..., where a is the channel average

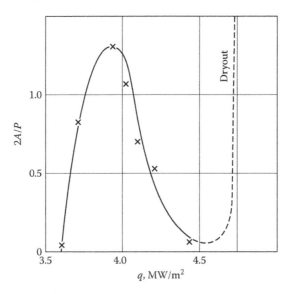

FIGURE 6.1

Variation of the relative peak-to-peak pressure oscillations with the increasing heat flux. Tube 6 × 1 mm, tube length (H) = 1.2 m; the heated section length (H_h) = 0.2 m; the heated section center coordinate (Zh/H) = 0.7; the pressure gauge location coordinate (Zp/H) = 0.25; ρW = 435 kg/m²s; P = 0.75 MPa; T_{in} = 25°C.

sound velocity and H is the distance between the acoustically open ends of the heated channel.

In an ideal case, where thermophysical parameters of coolant are constant along the length, the diagrams of the fluctuating velocity standing wave amplitudes (δW) and pressure (δP) for the first four harmonics are as shown in Figure 6.2 [108]. In the presence of TAOs, it is impossible to measure coolant flow velocity fluctuations because even the lowest frequencies usually have the order of several hundreds of hertz. In a real channel, standing waves were found to be asymmetric toward the channel exit because of flow inhomogeneity and change of compressibility. Figures 6.3 and 6.4 illustrate diagrams of pressure oscillations. The channel length (H) is 3000 mm, the heated section H_h is 410 mm long, and its center point Zh/H is located at 0.8 H from the channel inlet. Six pressure pickups were located along the length of the channel (tube: 6 mm diameter and 1 mm thick), allowing the use of eight points (including constant pressure points at two channel ends) for plotting the distribution of individual component amplitudes along the channel length.

Figure 6.3 shows relative amplitudes of the first and second harmonics for $\rho W = 875$ kg/m²s. The higher harmonic amplitudes are practically absent. Figure 6.4 presents the diagrams of the pressure oscillation first harmonic relative amplitude at different heat fluxes. The higher harmonics are also absent here. The results of the experiments show that with the heat flux variation, both oscillation amplitudes and their distribution along the channel length also varied. At lower heat fluxes (near the TAOs' initiation boundary), the pressure wave diagrams obey the harmonic distribution (e.g., the profile with $q = 1.65.10^6$ W/m² in Figure 6.4). With the increasing heat flux, the wave antinode shifts toward the higher coolant compressibility.

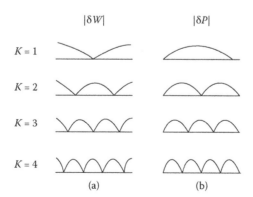

FIGURE 6.2
Standing wave diagrams |δW| and |δP| for the first four harmonics (two-end open tube): δW: velocity disturbance; δP: pressure disturbance. (a) Orifice plates-bounded tube; (b) a tube acoustically isolated by constant pressure points.

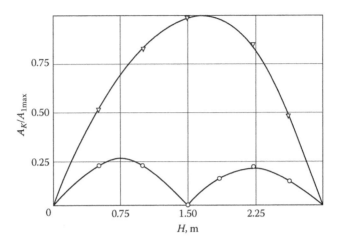

FIGURE 6.3
Distribution of the first and second harmonics of pressure fluctuations in the channel at $q = 1.28 \ 10^6$ W/m²; $\rho W = 875$ kg/m² s; $T_{in} = 25°C$; clear reversed triangles: first harmonic; clear circles: second harmonic.

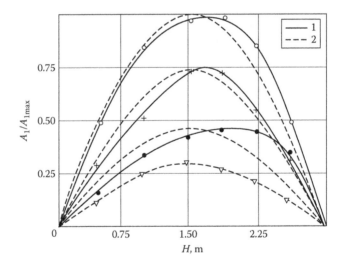

FIGURE 6.4
Distribution of the first harmonic of channel pressure fluctuation; $\rho W = 1010$ kg/m²s; $T_{in} = 25°C$; $q \ 10^6$, W/m²: ∇: 1.65; × = 1.736; + = 1.89; o = 2.093; solid lines: real pressure distribution; dashed line: sinusoidal pressure distribution.

By now, several possible mechanisms of TAOs' initiation and development have been proposed. Here, the TAOs' development description is based on the ideas expressed in Gerliga and Prokhorov [106] and Hayama [109].

Acoustically, a hydraulic channel represents an oscillatory system. This is determined by coolant compressibility and is expressed by the flow acoustics equations. Natural hydraulic channel pressure fluctuations are similar to those resulting from the deviation from the equilibrium of the string stretched between the channel ends (Figure 6.2).

In a boiling channel, turbulent disturbances and those resulting from boiling initiate low-intensity oscillations (at the level of hydraulic noises) that are close to natural harmonics. Further, amplitudes of these harmonics will be increasing if heat or mass input/output to/from coolant flow exists with the pressure rise/drop, respectively. In fact, it is the Rayleigh criterion of TAOs' initiation [4,108] and its essence is as follows: The in-phase disturbances of the rate of mass/heat input to the channel working fluid and those of the channel pressure facilitate TAOs' initiation. This mechanism of oscillations' initiation in the channel with surface boiling has some specific features.

Let us first consider the condensing bubble behavior in the pressure sound field. The work performed by the bubble during the oscillation cycle θ, if the bubble life is substantially longer than the oscillation cycle, is written as

$$A = \int_t^{t+\theta} \frac{d\delta V}{dt} \delta P dt, \tag{6.1}$$

where δV is the bubble volume disturbance due to the δP pressure acoustic fluctuation, and θ is the oscillation cycle.

If we assume that $\delta P = a_p \sin \omega t$ and $\delta V = a_V \sin (\omega t + \phi)$, then

$$A = -\pi a_p a_V \sin\phi, \tag{6.2}$$

where ϕ is phase shift between the pressure and bubble volume fluctuations.

The work performed by the bubble in the pressure sound field during the oscillation cycle will be positive—that is, will be performed on the surrounding coolant flow with

$$-\pi < \phi < 0. \tag{6.3}$$

Additionally, if we take into account that $E = \dfrac{d\delta R}{dt} \delta P$ is the acoustic energy flux per bubble unit surface [4], expression (6.1) will be the amount of acoustic energy radiated by the bubble during the oscillation cycle:

$$A = \int_t^{t+\theta} F_G \frac{d\delta R}{dt} \delta P \, dt = \int_t^{t+\theta} F_G E \, dt,$$

where F_G is the bubble surface.

Further, we confine ourselves to revealing the main relations in the TAOs' initiation mechanism, and bubble growth will be described in a simplified form.

Let us consider feedbacks replenishing the oscillatory system with energy.

When analyzing the pressure disturbance feedback, it may be assumed that the bubble is in the pressure harmonic antinode ($\partial \delta P / \partial z = 0$); that is, the bubble is not influenced by the pressure disturbance gradient. When the channel pressure increases, the bubble volume decreases together with the mass transfer surface. Let the bubble be located in water below T_s. Then, with the pressure rise, steam condensation drops (i.e., the steam outflow from the bubble decreases as compared to the bubble's undisturbed state). If the bubble's undisturbed state is assumed as neutral (zero), it may be said that with the increasing pressure, steam flow "rushes" into the bubble (condensation reduces due to the decreased bubble surface). When the pressure reduces, the bubble surface increases, as well as the steam outflow from the bubble. Thus, with the increasing pressure, steam "flows into" the bubble and contributes to the further pressure rise. With the reduced pressure, steam "flows out" from the bubble, as compared with the undisturbed state, thus promoting the further pressure drop. In accordance with the Rayleigh criterion, this pressure disturbance feedback results in the initiation of oscillations.

With the growing bubbles (superheated fluid), the disturbance of steam "feed" to the bubble is in antiphase with the pressure disturbance and no oscillations are induced. The bubble surface decreases as the pressure grows; steam input reduces and prevents the further pressure rise. As the pressure drops, the bubble surface increases, as does the steam input, thus preventing the further pressure drop.

Thus, the channel pressure harmonics induced by different disturbances (due to the channel behavior as an acoustic oscillatory system) govern the steam mass feed into the channel. Since $\rho_L \gg \rho_G$, it may be considered in the first approximation that the disturbance of steam outflow from the bubble passing through the given channel section is equal to the disturbance of steam outflow from the channel in the given section area.

The gain of the considered pressure disturbance feedback somewhat decreases due to the dependence of steam–fluid interphase temperature on pressure. With the increasing/decreasing pressure, T_s is found to increase/decrease, thus contributing to the weakening of the previously considered feedbacks concerning steam inflow/outflow into/from the bubble.

The considered TAOs initiation mechanism agrees with (6.2) and (6.3). Indeed, let the bubble volume stay constant under the steady-state condition during time θ. In this case, the bubble volume change will be in antiphase with the pressure change ($\phi = -\pi$). If pressure increases, the condensing bubble will contain more steam as compared to that in the undisturbed state, and vice versa when pressure drops. This will lead to a somewhat delayed change of δV for the condensing bubble as compared to the neutral bubble (i.e., $\phi > -\pi$).

The preceding may be proved by a simple example. Let the bubble volume growth be described by

$$\frac{d\rho_G V_G}{dt} = h_m F_G \Delta T ,$$ (6.4)

where ΔT is the difference between T_L and T_s and h_m is mass transfer coefficient. In terms of disturbance, Equation 6.4 will be

$$\frac{d\delta V_G}{dt} = \frac{h_m}{\rho_G} \cdot \frac{dF_G}{dV_G} \cdot \Delta T \delta V_G - \frac{V_G}{\rho_G} \frac{\partial \rho_G}{\partial P} \frac{d\delta P}{dt} .$$ (6.5)

When the inlet harmonic signal ($\delta P = a_p e^{j\omega t}$) of the linear system is used, the outlet signal may be written as

$$\delta V = a_v (j\omega)\, e^{j\omega t},$$

where $a_v (j\omega)$ is the complex amplitude.

The phase shift between δV and δP will be equal to the argument of $a_v(j\omega)$ quantity. In this case, the phase shift ϕ will be in the range of (6.2), if Im $a_v(j\omega) < 0$. After the substitution of expressions for δP and δV into (6.5), we have

$$\frac{a_v(j\omega)}{a_P} = \frac{j\omega \cdot \dfrac{V_G}{\rho_G} \dfrac{\partial \rho_G}{\partial P}}{\dfrac{h_m}{\rho_G} \dfrac{dF_G}{dV_G} \Delta T - j\omega} ,$$

from which the following expression for the imaginary part $a_v(j\omega)$ is written as

$$\text{Im}\left\{\frac{a_v(j\omega)}{a_P}\right\} = \frac{h_m \dfrac{dF_G}{dV_G} V_G \Delta T \dfrac{\partial \rho_G}{\partial P} \cdot \left(\dfrac{1}{\rho_G}\right)^2 \omega}{\left(\dfrac{h_m}{\rho_G} \dfrac{dF_G}{dV_G} \Delta T\right)^2 + \omega^2} .$$

The imaginary part is negative if $\Delta T < 0$ (i.e., $T_L < T_s$). Consequently, the condensing bubbles perform work on the surrounding flow (return positive acoustic energy) during the oscillation cycle. It follows from this simplest consideration that if a nondisturbed bubble is growing with time ($dV_G/dZ > 0$), no oscillations are initiated, and vice versa with $dV_G/dZ < 0$. The preceding analysis of the mechanism of thermal-acoustic instability with the pressure disturbance feedback has been performed for a more vivid understanding of the phenomenon in question. In doing so, the pressure disturbance gradient feedback was not taken into account.

A detailed analysis of the steam bubble behavior in the pressure acoustic field shows that the pressure disturbance gradient feedback ($\partial \delta P / \partial Z$), which causes disturbance in the bubble and surrounding fluid velocities, may be a more important feedback governing the steam mass feed into the bubble.

If the condensing steam bubbles are located beyond the antinode of pressure fluctuation modes, they are influenced by both the pressure disturbance and pressure disturbance gradient. This gradient, taken with the reverse sign and multiplied by the bubble volume $\left(-\dfrac{\partial \delta P}{\partial Z} \cdot V_G \right)$, is in fact a peculiar buoyancy force. Since the pressure disturbance ($\delta P = a_p(z) \sin \omega t$) and hence pressure disturbance gradient $\left(\dfrac{\partial \delta P}{\partial Z} = \dfrac{\partial a_p}{\partial Z} \sin \omega t \right)$ are the alternating quantities, the buoyancy force will also be an alternating quantity. The alternating behavior of $\left(\dfrac{\partial \delta P}{\partial Z} \right)$ is defined not only by $\sin \omega t$, but also by the sign of the amplitude gradient $\left(\dfrac{\partial a_p}{\partial Z} \right)$.

Let us further consider specifics of bubble behavior depending on its location using excitation of the first mode (harmonic) of pressure fluctuation as an example. Let the bubble be in the second half of the acoustically isolated channel (Figure 6.5). The buoyancy force is directed toward decreasing the pressure disturbance. In the given case with P increasing with time ($\delta P > 0$),

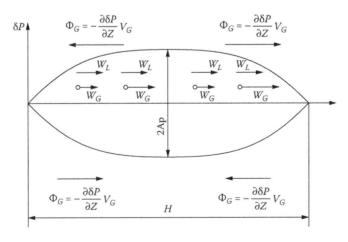

FIGURE 6.5
Effect of pressure wave on bubble slip.

the buoyancy force $\phi_G = V_G \cdot \left(\dfrac{\partial \delta P}{\partial Z}\right)$ accelerates the bubble and reduces the module of the bubble and fluid velocity difference ($|W_G - W_L|$) when $W_G < W_L$. It increases the module of velocity difference when $W_G > W_L$. In the first case, heat and mass transfer are reduced, while in the second case they are increased. In other words, in the first case with the increasing δP, steam condensation decreases due to the pressure disturbance gradient and, in a manner of speaking, the bubble receives an additional amount of steam. In the second case, with the increasing P, steam condensation increases and, due to $\partial \delta P / \partial Z$, an additional amount of steam outflows from the bubble, as compared to the undisturbed state.

The buoyancy force changes its direction when pressure is reducing. Now, with $W_G < W_L$, the velocity difference module becomes even larger with the resultant increased steam condensation; with $W_G > W_L$, the module $|W_G - W_L|$ becomes smaller and steam condensation decreases. Thus, in the first case, the pressure drop due to the pressure disturbance gradient leads to an even larger pressure drop, while in the second case, the disturbance of steam feed to bubbles is in antiphase with the pressure disturbance, thereby preventing further pressure drop.

It may be concluded from the preceding that the condensing bubbles located on the downward part of the harmonic induce TAOs when $W_G - W_L < 0$ and stabilize the process when $W_G - W_L > 0$. The first case relates to the initial stage of bubble motion, at which their velocity is less than that of the surrounding fluid due to the apparent mass. With time, due to the $\left(-\dfrac{\partial P}{\partial Z}\right) \cdot V_G$ force, the bubble velocity becomes higher than that of the fluid.

If the bubble is located in the first half of the channel (Figure 6.5), the buoyancy force changes its sign to the opposite, as compared to the previously considered cases, and thus produces a principally different effect on the disturbance of steam feed to the oscillatory system.

During the first half of the bubble motion, when $W_G < W_L$, the condensing bubble stabilizes the process. During the second half, when $W_G > W_L$, the condensing bubble promotes the TAOs' initiation.

It should be kept in mind that a certain phase shift may exist between $\partial \delta P / \partial Z$ and the velocity difference disturbance $\delta(W_G - W_L)$. Consequently, the condensing steam flow disturbance due to the bubble apparent mass and unsteady heat transfer between the bubble and surrounding fluid neglected in previous considerations.

A detailed numerical analysis of the behavior of condensing bubbles moving with a slip in the pressure sound field confirms the disclosed initiation mechanism of oscillations with the slip disturbance feedback. Figure 6.6 shows qualitative predictions of the work performed by a bubble when the latter is affected by the first mode of pressure fluctuations. When writing the equation of bubble motion, the apparent mass [111], the friction force, and

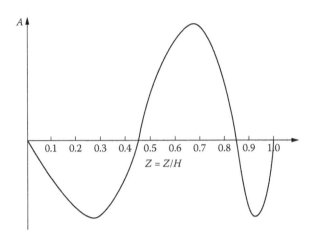

FIGURE 6.6
The work performed by a bubble depending on its coordinate along the channel length ($W_G < W_L$).

a force induced by the effect of pressure gradient on the bubble were taken into account. Heat transfer was calculated according to recommendations given in Kroshilin, Kroshilin, and Nigmatulin [112].

Positive work is performed by the bubbles when they are located in the second half of the channel, thus confirming the previous reasons concerning the mechanism of oscillation initiation.

Recently, one of the authors has performed a comparison of predicted and experimental stability boundaries that showed a good agreement between them. The predictive technique is based on the preceding mechanism of acoustic oscillations' excitation.

It is generally accepted by experimentalists that the cause of TAOs' initiation is the process of condensing bubbles' collapse. In contrast to this, it should be noted that the collapse is asymmetric (channel flow), thus reducing hydraulic shock with the bubble disappearance; that acoustic oscillations in a channel have a harmonic nature (collapse is not a harmonic process); and, finally, that the authors observed TAOs' initiation in heated short transparent channels without bubble collapse.

The bubbles growing on the wall may play a dual role in TAOs' initiation. In the case when the time of bubble growth is much longer than the oscillation cycle, pressure fluctuations lead to fluctuations in the height of the bubble growing on the wall. When the pressure increases, the bubble compresses as compared to its undisturbed state and "transits" into a hotter near-wall layer. Condensation (if any) at the bubble tip decreases and the bubble receives a larger amount of steam as compared to its undisturbed state, causing a further pressure rise. The picture is opposite with the decreasing pressure. The bubble volume grows; condensation increases and causes a further pressure drop. All this proceeds according to the Rayleigh criterion.

In the case when the time of bubble growth is much shorter than the oscillation cycle, the steam "layer" formed by the bubbles at the wall will be changing its volume in antiphase with pressure fluctuations, thus stabilizing the oscillatory system.

The intermediate ratios between the time of bubble growth on the wall and the oscillation cycle will yield a more complex picture that will be hardly predictable quantitatively.

It is supposed that the number of bubbles moving in the flow (α, the exit bulk steam quality at the onset of TAOs, is about 0.1–0.5) is much larger than the number of bubbles growing on the wall. This permits disregarding the latter bubbles, which is the case with the previously mentioned predictive technique.

It should be noted that in the presence of developed self-oscillations, the dependence of departure frequency, density of nucleation sites, and near-wall bubbles' behavior on the pressure change may have a significant effect on oscillation amplitudes.

The mechanisms of TAOs' initiation in the case of spot heating are considered in Hayama [109], where everything is reduced to the pressure feedback only. Since initiation of TAOs in real systems is largely dependent on the parameters' distribution, Hayama's results are of no interest for our case.

6.1.2 Effect of Flow Parameters on Oscillation Characteristics

The experiments performed with a test channel with transparent walls showed that TAOs appeared after the heated rod was covered with a layer of scale. This was achieved by a preliminary rod treatment by means of boiling water on it. After that, the channel was filled with smaller bubbles, which were large in number.

During distilled water boiling in a clean tubular channel for as long as 10 h, TAOs may not appear. If distilled water is replaced with salt-rich water, oscillations of considerable amplitude occur practically instantaneously. With the increasing boiling time, fluctuation amplitudes rise, and the onset of fluctuations shifts toward lower heat fluxes. The experiments with the channel with transparent walls showed that the observed specific features are due to scale on the heater surface. As can be seen from Figures 6.7 and 6.8, the amplitudes and boundaries of TAOs' existence strongly depend on the scaling thickness [104].

The experiments were performed using a 6 mm diameter, 1 mm thick tube with the total length of $H = 1.65$ m and the heated section $H_h = 0.41$ m. The pressure pickup was located at a distance of 0.26 m from the channel inlet. Heat fluxes varied from 0 up to $5.25 \cdot 10^6$ W/m^2. The heated section was made replaceable. Five different heated sections were tested. The scaling thickness was specified by different time of preliminary boiling of water with the total salt content of 250 mg/1 (hardness salts: 2750 μg-eqv/1; iron

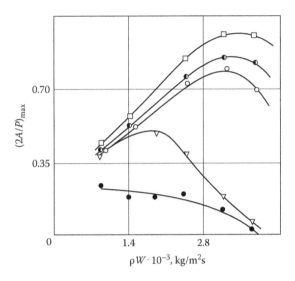

FIGURE 6.7
Effect of scale on the relative peak-to-peak pressure fluctuations. Dark circle: channel no. 1; triangle: channel no. 2; clear circle: channel no. 3; semidark circle: channel no. 4; square: channel no. 5.

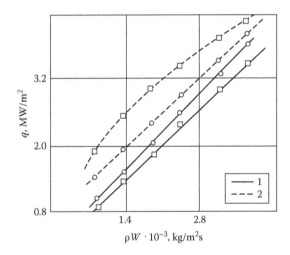

FIGURE 6.8
Effect of scale formation on the region of oscillations' existence. 1: oscillations' onset ($q_{Bn.l}$); 2: oscillations' termination ($q_{Bn.h}$); clear circle: channel no. 1; square: channel no. 5.

salts: 920 µg-eqv/1, pH = 7.4). The times for the channels were as follows: no. 1: 0 h, no. 2: 0.5 h, no. 3: 1 h, no. 4: 1.5 h, and no. 5: 2 h. TAOs' characteristics were investigated using distilled water with the total salt content of 3–5 mg/L (hardness salts: 60 µg-eqv/L, iron salts: 150 µg-eqv/L, pH = 7.4) with P = 2 MPa, T_{in} = 25°C, and ρW = 400–4000 kg/m²s. The scale thickness was assessed by x-ray on the cutout specimens and was found to have changed for about 20 μm.

With the beginning of scale formation, the lower TAOs' boundary ($q_{bn.1}$) shifts toward lower heat fluxes, while the upper boundary ($q_{bn.h}$) shifts toward higher heat fluxes.

At the same time, it should be noted that preliminary distilled water boiling in a channel will lead over time at significant subcooling to TAOs' boundaries and amplitudes that will be stable and practically unchanged throughout the channel operation time. In this case, operation of the channel in the boiling regime for dozens of hours does not cause the oscillation characteristics to change. On the other hand, the time of channel operation in the boiling regime with small subcooling and high pressure—when the region of oscillations' existence is narrow and the maximum amplitudes are of the order of fractions of an atmosphere—substantially influences TAOs' characteristics. Boiling within 20–30 h may lead to oscillations' degeneration. Probably, the roughness that increases with time on the heated surfaces leads to higher dissipation losses.

The dependence of standing wave amplitudes on the channel length coordinate complicates their experimental investigation. It is desirable to have several pickups along the channel that would be recording the profile of oscillation amplitudes' distribution over the channel length. This is not always possible; therefore, only one pickup is usually used for the purpose. If a TAO consists mostly of the first mode, the readings of a single pickup will sufficiently fully characterize the pressure profile. When one pickup is used for recording multimode oscillations or single-mode ones with the second and third harmonics, the general picture of pressure distribution along the channel levngth is not clear. Usually, experimentalists use one pickup that allows them to establish qualitative dependence of TAOs' parameters on flow characteristics.

In the majority of our experiments (P = 2 MPa, T_{in} = 25°C, water as coolant), the frequency spectrum was noted to develop from higher to lower harmonics when the heat flux was increasing. As a rule, lower harmonics have larger amplitude maxima. In terms of size, the first, second, and third harmonics are the main ones. A result of research on oscillations' excitation is presented in Figures 6.9 and 6.10 (H = 3 m, H_h = 410 mm, a pickup is located at 0.25 H from the channel inlet).

At first, the higher harmonics were excited. The amplitudes of separate harmonic components were found first to increase (along with the increasing q) and, then, having reached their maximum, to decay. With regard to the heat flux, there are regions in the channel where only one harmonic exists. With the decreasing mode number, such regions are shifted toward higher heat fluxes. At pressure rise, excitation of the first harmonic prevails.

When the heat flux is varied, the pattern of change in pressure fluctuation amplitudes in the channel remains similar at different velocities. Figure 6.11 presents the dependence of the relative pressure fluctuation peak-to-peak

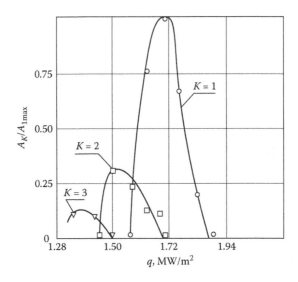

FIGURE 6.9
Harmonics excitation with the heat flux variation. $\rho W = 1272$ kg/m²s; $P = 20$ MPa; $Z_n = 0.85$ H; $Z_p = 0.25$ H; $H = 3$ m; clear circle: fundamental tone; clear square: second harmonic; clear reverse triangle: third harmonic; A_k: the k harmonic amplitude.

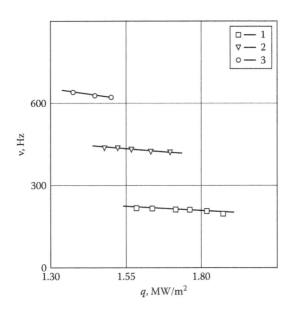

FIGURE 6.10
Frequencies of harmonics depending on the heat flux; flow parameters as in Figure 6.9.

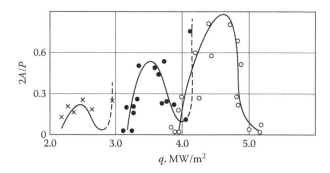

FIGURE 6.11
Relative peak-to-peak pressure oscillations depending on q at different coolant flow rates.
x: $\rho W = 3160$ kg/m²s; dark circle: $\rho W = 5153$ kg/m²s; clear circle: $\rho W = 6715$ kg/m²s; $P = 2.0$ MPa;
$Z_h = 0.7$ H; $Z_p = 0.25$ H.

amplitude on three velocities. With the increasing G, the maximum amplitudes of fluctuations are found to rise.

The TAOs excitation in gas flows is known to depend on the location of
the flow heat input section between the equal pressure points [114]. Similar
influence should also be expected for TAOs occurring in tubular channels with boiling water. The experiment on revealing the effect of location
of the surface boiling section was performed using distilled water with
salt content of 3 mg/L. The channel was pretreated by boiling until the
appearance of stable fluctuations. The heated section traveled along the
channel (H = 2.55 m, H_h = 410 mm) and peak-to-peak fluctuations; the region
of fluctuations' existence (with respect to the heat flux) and fluctuations' frequency depending on mass velocity were measured at each location of the
heated section.

An experiment with a tube 6 × 1 mm and mass flow rate of 280–
1420 kg/m²s showed that the fluctuations' amplitude has two maxima.
The first maximum was observed when the middle part of the heated section (H_h) was located in the antinode region of the second mode of fluctuations (H_h = 0.25–0.35 H), while the second maximum was observed at
H_h = 0.75 H. The outlet section was not investigated due to finite length
of the heater. This has also shown the practical absence of fluctuations of
the first and second modes when the heater was located in the middle of
the channel (H_h = 0.4–0.6 H). Sometimes, the third and higher modes with
insignificant amplitudes of the order of hydraulic noise appeared with the
same position of the heater. The performed special experiments proved that
these results are not due to specific distribution of scale thickness along the
channel length.

Figures 6.12 and 6.13 [104] illustrate the results of experiments on
establishing the effect of heated section location (the surface boiling section)
on the oscillations' initiation. A pressure pickup was placed at 0.25 H from
the channel inlet, thus making it possible to record amplitudes of the first

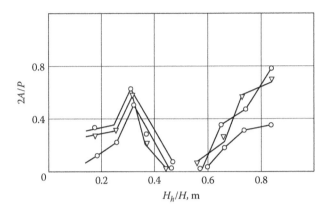

FIGURE 6.12
Relative peak-to-peak pressure fluctuations depending on the heater location. Clear circle: $\rho W = 600$ kg/m²s; clear triangle: $\rho W = 1060$ kg/m²s; $P = 2$ MPa; $T_{in} = 25°C$.

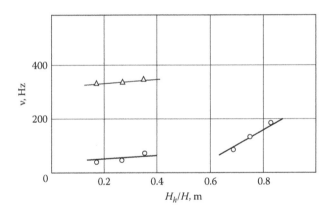

FIGURE 6.13
Frequencies of the first and second harmonics depending on the heater location. Clear triangle: first harmonic; clear circle: second harmonic.

three harmonics. The coordinates of the heated section center were as follows: 0.175, 0.25, 0.315, 0.36, 0.44, 0.49, 0.57, 0.65, 0.7, 0.73, 0.76, and 0.83. The third mode had an amplitude one order lower than those of the first two modes.

Figure 6.13 presents the frequencies of harmonic components after averaging with respect to velocity and heat flux. The frequencies were found to increase when the heater was shifted toward the outlet (i.e., the two-phase section length was decreased). Thus, the first harmonic appears when surface boiling takes place in both the first and second half of the channel, while the second one occurs when surface boiling occurs in the first half, with exception for the middle portion.

Let us first discuss excitation of the first mode (see Figure 6.5). The antinode of the steady-state pressure wave of the first harmonic is located in the channel's central region. The maximum pressure amplitudes are located here and the pressure disturbance gradient is close to or equals zero. When the surface boiling section is located within the pressure wave antinode, excitation of the fluctuation mode does not occur. This proves the fact that the earlier described pressure disturbance feedback is of secondary importance. In addition, the bubbles start performing work when they leave the area heated by the incoming liquid (i.e., practically the surface boiling section itself). With the location of the surface boiling section in the second half of the channel (i.e., in the region of negative $\partial\delta P/\partial z$ gradients), the onset of TAOs occurs.

This fact proves the primary role of the pressure gradient feedback for the mechanism of oscillations' development. In general, when the surface boiling section is located in the channel's middle portion, bubbles continue condensing in the second half of the channel too. In this case, the velocity of the bubbles arriving at the region of large pressure disturbance gradients becomes greater than that of liquid and, as shown earlier, the location of such bubbles in the second half of the channel stabilizes the process and excitation of the first mode does not occur. When the surface boiling section is located in the first half of the channel, the excitation of the first mode may be explained as follows: Bubbles get into the subcooled middle of the flow while in the antinode area or past it. This causes oscillations to appear.

Now, let us consider excitation of the second mode. The reasoning by analogy with Section 6.2.1 (that discusses the effect of $\partial\delta P/\partial z$ feedback on the first mode excitation) yields a conclusion that the condensing bubbles induce the second mode if they are located in the second and fourth quarters of the channel. It does not contradict the experimental data presented in Figures 6.12 and 6.13. When the heated section (surface boiling) is located in the first half of the channel, the maximum amplitude of the second mode is found in the beginning of the second quarter. It should be kept in mind that the earlier simplified presentation of TAOs' initiation mechanism cannot embrace the entire multiaspect nature of this phenomenon.

The experimental investigations of the possible appearance of TAOs within the range of flow parameters' characteristic of VVERs are of great interest. Therefore, ρW from 500 to 3000 kg/m²s, T_{in} from 100°C to 280°C, and P from 8 to 16 MPa were investigated. The experiments employed a tube 6 mm in diameter and 1 mm thick with the heated section 940 mm long.

Figure 6.14 presents the regions of TAOs' existence in the $q–T_{in}$ coordinates for P = 10 MPa at ρW of 500, 1000, 2000, and 3000 kg/m²s.

The increasing inlet temperature narrows the TAOs' region—that is, decreases the difference between heat fluxes at the upper and lower boundaries ($q_{Bn.h}$; $q_{Bn.1}$) and reduces values of the boundary heat fluxes. The ρW reduction leads to the same effect. For each ρW value there exists an inlet temperature boundary value, above which oscillations are absent (the dashed

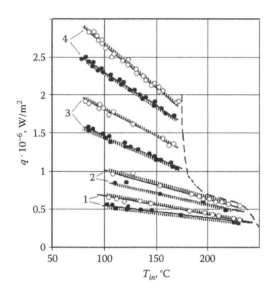

FIGURE 6.14

Regions of TAOs existence. open circle – upper boundary; solid circle – lower boundary. ρW, kg/m²s: 1 – 500; 2 – 1000; 3 – 2000; 4 – 3000.

line in Figure 6.14). When approaching the dashed line, TAOs' amplitudes become insignificant and comparable with fluctuations of hydraulic origin. Qualitatively, the oscillations' region location remains like that for other pressures, too.

Figure 6.15 shows the dependence of the lower boundary of oscillations' existence in the same q–T_{in} coordinates at 8, 10, 12, and 13 MPa. The stability boundaries represent straight lines, which are practically parallel to each other. With increasing T_{in} of water, $q_{Bn.l}$ is found to decrease. Similar behavior was observed for the upper boundaries ($q_{Bn.h}$).

Figure 6.16 shows the dependence of the oscillations' peak-to-peak amplitude (double amplitude) on the specific heat flux at 10 MPa (the upper portion of the figure) and 13 MPa (the lower portion of the figure). T_{in} varied from 100°C to 170°C. It is obvious from the figure that the dependencies A(q) approach vertical lines in the $q_{Bn.l}$ and $q_{Bn.h}$ regions. With a smooth heat flux increase, a transition from the stable to unstable region proceeds as follows. When $q = q_{Bn.l}$, the amplitude of pressure fluctuations increases spontaneously up to a certain value marked by the first experimental point in the curves given in Figure 6.16. Further, a certain relation is established between the peak-to-peak fluctuations and specific heat flux, because each variation in q corresponds to a certain fluctuation amplitude change.

If the maximum fluctuation amplitudes are small, there is no moment of spontaneous rise and drop of amplitudes at $q = q_{Bn.l}$ and $q = {}_{qBn.h}$, respectively.

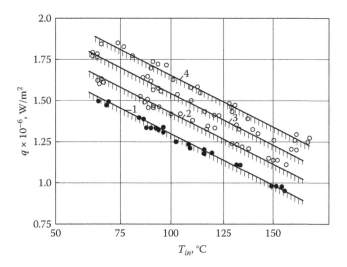

FIGURE 6.15
The TAOs' existence region lower boundary at $\rho W = 2000$ kg/m²s. P, MPa: 1: 8, 2: 10, 3: 12, 4: 13.

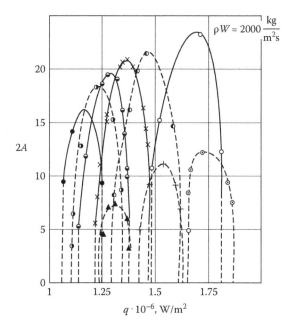

FIGURE 6.16
Dependence of peak-to-peak pressure fluctuations on specific heat flux at $\rho W = 2000$ kg/m²s. Inlet temperatures at P = 10 MPa: ○: 100°C; ◑: 120°C; X: 140°C; ◒: 150°C; ◐: 160°C; ●: 170°C; inlet temperatures at P = 13 MPa: ⊙: 102°C; +: 140°C; ▲: 170°C.

With the increased water subcooling below the saturation temperature at the channel inlet ($T_s - T_{in}$), fluctuation amplitudes increase and the regions of fluctuations' existence expand ($q_{Bn.h} - q_{Bn.l}$). The channel pressure rise leads to a decrease in fluctuation amplitudes. At pressure over 16 MPa and up to critical, there were practically no TAOs in the considered region of parameters.

Figure 6.16 shows that the total fluctuation amplitudes consist of three modes ($\omega_1 = 500$ Hz, $\omega_2 = 1000$ Hz, and $\omega_3 = 1500$ Hz), of which the second and third modes are insignificant and amount to several percent of the first mode amplitude.

TAOs' excitation at both large and small subcritical pressures depends to a great extent on the channel surface condition. In a new channel with the degreased heated surface, oscillations occur after 2–4 h of preliminary boiler water boiling, and then their boundaries of existence and oscillation amplitudes remain unchanged during a long period of time.

In the case of water's insignificant subcooling below the saturation temperature at the channel inlet, where double amplitudes of oscillations may be rather large (over 5% of the average pressure), TAOs' characteristics could not be significantly reduced at the expense of a preliminary boiling period that totaled over 100 h in experiments.

The regions of TAOs' existence with pressure fluctuation amplitudes of the order of 1% of the channel average pressure (at low subcooling below T_s) change their characteristics substantially as preliminary boiling progresses. A special experiment was aimed at establishing the dependence of peak-to-peak amplitudes and width of the fluctuations' existence region ($q_{Bn.h} - q_{Bn.l}$) on the channel operation time at constant $P = 10$ MPa, $\rho W = 2000$ kg/m^{2}s, $T_{in} = 150°C$. After the onset of TAOs, there was a time period during which the amplitudes and boundaries of fluctuations' existence practically did not change. It is at this particular moment that all data on TAOs were registered. Later, the existence region narrows down, oscillations lose stability, and their amplitudes decrease down to zero. The further channel boiling does not lead to appearance of oscillations.

The previously described peculiarities of TAOs are associated with changes in properties of the channel inner surface—that is, with the increased surface roughness. This was proved by hydraulic testing of the channel before and after the experiment. Hydraulic friction in the described channel was several percent higher than that in the initial one.

It has been shown [113] that in the presence of acoustic oscillations in the channel, the boundary dynamic layer is thin, while the flow core oscillates as an entity. The thickness of the oscillating dynamic boundary layer is described by $\delta_K \sim \left(\dfrac{2v}{\omega} \right)^{1/2}$.

The higher the oscillation frequency is, the thinner is the dynamic boundary layer. It may be supposed that with increasing channel operation time, scale hillocks become larger than δ_K and interact with the fluctuating flow core, thus promoting dissipation of energy of the fluctuating flow. The larger the boiling time is, the higher the scale hillocks are and the lower is the TAOs' amplitude.

All the preceding experimental results (high pressures) were obtained when the channel was heated along its entire length. Additionally, in some experiments the channel was only partially heated (e.g., one-fifth of the total length, $H_h = 180$ mm). All in all, eight different locations of the heated section were tested at $\rho W = 2000$ kg/m²s and $P = 10$ MPa; the oscillation frequencies, amplitudes, and existence regions were measured.

The differences in the heated section location were found to influence the initiation of TAOs in a different way in comparison with the results earlier obtained at 2 MPa.

When the heated section was located at the channel inlet region, oscillations (of relatively small amplitude) appear at lower specific heat fluxes than those when the heated section was located in the channel outlet region, where larger oscillations were observed. Probably, this is due to the fact that at identical heat fluxes, the channel contains more bubbles in the first case (hence, a larger acoustic energy is generated when bubble condensing occurs), as well as due to the redistribution (as the bubbles' pressure grows) of priorities between the earlier mentioned feedbacks responsible for TAOs' excitation.

Among all the factors influencing TAOs, of most importance is the concentration of gas dissolved in water. At first, it was established experimentally [115] and then shown theoretically [104]. The experiment employed a heated tube 6 × 1 mm in diameter, with $H = 1.2$ m, $H_h = 0.65$ H, $H_p = 0.25$ H, $P = 2$ MPa, $T_{in} = 25°C$, $\rho W = 1000$–5000 kg/m²s, and $q = (0$–5.6$)$ 10^6 W/m². Water was saturated with air by bubbling at 0.1, 0.26, 0.4, and 0.6 MPa. The respective weight concentration was $0.19 \cdot 10^{-4}$, $0.46 \cdot 10^{-4}$, $1.39 \cdot 10^{-4}$, and $1.98 \cdot 10^{-4}$.

With an increase in the dissolved gas concentration, the regions of TAOs' existence decrease (Figure 6.17), the difference between the upper ($q_{Bn.h}$) and lower ($q_{Bn.l}$) boundaries diminishes, and the onset of instability in the ρW parameter range shifts toward larger ρW values.

The amplitude of pressure fluctuations (Figure 6.18) decreases with the increasing air concentration, and at a certain amplitude value no oscillations are excited at all. In the considered case with $C > 1.98 \cdot 10^{-4}$, no TAOs appeared. A theoretical analysis [104] of the work performed by a steam–air bubble during the oscillation cycle showed that an increased concentration of the dissolved air decreases the amount of work performed by condensing bubbles. This decrease is associated with the $|dV_G/dz|$ module decrease (in the undisturbed state), which is determined by the input of air released from water into the bubble.

This chapter provides only a general picture of such a complex phenomenon as TAOs in heated channels at coolant subcritical pressures. A detailed knowledge of the phenomenon requires further investigations.

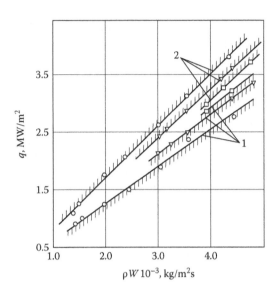

FIGURE 6.17
Regions of oscillation existence at different concentrations of dissolved air: $\bigcirc$: 0.19 10^{-4}; ∇: 1.39 10^{-4}; $\square$: 1.98.10^{-4}; 1: lower boundary; 2: upper boundary.

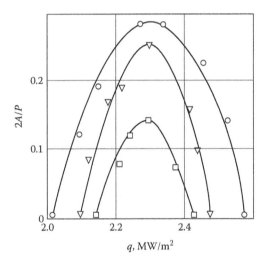

FIGURE 6.18
Relative maximum amplitudes of oscillations at different air concentrations in water at $\rho W = 7020$ kg/m²s. Weight concentrations: $\bigcirc$: 0.19·10^{-4}; ∇: 1.39·10^{-4}; $\square$: 1.98·10^{-4}.

6.2 TAOs at Supercritical Pressures

6.2.1 Oscillations' Development Pattern: A Concept of Oscillations' Initiation Mechanism

The interest in coolants used at $P > P_{s.cr}$ is due to rocket engineering needs [105] and the possibility of using supercritical pressure coolants in nuclear reactors.

The processes occurring in channels at supercritical pressures differ greatly from those occurring at subcritical ones. The main difference is in the existence of boiling at $P < P_{s.cr}$, while at $P > P_{s.cr}$ there is the so-called quasiboiling. The mechanism of quasiboiling is most completely presented in Kafengaus [117] using the experimental data. When $T_L \ll T_m$, and $T_w > T_m$ (T_m is the temperature of heat capacity (C_p) maximum), the core of a turbulent flow in a tube is cool liquid, while a thin near-wall layer is hot gas. Turbulent eddies destroy the near-wall gas layer and entrain some volumes of gas into the cool core. There, under the effect of surface tension arising from a large difference in densities, these gas volumes acquire the shape of a bubble ("pseudobubble" or quasibubble). Quick cooling and compression of bubbles create additional "turbulization," thereby enhancing heat transfer. An inverse process of getting separate volumes of cool liquid from the flow core onto the hot wall causes formation of droplets, a rapid heating and expansion of which promotes additional turbulization. It may expected that (similarly to $P < P_{s.cr}$) the compression and expansion of pseudobubbles will generate TAOs. It should be supposed that the differences in peculiarities of boiling and quasiboiling will induce their own specific differences in TAOs characteristics at sub- and supercritical pressures.

By now, a wealth of facts on TAOs at $P > P_{s.cr}$ has been accumulated (e.g., references 105, 117–122), with a comprehensive overview offered in Kafengaus [102].

Due to problems arising in high-pressure experimentation, most investigators used to choose coolants with low $P_{s.cr}$ values. Most experiments employed hydrocarbon coolants, and only a few used water. The analysis of their results shows that the results for water and hydrocarbon coolants are qualitatively consistent, provided no hydrocarbon coolant decomposition occurs.

Let us consider the behavior of pressure in an experimental tubular channel with flowing coolant with a heat flux (heating by passing electric current through the channel walls) increasing from zero to the level at which the channel reddens.

The experimental channel represents a tube linking two vessels with capacities ensuring constant pressure in the vessels during pressure fluctuations in the channel, so the experimental channel is dynamically (acoustically) decoupled from other test facility components (i.e., it is a channel with acoustically open ends). Otherwise (as in the case with $P < P_{s.cr}$), pressure fluctuations of the acoustic frequency generated in the tube will penetrate into the main components of the test facility. This will yield a complex picture of reflected and superimposed acoustic waves characteristic of the given test facility only. It should be stressed that in the absence of acoustic isolation of the heated channel, the obtained data may relate to the stability boundaries only, and they will be mostly qualitatively consistent with those obtained for the dynamically isolated channels. The following results have been obtained employing the dynamically isolated heated channels.

At $q_{bn.1}$, high-frequency fluctuations of working parameters appear in the channel. Repeated oscillography of the fluctuations' onset showed that the excitation of TAOs at supercritical pressures (similar to those at subcritical ones) proceeds in a smooth manner. A gradual power increase brings the channel to the stability boundary at $q = q_{bn.1}$ and spontaneous exponential increase in the amplitude of high-frequency fluctuations starting from zero up to a steady-state value is observed in the channel. With a further increase of the heat flux, a clear relation between the heat flux value and pressure fluctuation amplitude becomes established. As in the case with $P < P_{s.cr}$, it is the peak-to-peak amplitude (double amplitude 2A) rather than the fluctuation amplitude, since fluctuations with large amplitudes become asymmetric with respect to the average pressure.

The dependence of fluctuations' peak-to-peak amplitude on specific heat flux, 2A(q), behaves in a different manner depending on the channel heated length, coolant inlet temperature, pressure, etc. Three variants of 2A(q) curves are observed:

1. Peak-to-peak amplitudes reach their maximum and sharply (stepwise) decrease with a further increase of the heat flux, and fluctuations with higher frequencies appear in the channel. A further increase of the heat flux causes a repeated increase in peak-to-peak amplitudes until the reddening of the channel wall. This is illustrated in Figure 6.19, with respective spectral analysis given in Figure 6.20. Sharp changes in pressure fluctuation amplitudes are apparently associated with generation or decaying of separate harmonic components (Figure 6.21). The results have been obtained employing an electrically heated tube 4 mm diameter, 0.5 mm thickness of the tube, and 1 m long.

2. As the heat flux grows after the transit across the stability boundary (qBn.1) and establishment of 2A(qBn.1), peak-to-peak amplitudes keep increasing and reduce their growth rate when the heat flux reaches its critical value (the channel wall reddens). The 2A(q) curve may either not reach its maximum or pass through it;

3. With the increasing heat flux, the 2A(q) curve reaches its maximum and starts to decrease smoothly, as q increases further. At q = qBn.h, the peak-to-peak amplitude equals zero. This boundary is usually called the upper TAOs' boundary.

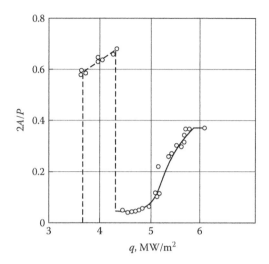

FIGURE 6.19

Dependence of the pressure fluctuations' amplitude on the heat flux. $\rho W = 0.466 \ 10^4 \ kg/m^2s$; $P_k = 3.9$ MPa; $T_{in} = 288$ K; working fluid: T-7 kerosene.

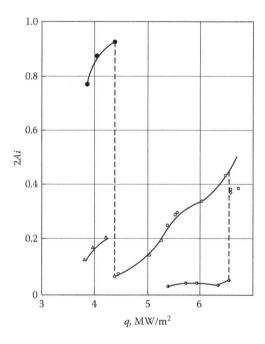

FIGURE 6.20

Amplitudes of harmonic components of pressure fluctuations depending on the heat flux. Working fluid: T-7 kerosene. ν, Hz: ●: 960, △: 1950, ○: 2750, ◑: 5500.

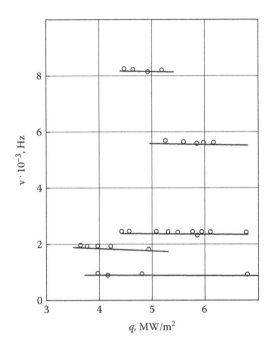

FIGURE 6.21
Effect of the heat flux on oscillation frequency.

Many researchers have proved that at $P > P_{s.cr}$, in the acoustically isolated channel, TAOs represent standing waves of different multiple frequency [102,118].

At sub- and supercritical pressures, temperature conditions of heated channels share a certain similarity (i.e., an improved heat transfer that occurs at $P > P_{s.cr}$ when the channel wall temperature exceeds the maximum C_p temperature) similar to surface boiling at $P < P_{s.cr}$. Goldman [119] was among the first to report the fact and proposed the following mechanism of fluctuations' initiation. Large groups of liquid molecules contact the heat transfer surface and break, thus forming gas molecule cavities within the liquid. It is further supposed that the growth and subsequent breakup of such cavities in a low-density liquid may be so energetic that the channel pressure fluctuations will appear as in the case with $P < P_{s.cr}$.

It has been stated [105] that pressure fluctuations appear due to a strong dependence of dynamic viscosity on temperature at pressures close to critical. The largest temperature changes near the critical point may cause substantial changes in viscosity. A random slight wall temperature rise may cause a noticeable thinning of the laminar near-wall layer with the resultant wall temperature drop and respective viscosity increase, which again leads to the growth of the laminar near-wall layer. The latter causes the wall temperature to rise and the process repeats itself. The proposed mechanism should be treated as a secondary one, because estimates show the location of fluctuating dynamic and thermal boundary layers within the near-wall laminar layer in the presence of TAOs. In this case, the fluctuating change of the heat flux from the wall to coolant will be governed not by the laminar sublayer thickness, but rather by the fluctuating thermal boundary layer and fluctuating temperatures ensured by fluctuation of physical parameters—energy equation constituents.

Of interest are the results of Aladiev et al. [120], where high-speed filming of pseudoboiling in the annular channel with transparent outer wall has shown an improved heat transfer in pseudoboiling caused by TAOs. Bubbles with a diameter from 0.01 to 0.1 mm were observed. Outwardly, the process of pseudoboiling was similar to that of subcooled liquid boiling at subcritical pressure, with the only difference that, in pseudoboiling, bubbles formed not on the wall, but rather from turbulent eddies in the homogeneous flow, in the presence large transverse temperature and density gradients. In other words, the observed bubbles resulted from the unstable slip motion of two layers of liquid with sharp differences in velocities and thermophysical properties. The wave crests, torn off from the layer superheated with respect to T_m, get into the cold layer (flow core) and acquire the shape of bubbles.

With the initiation of TAOs, the process of pseudoboiling changes qualitatively: The bubbles appear and disappear simultaneously across the entire surface at a frequency equal to the fundamental TAOs' frequency. The standing pressure wave intensifies the bubbles' formation and collapse. Further, the concepts expressed in Aladiev et al. [120] find their development in Beschastnov and Petrov [121], where it is shown that the frequency of pseudobubble volume oscillations coincides with the frequency of pressure oscillations with the maximum amplitude, and a one-fourth cycle phase delay. This speaks in favor (see Equation 6.2) of positive work done by quasibubbles in the surrounding fluid during the oscillations' cycle.

Such a work was numerically determined for a single pseudobubble in Gerliga and Vetrov [123] by considering the unsteady problem formulated in spherical symmetry. Outside the bubble, $T < T_m$, while inside $T > T_m$. No positive value was found for the work considering only molecular thermal conductivity and without taking possible eddies into account. This contradicts the experimental results presented in Beschastnov and Petrov [121]. If one applies the steam bubble analogy ($P < P_{s.cr}$), then the higher rate of mass transfer between the quasibubble and surrounding fluid promotes destabilization of the process in the flow. As a consequence, the neglect of turbulent mass transfer and the possible quasibubble rotation may be the cause of the discrepancy in the results of Beschastnov and Petrov [121] and Gerliga and Vetrov [123].

The mechanism of TAOs' initiation at supercritical pressures is as follows. When $T_w > T_m$, the flow section may be divided into two regions: the core with $T < T_m$ and the near-wall region with coolant low density and $T > T_m$. Let us assume that heat is removed from the low-density layer to the flow core by turbulent thermal conductivity proportionate to $C_p\rho$. When pressure in the heated channel increases, the maximum of T_m and, hence, that of $C_p\rho$ is shifted toward the region of higher temperatures because of their peculiar dependence on P [124]. This leads to a decreased turbulent thermal conductivity at a layer boundary and, hence, to a decreased heat removal from the superheated layer. From this it follows that with the increasing pressure, heat removal from the region of low density to that of high density (flow core) decreases. In other words, the rate of heat accumulation in the channel near-wall region with low coolant density increases. In addition, the low-density layer compresses, thereby increasing heat removal from the wall.

As the pressure in the channel decreases, coolant density near the wall also decreases, as well as thermal conductivity and heat removal from the wall. On the other hand, the decreasing pressure shifts the $C_p\rho$ maximum toward the flow core, increases $C_p\rho$ in the $T < T_m$ region, and decreases it in the $T > T_m$ region. As a result, pressure reduction in the channel corresponds to a reduced rate of heat accumulation in the near-wall region with low coolant density (turbulent thermal conductivity from the outside of the near-wall layer increases). The higher the gradient between the channel wall and flow core temperature is, the more intensive is the described mechanism of fluctuations and heat accumulation in the near-wall regions superheated with respect to the T_m region.

Since the flow core density dependence on temperature is weak, the disturbance of the heat flux from the low-density layer to the flow core will mostly influence the change of the near-wall region volume, because here the modular derivative $\partial\rho/\partial T$ is considerably greater than in the flow core.

Therefore, the average density oscillations in the flow section will depend on density oscillations in the near-wall region, while the phase shift between the heat transport disturbances in this layer and the flow section pressure will correspond, according to the Raleigh criterion, to the condition of TAOs' initiation.

The presented mechanism of TAOs' initiation at coolant supercritical pressure was numerically proved in Vetrov [125], where it is shown that the work of the section-averaged specific volume during a cycle of high-frequency oscillations is positive when $T_N < T_m < T_w$ and the described inner heat transfer is taken into consideration.

It may be supposed that a similar principle of flow "swinging" is also characteristic of quasibubbles. If $T > T_m$ inside a quasibubble and outside $T < T_m$, the dependence of heat removal fluctuations on pressure fluctuations may be constructed in much the same way as it has been done before.

According to the foregoing, laminar thermal conductivity is insufficient for quasibubbles' swinging. Apparently, turbulent heat transfer from the flow core to a relatively cold surrounding liquid should be taken into account. The experiments by N. L. Kafengaus [121] showed that the mechanism of TAOs' maintenance at the expense of quasibubbles does exist. Quasibubbles appear after TAOs' initiation, and Beschastnov and Petrov [121] managed to trace the phase shift between pressure disturbances in the channel and the quasibubble volume, which corresponds to the range of (6.3).

It should be noted that the relation $T_N < T_m < T_w$ (T is the core temperature) is a necessary condition of TAOs' initiation at $P > P_{s.cr}$. A larger $T_w - T_N$ difference leads to destabilization of the process in the heated channel.

6.2.2 Effect of Flow Parameters on Oscillation Characteristics at Supercritical Pressures

Let us first consider the results obtained on a heated tubular channel 1125 mm long with a heated length of 960 mm, using water as coolant. The regions of oscillations' existence in the heat flux (q) – inlet temperature (T_{in}) system of flow parameters are illustrated in Figure 6.22.

A rise in water inlet temperature at a fixed mass velocity depresses the lower and upper boundaries. The reason is that, on the one hand, the maximum C_p (T_m) temperature is achieved at the wall at lower heat fluxes with the rise in T_{in}; on the other hand, the increasing T_{in} promotes an earlier reaching of the transverse temperature gradient limiting value, below which no oscillations are initiated. With further rise in T_{in}, the region of pressure fluctuations' existence is found to decrease until a complete disappearance occurs.

A higher flow rate tends to narrow the instability region and shift it, with respect to T_{in} and q, toward lower inlet temperatures and higher heat fluxes, respectively. Evidently, the condition of $T_w = T_m$ demands higher q when ρW is increasing. An increased velocity results in higher dissipative losses, and in this case the initiation of TAOs requires a more intensive heat transfer

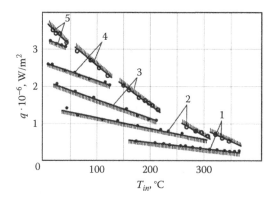

FIGURE 6.22

Region of pressure fluctuations' existence. $P = 24$ MPa; $d = 3.3$ mm; $H = 1125$ mm (channel length); $Z_{h.B} = 125$ mm (heating initiation coordinate); $H_h = 960$ mm (heated section length); $Zp = 570$ mm (pressure gauge location coordinate). ρW, kg/m²s: 1: 560; 2: 1000; 3: 2000; 4: 2500; 5: 3000. Coolant: water.

from the near-wall region, superheated relative to T_m (i.e., the flow core shall be cooler).

It should be noted that at the decrease of ρW (see Figure 6.22), the upper boundary of TAOs' existence ($q_{Bn.h}$) is reached only at high inlet temperatures. Shading faces stability regions. When T_{in} is decreasing, $q_{Bn.h}$ increases and exceeds critical heat fluxes, causing channel reddening.

The same reason limits the lower boundary ($q_{Bn.l}$) of oscillation region from the side of low inlet temperatures.

No considerable effect of pressure on the lower boundary and also on the oscillations' existence region was observed in the investigated pressure range from 22.5 to 28.0 MPa. The reason is that the change of T_m is small in the considered pressure range.

The upper boundary of thermoacoustic instability could not be reached on the channel with a diameter of 4.4 mm and the heated length of 1 m, since the channel wall temperature was increasing up to the critical value. The upper boundary of the oscillations' region was determined for that channel with the heated length below 0.5 m. Experiments with a small heated length were conducted also for determining the effect of heated section location on oscillations' initiation.

The heated section location along the channel has practically no effect on the oscillations' onset boundary, while at subcritical pressures, this effect is substantial, especially at relatively small pressures. As was stated previously, the heated section location at $P < P_{s.cr}$ significantly influences the oscillations' initiation boundary if the pressure disturbance gradient along the channel length plays an essential role in the "swinging" mechanism. The gradient disturbance causes a disturbance of slip between the light and heavy phases. The obtained experimental results indirectly indicate that the major role in the TAOs' initiation mechanism at $P > P_{s.cr}$ is played rather by the pressure disturbance than by the gradient disturbance (velocity disturbance) feedback.

A reduction in the tube heated length leads to a higher $q_{Bn.l}$ and the oscillations' amplitude growth. Indeed, in the case of a smaller heated length, a higher specific heat flux is required for achieving $T_w = T_m$. Also, the near-wall transverse temperature gradient is increasing in this case and enhancing heat and mass transfer between the relatively superheated near-wall layer and flow core. The latter increases the amount of work per oscillation cycle performed by both the near-wall region and the formed bubbles.

The general picture of changes in the oscillations' amplitude depending on the heat flux growth is similar to that observed at $P < P_{s.cr}$. At the beginning of the oscillations' region, the amplitude increases sharply; then, it passes its maximum and keeps decreasing until it disappears at the upper boundary of the instability region. Naturally, under conditions when the upper boundary was not reached, the general form of this dependence was not obtained.

The maximum oscillation amplitudes were reliably registered when there was an upper stability boundary with respect to heat flux. When the water inlet temperature is increasing, the maximum oscillation amplitudes decrease. Coolant flow rate variation has an ambiguous effect on the change of amplitudes. The experiments were performed at ρW = 560, 1000, 2000, and 2500 kg/m²s. The highest oscillation amplitudes were observed at ρW = 2000 kg/m² s (see Figure 6.23).

At ρW = 2000 and 2500 kg/m² s, only the first mode was excited, while with ρW = 560 and 1000 kg/m²s, the first and second modes were excited. The amplitudes of the second mode of oscillations were small.

TAOs' characteristics in annular channels have been investigated [122]. The test section was represented by a coaxial channel, whose outer and inner surfaces were formed by 13 × 1.5 mm and 6 × 1 mm tubes, respectively, with hydraulic and heated lengths of 385 and 300 mm, respectively. Only the lower thermoacoustic stability boundaries were registered with both external and internal heating. The upper boundaries of the oscillation region were not attained because the heated surface wall temperature reached 600°C before that. With internal heating, TAOs initiated at somewhat higher specific heat fluxes as compared to the case of external heating. This may be explained by a smaller area of the annular channel inner surface. The type of heating is of insignificant effect on the oscillation amplitudes. The influence of flow rate and water inlet temperature on the TAOs' lower boundary location does not qualitatively differ from the case of round heated channels.

Gerliga and Vetrov [118] present the results of TAO investigations that employed a 9 × 0.5 mm heated tube with 1 m spaced acoustically isolated ends and the heated length of 0.2 m. T-7 purified kerosene ($P_{cr.}$ = 2.2 MPa, $T_m \approx 690$ K) was used as coolant. The relative range of some flow parameters' variation was larger as that for water. For instance, pressure varied from one to three $P_{s.cr}$ and ρW from 1000 up to 9000 kg/m²s. Large subcooling and

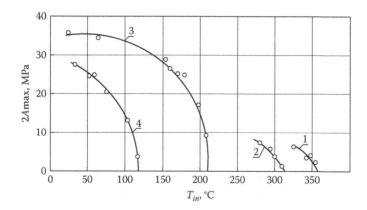

FIGURE 6.23

Peak-to-peak pressure oscillations' depending on the channel inlet coolant temperature. d = 3.3 mm; H = 1125 mm; $Z_{h.B}$ = 125 mm; H_h = 960 mm; Z_p = 570 mm; P = 24 MPa; ρW, kg/m²s: 1: 560; 2: 1000; 3: 2000; 4: 2500. Coolant: water.

shortness of the heated section predetermined peculiarities of oscillations' development.

A general picture of TAOs' initiation and development is presented in Figure 6.19. With a smooth rise of heat flux, oscillation amplitudes increase spontaneously up to steady-state values having reached the lower stability boundary ($q_{bn.r}$). Further, amplitudes keep growing along with the increasing q, and at a certain heat flux, limiting level oscillation amplitudes of the given frequencies drop to zero and higher frequency oscillations are excited. Finally, the process comes to an end with channel reddening. Spectral analysis of the curve is given in Figures 6.20 and 6.21.

When pressure is increasing, the amplitude of oscillations shifts toward higher heat fluxes. Despite a strong reduction of oscillation amplitudes at the increase of T_{in}, the lower oscillations' boundary (the upper has not been reached) changes—but insignificantly, as in the case with water. Generally speaking, investigations of the effect of all flow parameters on TAOs for organic coolant yield no qualitatively new data as compared to water.

6.2.3 Effect of Channel Design Specifics on Heat Removal and TAOs' Characteristics

1. Artificial heat transfer enhancement in a tubular heated channel due to smooth annular ribs on the inner surface was recommended for $P < P_{s.cr}$ in Bogovin [126]. Such cross ribbing of the tube significantly influenced both heat transfer and TAOs at $P > P_{s.cr}$. The tube with annular depressions (knurls) had the following parameters: spacing of 6 mm and ratio between the annular inner ribs' diameter and the channel diameter of 0.9.

Figure 6.24 shows the dependence of specific heat flux at the TAOs' onset on mass velocity. Stability regions are located on the shadowed side of straight lines. Artificial turbulization of the near-wall layer caused an appreciable increase in specific heat flux at the onset of pressure fluctuations. Oscillation amplitudes also decrease strongly: The larger the value of ρW is, the smaller the maximum oscillation amplitudes are (Figure 6.25). Artificial turbulization shifts the moment when the wall temperature reaches the T_m value to the region of high heat fluxes and reduces the transverse temperature gradient. According to the proposed mechanism of TAOs' initiation, this leads to decreased oscillation amplitudes. The most intensive amplitude reduction was observed at high ρW values, since in this case the channel inner ribs cause the highest flow turbulization. This leads to a more intensive equalization of channel temperature profile and a higher dissipation of the coolant velocity longitudinal acoustic waves. Figure 6.26 shows the $T_w(q)$ dependence for a smooth channel and that with artificial turbulizers.

The regions of improved heat transfer in both tubes with inner annular ribs and smooth ones appear with the initiation of TAOs, though in the case of smooth tubes the better heat transfer section is more apparent.

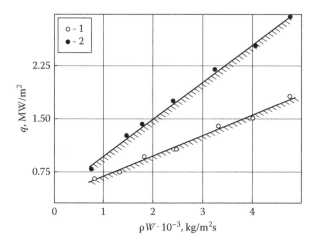

FIGURE 6.24
Regions of pressure oscillations' existence in knurled and smooth channels. P = 4.5 MPa; 1: smooth channel; 2: channel with turbulizers. Coolant: toluene.

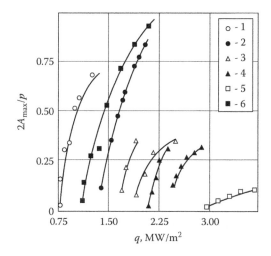

FIGURE 6.25
Dependence of the pressure oscillations' relative amplitude on q at different mass flow rates (knurled channel). P = 4.5 MPa; ρW, kg/m²s. 1: 760; 2: 1,740 3: 2,440; 4: 3,330; 5: 4,850; 6: 2,440 (smooth channel). Coolant: toluene.

It is obvious from the given figure that the artificial flow turbulization substantially improves heat transfer. Even for the improved heat transfer section (in presence of TAOs) of a smooth channel, the heat transfer coefficient is 1.5 times smaller than in a channel with inner ribs and with no oscillations.

At q = 1.4 MW/m², an improved heat transfer regime is attained (T_W exceeds T_m by 25°C in the considered section) in a smooth channel. TAOs' initiation

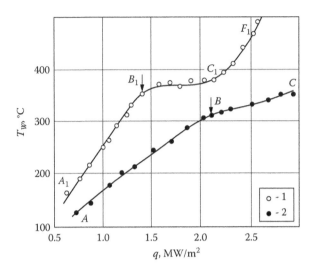

FIGURE 6.26

$T_w = f(q)$ dependence for a smooth channel and that with artificial turbulizers. $\rho W = 3330\,kg/m^2s$; $P = 4.5\,MPa$; $Z/d = 94$; 1: smooth channel; 2: knurled channel. Coolant: toluene.

corresponds to this moment, and T_w is found to be almost constant (B_1C_1 section) with the further increasing heat load. The $T_w = f(q)$ curve for a channel with turbulizers lies considerably lower.[*]

Under the conditions of TAOs' initiation in a smooth channel, T_w in a knurled channel is 120°C lower for the considered sections, and oscillations appear only when heat fluxes correspond to those at the beginning of the section with the worsened heat transfer in a smooth tube (C_1F_1 section in Figure 6.26). Within the considered range of q, only two heat transfer sections were found in the knurled channel: the normal heat transfer section AB and the poor heat transfer section BC. The limiting value of T_w (500°C is the temperature of coolant decomposition) in the channel exit section was achieved at heat fluxes 1.7 times higher than those for smooth channels.

In contrast to the smooth channel, the pressure drop across the test channel was changing with the increasing heat load in the following way: At first, it was decreasing along with the increasing heat flux ($\rho W = const.$), and then, with the appearance of TAOs, it slightly raised, but never exceeded the pressure drop across the nonheated channel. At large ρW values (4000–5000 kg/m²s), the pressure drop across the channel was monotonously reducing in the entire range of heat loads and finally became 15%–20% lower as compared with the initial value.

---

[*] The results presented in Sections 6.2.3 and 6.2.4 were obtained for toluene (C7H8 with $P_{s.cr} = 4.05\,MPa$, $T_m = 320.8°C$). The test tube has inner diameter of 4 mm and wall thickness of 0.5 mm. Until otherwise specified, H = 1 m, heated length = 416 mm, and H_h = 500 mm. Channel inlet temperature was 20°C. In figures, x/d = coordinate of thermocouple location starting from tube exit section.

A comparison of TAOs and heat transfer characteristics for tubes with smooth inner surface and those with artificial turbulizers in the form of protruding transverse rings yields a conclusion that forced deterioration of the interphase between the "hot" layer and "cold" flow core leads to a significant decrease of the region of TAOs' existence, of TAOs' amplitude, and an appreciable heat transfer improvement.

2. TAOs' characteristics greatly depend on the length of the dynamically (acoustically) isolated pipeline. It was used in Kalinin et al. [127] for solving the problem of the effect of oscillations on heat transfer. Two channels (two 6×1 mm tubes, 1 and 31 m long) were considered. In both cases, the heated section was 416 mm long and was located at the channel outlets. The long channel was intended to increase dissipative losses and, hence, narrow down the region of TAOs' existence in comparison with the short channel instability region. By using sufficiently high-capacity volumes, the channels were acoustically isolated from both upstream and downstream components of the test facility. The experimental results showed that, in short channels, TAOs initiated at any mass velocities, while within the range of mass flow rates from 900 to 1500 kg/m²s in long channels, pressure fluctuations were observed to be not in the entire range of heat load variations. The absence of oscillations substantially influenced heat transfer. The results of heat transfer tests using toluene and presented as $T_w = f(q)$ at $\rho W = 1260$ kg/m²s in two pairs of similar sections of the previously mentioned tubular channels are shown in Figure 6.27.

For different channels, the $T_w = f(q)$ dependencies differ greatly starting with TAOs' initiation. For a long channel, T_w increases linearly with the increasing heat load (BC and B_1C_1 sections), while in the case with the first

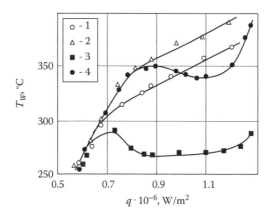

FIGURE 6.27

Dependence of the wall temperature on the heat flux density without TAOs and with pressure fluctuations. P = 4.5 MPa; ρW = 1260 kg/m²s; 1: Z/d = 46; H = 31 m; 2: Z/d = 85; H = 31 m; 3: Z/d = 46; H = 1 m; 4: Z/d = 85; H = 1 m. Coolant: toluene.

channel, a section of first improved and then deteriorated heat transfer sections appeared.

The shape of the $T_w = f(q)$ curve for the long channel (without TAOs) qualitatively corresponded to the behavior of the temperature for a wire with natural circulation heat transfer [128].

The curves of T_w behavior along the heated section length significantly differed for the channels in question (see Figure 6.28). The difference between the considered flow conditions is only in the TAOs' presence in the short channel and their absence in the long one.

The presented data prove another time that the pressure TAOs in channels with smooth inner surface are the cause of improved heat transfer at supercritical pressures. A somewhat improved heat transfer without TAOs (BC section in Figure 6.27) may be due to the quasiboiling observed visually in Aladiev et al. [120].

3. In the earlier described experiments with toluene, the first and second harmonics of pressure oscillations were excited simultaneously in the channel, with no pressure unit in the heated section in the second half of the channel. Maybe that was the reason why no T_w jump was observed (e.g., in Kalbaliev [129]). The channel inlet or exit throttling resulted in the redistribution of pressure oscillation modes in the presence of TAOs. In this case, changes in TAOs' regions of existence and of amplitudes should be expected. The results of the experiments showed that in the case of the channel exit section heating, the installation of local inlet resistances produced insignificant effect on the fluctuation amplitude characteristics and heat transfer. As a result of exit throttling using an orifice, the peak-to-peak amplitudes of pressure fluctuations decreased three times and q at the oscillations' onset somewhat increased. When the channel was acoustically

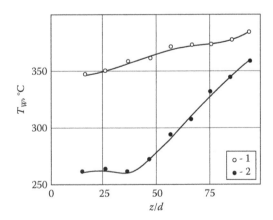

FIGURE 6.28

Wall temperature conditions for channels of different length. P = 4.5 MPa; ρW = 1260 kg/m²s; q = 1.17 MW/m²; 1: H = 31 m; 2: H = 1 m. Coolant: toluene.

closed at the outlet, the wall temperature conditions changed greatly along the heated section and heat transfer deteriorated locally at any mass velocity. An approximate location of pressure fluctuation modes is illustrated in Figure 6.29.

In the case of a prevailing second mode, the clearly expressed maximum of the wall temperature was observed closer to the beginning of the heated section. An improved heat transfer in the case of excitation of the third mode only is illustrated in Figure 6.30. At the initiation of TAOs, the first harmonic existed at the start. As the heat flux increased, the first harmonic of pressure fluctuations disappeared and the second harmonic appeared. In this case, the wall temperature increased in the heated section in the region of the pressure node location. At $q = 1.56$ MW/m, the coordinate of T_w maximum coincided with the coordinate of the third harmonic node of pressure fluctuations. The further rise of q caused the shift of T_w maximum toward the exit, with the simultaneous shift of the node to the exit [126].

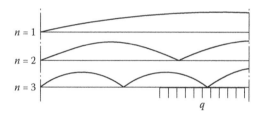

FIGURE 6.29
Profiles of pressure standing waves for the first three harmonics (single-end open tube).

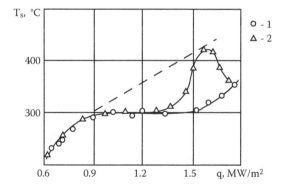

FIGURE 6.30
The wall temperature variation depending on the heat flux density for the acoustically open channel and that with closed outlet. $P = 4.5$ MPa; $Z/d = 66$; 1: conventional channel; 2: channel with a hydraulic resistance at the outlet; dashed line: trend of $T_s = f(q)$ dependence without TAOs. Coolant: toluene.

6.2.4 Effect of Dissolved Gas on TAOs' Characteristics

Under both supercritical and subcritical pressures, the presence of dissolved gas in the coolant greatly affects the TAOs' regions of existence and amplitudes. Figure 6.31 shows the dependence of specific heat flux at the onset of pressure fluctuations on the mass velocity at different concentrations of N_2 dissolved in the coolant (toluene). With the increasing concentration, initiation of oscillations requires higher heat fluxes ($q_{Bn.l}$). When coolant is heated, the dissolved gas starts depositing in the form of bubbles on the channel wall [128]. Further, the detached bubble turbulizes the flow and diminishes the temperature difference between the wall and flow core. Higher heat fluxes are required for the initiation of oscillations. The dissolved gas concentration will keep increasing until a moment when the TAOs' lower boundary cannot be obtained [126] because of the channel walls' reddening. This is probably due to separation of the coolant from the wall by a gas layer. At concentrations given in Figure 6.31, TAOs' boundary at $\rho W = 2500$ kg/m²s was concentration independent. This may be explained by the fact that the near-wall temperature and stress gradients are large in high-velocity flows, preventing bubble existence at weak forces of the surface nature.

Figure 6.32 shows the dependence of the relative maximum peak-to-peak amplitude of oscillations on the heat flux at different nitrogen concentrations for two mass velocities. The ranges of heat fluxes corresponding to the TAOs' existence region sharply reduce with an increase in the dissolved

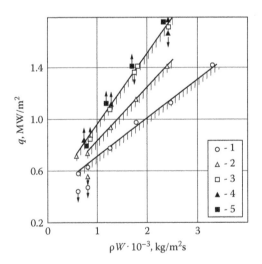

FIGURE 6.31
TAOs' origination boundary depending on mass velocity at different gas contents. P = 4.5 MPa; T_{bn}^{in} = 20°C; C, n mol N_2/kg. 1: 0; 2: 1,250; 3: 2,450; 4: 4,050; 5: 5,500; ↓: unstable oscillations; ↑: specific heat flux at which channel reddening is observed at a given concentration. Coolant: toluene.

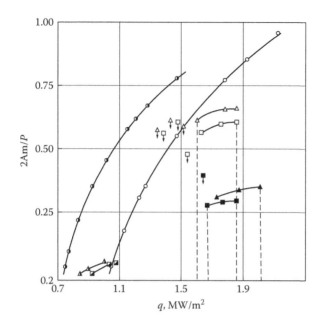

FIGURE 6.32

Relative peak-to-peak amplitude of pressure fluctuations depending on the heat flux density for two mass velocities and different gas contents. P = 4.5 MPa; C, n mol N_2/kg for ρW = 1260 kg/m²s. 1: 0; 2: 1,250; 3: 2,450. C, n mol N_2/kg for ρW = 2440 kg/m²s. 4: 0; 5: 1,250; 6: 2,450; 7: 4,050; 8: 5,500; ↓: unstable oscillations. Coolant: toluene.

gas concentration. With the increasing q, the growth of curves presented in Figure 6.32 is limited by the onset of burnout (channel reddening).

A pressure drop across the heated section is associated with pressure fluctuations and their amplitude (Figure 6.33). Until the initiation of TAOs, pressure drop practically did not change at the increase in heat flux density. After the excitation of high-frequency oscillations, the pressure drop across the channel between the volumes with equal pressure increased in proportion to the rise of the oscillation amplitude, either smoothly or in jumps. In the latter case, oscillation amplitudes also increased jump-wise at the moment of TAOs' initiation. The growth of hydraulic friction may be due to the appearance of secondary near-wall eddies induced by acoustic oscillations [130].

At ρW > 2500 kg/m²s, high-frequency pressure fluctuations were excited irrespective of gas concentration. This may be due to the fact that the working fluid passes the heated section promptly and gas is released in smaller amounts than at lower velocities. It should be noted that this relates to fluctuation-free regimes close to the stability boundary. The oscillation amplitude depends on gas concentration. The maximum saturation of coolant with gas sometimes makes it possible to reduce fluctuation amplitudes several times.

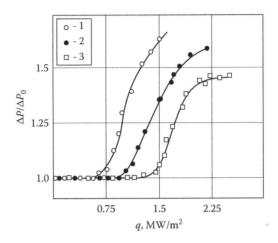

FIGURE 6.33
Dependence of the relative pressure drop across the channel on the heat flux density. $P = 4.5$ MPa; $T_{in} = 20°C$; ρW kg/m²s. 1: 1,260; 2: 2,440; 3: 3,330; $\Delta P_{c.o}$: pressure drop in the channel prior to oscillations' onset. Coolant: toluene.

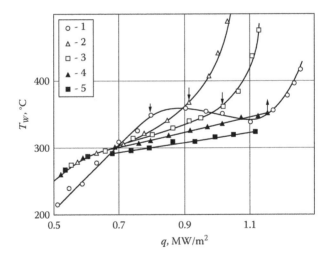

FIGURE 6.34
Wall temperature dependence on q at different gas contents: $P = 4.5$ MPa; $\rho W = 1260$ kg/m²s; $Z/d = 85$; C, n mL N_2/kg. 1: 0; 2: 1,250; 3: 2,440; 4: 4,050; 5: 5,500; ↓:start of oscillations: ↑: burnout onset. Coolant: toluene.

The effect of dissolved gas on the intensity of heat transfer with turbulent toluene flow under supercritical pressure is characterized by the $T_w = f(q)$ dependence at different gas contents and is illustrated in Figure 6.34 for $\rho W = 1260$ kg/m²s. The behavior of T_w curves at lower ρW values (e.g., at 760 kg/m²s) is qualitatively similar. When the wall temperature is below pseudoctritical, heat transfer becomes impaired: When coolant is saturated with gas, the wall temperature may be higher than T_w when concentration of the gas dissolved in coolant equals zero. This is in agreement with the results obtained for the case of free-convective heat transfer from a horizontal wire [128] and also may be due to a lower coefficient of thermal conductivity as a result of gas release.

Several dozens of degrees below T_m, heat transfer with gas-saturated liquids becomes improved. Probably, intensive release of gas bubbles occurs, which enhances heat transfer. On the one hand, such "gas boiling" causes a reduction of wall temperatures and TAOs' initiation is shifted toward higher heat fluxes. It should be noted that T_w at the onset of TAOs in coolant with $C \neq 0$ is somewhat higher than T_w at $C = 0$.

On the other hand, gas boiling leads to a decreased transverse temperature gradient in liquid. It causes a reduction of work per cycle of quasibubbles' oscillations and of the near-wall coolant layer superheated with respect to T_w. This makes oscillation amplitudes at $C \neq 0$ several time lower than in the case of $C = 0$. At high concentrations of dissolved gas ($C = 4050$ and 5500 n mL N_2/kg, as in Figure 6.34), higher heat fluxes do not lead to TAOs' initiation due to channel reddening. Burnout evidently occurs because of the appearance of the near-wall film formed by the released gas.

At $C \neq 0$, the initiation of TAOs due to small amplitude causes no enhanced heat transfer. At a higher flow velocity, the process of heat transfer and TAOs undergo a change. A series of experiments performed at $\rho W = 2400$ kg/m²s and different concentrations of dissolved nitrogen showed that, under the considered conditions, this regime can be called boundary to some extent. At the given and higher velocities, no heat transfer worsening at a wall temperature below pseudocritical was observed, and the $T_w = f(q)$ dependence almost coincided for all gas concentrations.

At high ρW and wall temperatures above the pseudocritical one, the $T_w = f(q)$ curves change. The maximum heat transfer coefficient was observed at $C = 0$ when TAOs' amplitudes were maximum. At higher concentrations, the range of heat fluxes at which TAOs exist and oscillation amplitudes decrease, thus leading to an increased $T_w = f(q)$, shortening of the section with improved heat transfer, and, further, to its complete degeneration.

7

Instability of Condensing Flows

7.1 Introduction

Two-phase condensing tube and channel flows are widely employed in various devices of many branches of engineering (e.g., in nuclear power plants and space nuclear power plants, in solar energy converters and chemical equipment). Some of these devices are as follows:

1. Cooling radiators of space nuclear plants [131] (In this case, the waste steam is fed into the tubular condensing radiator after it has passed the turbine. Heat is removed by radiation into space from the ribs connecting the tubes along their length.)

2. Reactor cooldown systems (Steam is condensed in a tubular heat exchanger, while condensate is directed by natural circulation to the intermediate heat exchanger for heat removal from the reactor.)

3. Steam separator-superheaters (SSH) used at nuclear power plants

4. Bubbling condensers

5. Hot-air heaters for cold air preheating of gaseous and liquid fuel preheaters prior to their compression [132,133]

Steam separator-superheaters SPP-220, SPP-500-1, and SPP-1000 have a similar design. Most completely, stable operation has been analyzed for SPP-220 [134] and SPP-500-1 [135].

The commercially produced SPP-220 comprises a cylindrical vessel accommodating a chevron-type separator in its upper part. The lower part contains two superheaters made of longitudinally finned tubes arranged in assemblies.

From the high-pressure cylinder of the turbine, wet steam is supplied to the moisture separator, dried, and fed to the intertube space of the primary superheater arranged along the periphery of the SSH lower part. Near the cylinder lower bottom, the steam flow turns and travels upward via the assembly intertube space in the secondary superheater. The heating steam is supplied via two receiving chambers located on the casing and further flows via individual tubes per each assembly of the primary and secondary superheater to be condensed within the tubes. The drain water flows downward and is removed from the assemblies via the drain chambers (two per each section) to the hotwells. The chambers are connected with the assembly outlets by connecting tubes.

Tests [134] conducted at the Novo-Voronezh NPP (nuclear power plant), unit 4, showed that the system of two SSHs connected to one hotwell (HW) operates unstably under certain conditions. The secondary superheater was periodically flooded with condensate up to the level of 1500 mm (counting from the drain chamber axis); the drain water temperature was 50°C below that of saturation. The amplitude of the condensate level fluctuations in the SSH was 250 mm with a period of 20 s.

An analysis of the heating steam and condensate drain pipelines showed asymmetry, which caused hydraulic maldistribution.

Introduction of a steam pipe connecting the receiving chambers of the drain chamber connecting pipe and keeping the hotwell level constant stabilized SPP operation. All of the previously mentioned modifications were aimed at elimination of maldistribution.

An experimental investigation of flow instability using the model of the condenser assembly is described in Artemov et al. [134]. The limited amount of data obtained by the authors is due to incomplete modeling of parallel operation of both condenser assemblies and SSHs. It was shown in the paper that two kinds of self-oscillations are possible in the *condenser–HW* system. (The condenser inlet was connected to the hotwell by the auxiliary connecting pipe. The steam flow variation through that pipe permitted the condenser pressure drop to be regulated.) The low-frequency condenser flow parameter fluctuations with a period of 660 s existed in the absence of HW blowdown and with a backpressure of 0.03 MPa. At a lower backpressure and with the HW space steam blowing, the low-frequency oscillations were replaced by oscillations of comparatively high frequency with a period 10 to 15 s. When backpressure was zero, the condenser–HW system operated stably.

The test results of Desyatun et al. [135] for the primary and secondary superheaters of SPP-500-1 showed substantial influence of HW condensate level on stable operation of the SPP. The heat exchange surface was made of smooth tubes assembled in parallel-connected modules. The heating steam was condensed in the intertube space. At nominal power and the condensate level of 1200 (in HW-I) and over 1100 mm (in HW-II) of the primary and secondary heaters, periodic supercooling of condensate in all drain chambers was observed. In the secondary superheater, the temperature fluctuation period was 600 s, while the amplitude of temperature fluctuations in some SSHs reached 70°C.

The condensate temperature fluctuations indicate periodic flooding of superheater modules. The measurements showed that this was leading to respective fluctuations of the total turbine steam flow rate and of live steam pressure. After bringing the condensate level down to 900–800 mm in HW-I and to 1000 mm in HW-II, condensate depression in the drain chambers and instabilities disappeared.

It follows from this that, under certain flow parameters, a complex instability representing a combination of fluctuations with different frequencies and, probably, of different nature develops in SSH. In Artemov et al. [134] and Desyatun et al. [135], little attention was paid to the possible mechanisms of such self-oscillation development.

The SSH connection schemes are very diverse for different turbines [137], but nevertheless all SSHs may be considered as systems of condenser assemblies or modules placed in parallel.

Figure 7.1 illustrates the SSH as a system of parallel condensing channels for the K-220-44 turbine unit [137]. With regard to the K-500-65/300 turbine unit (SPP-500-1), the connection scheme is more complicated because of four SSHs.

In the case shown in Figure 7.1, one HW for the primary heaters and another one for the secondary heaters are used. The considered heating steam system of the figure is dynamically isolated by two units—namely, steam generator and deaerator, since these contain great steam spaces. When self-oscillations develop in steam superheaters, pressures in the steam generator and deaerator may be regarded as constant. Apparently, boundary conditions should be selected in this manner when performing the design analysis of the SSH dynamics. In our case, the qualitative analysis of stability in the SSH heating path permits some narrowing of the considered system: The pressure and

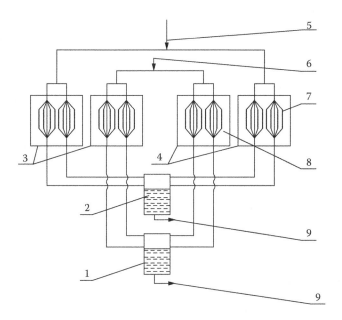

FIGURE 7.1

Connection of superheaters of two SSH of K-220-44 turbine unit to heating steam and condensate lines [137]. 1: HW-I; 2: HW-II; 3: first SSH superheaters; 4: second SSH superheaters; 5: live steam supply; 6: second extraction steam supply; 7: secondary superheater assembly; 8: primary superheater assembly; 9: condensate removal.

enthalpy in supplying steam lines are constant, and condensate levels in HW-I and HW-II are constant over time. Also, the condensate flow rate fluctuations at the HW outlet do not affect the processes occurring in HWs and above them.

If all components of the primary and secondary SSHs are considered to be identical, as well as the corresponding steam and fluid lines, SSHs may be treated as parallel-switched identical systems.

The instabilities developing in both primary and secondary steam superheaters may be either in phase or in antiphase. In the first case, the HW pressure will vary, while in the latter case it will be constant. This is true for linear systems (in the range of small deviations of SSH flow parameters from their steady-state values). When parameter fluctuations in steam superheaters are in phase, the latter should be considered jointly with HW as a united dynamic system; further on, its unstable operation will be classified as the system-wide instability.

By analogy with the analysis of multichannel steam generators, the linear representation reveals the following types of instabilities:

1. Each assembly is an entity of parallel-connected identical condensing tubes (modules). Hence, the assembly may exhibit the intertube instability.

2. If each assembly is considered as a channel, the interassembly instability may be supposed.

3. Both primary and secondary steam superheaters are respectively separated into two parts by the systems of heating steam supply and condensate removal (Figure 7.1). The parts are practically identical and constitute two pairs of specific reduced channels with common headers. In this case, instability may also occur in between two pairwise connected channels.

At small fluctuation amplitudes (at the initial phase of instability propagation) and with identity of the interconnected channel assemblies, these three kinds of instabilities do not cause HW pressure variation.

The measurements performed by the authors of references 134–136 made it possible to register only the system-wide instability and instability of groups of assemblies into which steam superheaters are separated.

7.2 Instability of Condenser Tube and Hotwell System

The basic relations in the mechanism of the system-wide instability initiation are those between the HW steam pressure and the supercooled condensate flow rate in the supplying pipes. The dynamics of the condensation proper in the assemblies is of secondary importance here.

Let us consider a system consisting of the supercooled condensate flow in the HW supplying pipes and in the HW steam space. This is the system with static instability in the small. Indeed, any accidental rise of the supercooled condensate flow rate in HW causes great steam condensation there. The HW pressure will drop and cause the subsequent increase in the condensate flow rate, and so on. When the flow rate decreases for the time first, the HW pressure increases as compared to its undisturbed value, thus causing further condensate flow rate decrease, and so on.

The conditions under which there will be no parameter spontaneous drift from the steady-state regime in the considered system can be obtained from simple mathematical expressions.

The equations for the condensate flow and pressure in the HW can be written as

$$\frac{dG_{con}}{dt} \int_0^H A_{con}^{-1} dZ + \Delta P_f(G_{con}) = P_G - P_{hw}, \tag{7.1}$$

$$P_{hw} = P_{hw}(G_{con}), \tag{7.2}$$

where G_{con} is the supercooled condensate flow rate, H is the condensate travel distance (from the steam superheater level to HW), A_{con} is the condensate flow cross-section area, ΔP_f is the pressure friction loss, P_{hw} is the pressure in the HW, and P_G is the pressure in the steam superheater.

Equation (7.2) takes into account the dependence of P_{hw} on G_{con} in an implicit form.

If the heating steam pressure is presumed constant, it can be found from (7.1) and (7.2), upon linearization, that

$$\frac{d\delta G_{con}}{dt} \int_0^H A_{con}^{-1} dZ + \left(\frac{d\Delta P_f}{dG_{con}} - \left| \frac{dP_{hw}}{dG_{con}} \right| \right) \delta G_{con} = 0,$$

from which it follows that the process is stable at

$$\frac{d\Delta P_f}{dG_{con}} - \left| \frac{dP_{hw}}{dG_{con}} \right| > 0, \tag{7.3}$$

That is, when the condensate flow rate is increasing, the friction losses should exceed the HW pressure drop due to increased steam condensation and the level controller triggering. The HW level controller performs its own functions and at the same time destabilizes the considered process. So, with no controller, the rise in supercooled condensate flow rate would cause a smaller HW pressure drop. The steam volume reduction at the expense of the HW condensate level rise would somewhat prevent the pressure drop.

Probably, the best way of eliminating the instability in question is either the HW level stabilization or prevention of condensate supercooling. An increase in hydraulic friction in the condensate line also contributes to process stabilization.

The HW blowdown stabilized pressure in it and therefore facilitates stable operation of the *SSH–HW* system [134].

The stabilizing effect of a lower level in HW [134,135] manifests itself due to the following circumstances. When the level lowers, the driving force influencing the condensate column in the drain line increases. The condensate flow rate rises and the condensate level in steam superheaters decreases. The degree of condensate depression reduces, thereby stabilizing the process, since the HW pressure becomes less sensitive to the flow rate disturbance of the condensate to be drained. Additionally, the dependence of P_{hw} on G_{con} becomes less strong because of the increased steam space.

Many authors have stated that one of the main causes of SSH unstable operation is the difference in hydraulic friction in steam and condensate lines. Periodic additional flooding of the heat transfer surfaces takes place, first of all, in devices with large hydraulic friction in live steam and condensate upstream and downstream lines [132–134]. This is probably due to the fact that a smaller amount of heating steam flows through the superheater with a larger total hydraulic friction. The condensate is drained from such a steam superheater at a higher degree of depression and it is the cause of instability initiation. Also, the appearance of the liquid section in condensing assemblies is, by itself, a cause of interchannel instability in the steam superheater [138–141].

The considered aperiodic departure of the SSH–HW system from the steady-state regime further transforms into self-oscillations. The work [134] offers characteristics of self-oscillations that occur in the test facility modeling the SSH–HW system. Physical SSH and HW models were interconnected by a parallel steam line.

The oscillation period in Artemov et al. [134] is divided into two parts, the boundaries of which happen at the moment of sharp rise in the condensate column H_{con} at t = 470 s. In the first part of the period (t < 470 s), the heating steam enters the test section from two sides: from the condenser steam inlet and from the HW side. In this case, H_{con} is found to rise gradually, as well as the condenser backpressure, along with the decreasing steam flow rate G in the parallel steam line. At t = 470 s, steam supply to the test section from below stops and the condensate is drained into the drain pipeline. The level of H_{con} is reduced but not significantly, since the condenser cross-section area essentially exceeds that of the drain pipeline.

During the second part of the period (t = 470–680 s), H_{con} tends to rise with simultaneous filling of the hydraulic lock riser. The hotwell backpressure is constant, G = 0. Further on, the period terminates in a sudden drop of backpressure and H_{con} down to zero. Condensate is fully drained because of the sharp HW pressure drop due to steam condensation caused by getting the supercooled condensate into HW.

7.3 Interchannel Instability in System of Parallel-Connected Condensing Tubes

Similarly to the parallel-connected steam-generating tubes, the interchannel instability may also develop in the system of parallel-connected condensing tubes. Theoretically, it is consistent with Gerliga and Dulevsky [36]. In this case, the interchannel stability boundaries in the small coincide with the boundaries in the small for the dynamically isolated condensing channel. The conditions of dynamic isolation are similar to those for steam generators:

$$t_{in} = \text{const.}; \quad P_{in} = \text{const.}; \quad P_e = \text{const.},$$

where the subscript index *in* relates to the inlet header, while *e* relates to the outlet header.

It follows from the published data that both static and oscillatory instabilities may occur in the condensing channel.

As is known, the static instability in the small is due to ambiguity of the hydraulic characteristic curve. Location of the working point on its dropping section causes uncontrolled static drift of flow parameters to the rising sections of the hydraulic characteristic curve. This may result either in a new steady-state regime or in self-oscillations.

It was predicted [133] that the condensing channels may have ambiguous hydraulic characteristics. Figure 7.2 presents the curves of pressure loss depending on the flow rate for the upflow and downflow condensing tubes. The curve behavior shows that the negative slope of the hydraulic

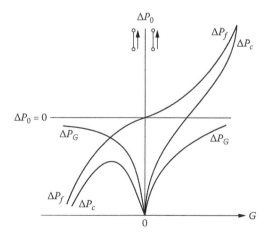

FIGURE 7.2

Hydraulic characteristic curves of the condensing tube [133]. ΔP_G: pressure drop gravity component; ΔP_f: friction pressure drop; ΔP_C: total pressure drop.

characteristic ($d\Delta P_C/dG < 0$) is typical of the upflow tubes at small values of the condensing steam flow rate. The horizontally located condensing tubes also feature sections of the hydraulic characteristic curve with a negative slope [133].

In vertical condensing components with internal steam condensation (SSH is an example), no static instability is observed according to Lokshin [133].

The predictions and experimental results of Wedekind and Bhatt [139] indicate that a small change in Freon-12 vapor flow rate at the tubular condenser inlet immediately causes a substantial short-term change in the supercooled outlet condensate flow rate. The horizontal condenser is made up of two copper coaxial tubes 5 m long. The condensing Freon-12 flows via the inner tube 8 mm in diameter. Figure 7.3 shows condensate flow rate variation at the 11.4% increase in steam flow rate. The peak of fluid flow rate disturbance is over 10 times higher than the steam flow rate disturbance. Physically, this phenomenon may be explained as follows. The increased steam flow rate causes a rise in the condensing steam pressure, which in turn results in an abrupt shift of the condensing section boundary to the outlet and hence in the increased condensate flow rate. The condensing surface expands until it reaches a value sufficient for condensation of the total steam in the condenser. With the increasing condensation, the tube pressure drops, as well as the condensate flow rate.

The condensate flow rate behavior at the stepwise inlet steam flow rate reduction is illustrated in Figure 7.4.

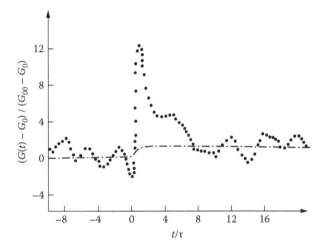

FIGURE 7.3

Characteristics of the liquid flow rate at the condenser outlet after increasing the inlet flow rate [139]. Dotted dashed line: inlet steam mass flow rate; dotted line: outlet liquid mass flow rate; $G(t)$: condensate flow rate over time; G_0: working medium flow rate before disturbances; G_∞: final working medium flow rate; working medium: Freon-12; $P = 620$ kPa; $i_{in} = i_G$; $\bar{\alpha} = 0.084$; $G_0 = 4.69$ g/s; $G_\infty = 5.22$ g/s; $\bar{q} = 13.0$ kW/m^2; $\tau = 0.65$ s; $d = 0.8$ cm; $\bar{\alpha}$ = length averaged quality; $\bar{q}$ = length averaged specific heat flux.

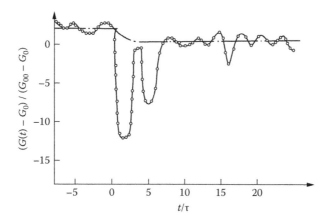

FIGURE 7.4
Characteristics of the outlet liquid flow rate after decreasing the inlet flow rate [139]. Dotted dashed line: inlet steam mass flow rate; O: outlet liquid mass flow rate. Working medium: Freon-12; P = 668 kPa; $i_{in} = i_G$; $\bar{\alpha} = 0.831$; Go = 5.82 g/s; $G_\infty = 4.96$ g/s; $\bar{q} = 17.4$ kW/m²; τ = 0.51 s; d = 0.8 cm.

The results of an experimental investigation aimed at determining the effect of liquid section throttling on the overshoot value reduction are presented in Wedekind and Bhatt [139]. Figure 7.5 contains the experimental data. When throttling is increased, overshoot reduces to zero. An almost instantaneous rise of the condenser steam pressure becomes insufficient to accelerate the liquid.

The same authors investigated stability boundaries of a horizontal tubular condenser located between two tanks [140]. The occurring self-oscillations had the constant frequency. Figure 7.6 shows stability boundaries in N_c and N_i coordinates,

$$N_c = \frac{\tau}{V_G \dfrac{d\rho_G}{dP} \dfrac{d\Delta P_G}{dG}} ; \quad N_i = \frac{A \dfrac{d\Delta P_L}{dG} \cdot \tau}{H},$$

where V_G is the tube steam volume, H is the length of the liquid section under the steady-state regime, A is the flow area, ΔP_G is hydraulic losses due to friction in the steam section, and ΔP_L is hydraulic losses due to friction in the condensate section.

The time constant τ is

$$\tau = A\bar{\alpha}\,\rho_G i_{LG} / \bar{q}\,\Pi$$

where $\bar{q}$ is the condenser length averaged heat flux, Π is the heated perimeter, $\bar{\alpha}$ is the condenser length averaged steam void fraction, and i_{LG} is the latent heat of evaporation.

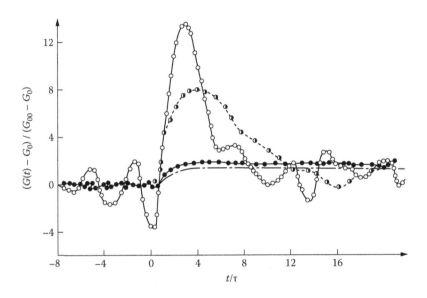

FIGURE 7.5
Characteristics of the outlet liquid flow rate with the outlet throttling [139]. Clear circle: $\Delta P_{e.t.} = 1.9$ kPa, negligible throttling; crossed circle: $\Delta P_{e.t.} = 6.0$ kPa, weak throttling; dark circle: $\Delta P_{e.t.} = 87.0$ kPa, strong throttling; dotted dashed line: inlet steam mass flow.

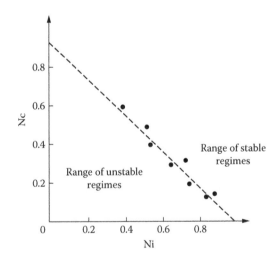

FIGURE 7.6
Condensate stability boundary [140]. Condensing flow: Freon-12; $P = 640$ kPa, inlet quality $x = 1$; $i_{in} = i_G$; $d = 0.762$ cm; $\bar{\alpha} = 0.83$; $\bar{q} = 7.56$ kW/m^2; ●: boundary experimental points.

From Figure 7.6 it follows that an increase of N_c and N_i stabilizes condensation; that is, attaining stable conditions in a horizontal condenser requires an increase of ΔP_L and a decrease of ΔP_G, H, $\bar{q}$.

Stability boundaries of a vertical 1.5 m long condenser (tube-within-a-tube design, inner tube diameter of 9 mm) have been investigated [138]. The condenser was located between the tanks, with the inlet steam and outlet liquid–gas blankets. The tanks ensured constant pressure before and past the condenser in its unstable operation.

The effect of the inlet (ΔP_{in}) and outlet (ΔP_e) throttling, of the working medium flow rate G, of the cooling fluid flow rate G_{cool}, of the condenser pressure P, and of the angle of condenser inclination have been investigated.

It was shown experimentally that in the considered ranges of flow parameters (G = 3.10^{-3}–6.10^{-3} kg/s; P = 0.25–0.8 MPa; G_{cool} = 0.05–0.1 kg/s, i = i_s, at the outlet) in the condenser, there always exist fluctuations with a frequency ν_1, the amplitude of which (equal to 5%–8% of the steady-state flow rate) always stays practically constant with ΔP_{in} and ΔP_e variation.

Visual investigations described in Gerliga and Shelkhovskoy [138] allow a supposition that the noticed fluctuations are due to condensate wavy motion. Their experiments showed that the existence of fluctuations with the ν_1 frequency does lead to a noticeable disruption of condenser operation and to the steam breakthrough at the condenser outlet. When the outlet pressure drop is significant (ΔP_e = 0.3 MPa) and the inlet throttling is absent, only fluctuations with the ν_1 frequency exist in the condenser. When the outlet pressure drop is decreasing, while other parameters are kept unchanged, fluctuations with the frequency ν_2 = 0.4 Hz appear in the condenser. The further decrease of ΔP_e down to 0.1 MPa (e.g., with a flow rate of 5.10^{-3} kg/s) causes the flow rate fluctuation amplitude with the frequency ν_1 to increase insignificantly. A ΔP_e below 0.1 MPa leads to a sudden rise in the condensate flow rate fluctuation. At the minimum possible values of ΔP_{in} and ΔP_e, the amplitude of flow rate fluctuations with the frequency ν_2 may reach the value of the initial steady-state flow rate. It should be noted that fluctuations with the frequency ν_1 are always present. (They are superimposed over fluctuations with the frequency ν_2.) The described picture of condensate fluctuations is presented in Figure 7.7.

Initiation and development of unstable condenser conditions is clearly demonstrated by the working medium flow rate behavior at the condenser outlet. This is the reason why this parameter (flow rate) was chosen as the determining one for the experiments.

The *open inlet/outlet* means the fully opened throttle at the inlet/outlet.

It is indicated in Wedekind and Bhatt [139] that the amplitude of supercooled fluid flow rate fluctuations with the frequency ν_1 may reach 5%–30% of G_0. Additionally, it was noticed that in the case of weak throttling at the condenser outlet, these random fluctuations transform into the system instability expressed by the ordered harmonic fluctuations. It is also said that, to a certain extent, random fluctuations initiate condensation instabilities.

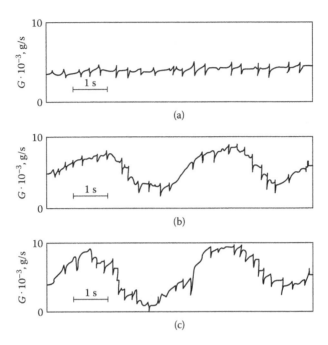

FIGURE 7.7
Condensate flow rate fluctuations development [138]. $G = 5.10^{-3}$ kg/s; $P = 0.4$ MPa; $G_{cool} = 0.1$ kg/s; (a) $\Delta P_e = 0.3$ MPa; (b) $\Delta P_e = 0.044$ MPa; (c) $\Delta P_e = 0.015$ MPa.

Figure 7.8 presents the $\overline{A}_G = f(\Delta P_e)$ dependence for a number of condensate flow rates. Here,

$$\overline{A}_G = \frac{A_G}{G}; \quad A_G = 0.5 \cdot (G_{max} - G_{min}).\tag{7.4}$$

The effect of the condenser inlet pressure drop (ΔP_{in}) on the condenser stable operation can be conveniently traced starting from the fully opened inlet/outlet. It has been stated previously that fluctuations with the frequency ν_2 exist in the condenser. When ΔP_{in} is increasing, the amplitude of fluctuations with the frequency ν_2 decreases and then fluctuations with a frequency ν_3 (variable frequency of 1.4–3.9 Hz) and maximum amplitude, depending on the flow rate, appear. As the condensate flow rate decreases, the relative value of the fluctuation amplitude A_G with the frequency ν_3 increases insignificantly, while the frequency is found to decrease. With the increasing inlet pressure drop, the fluctuation amplitude with the frequency ν_2 decreases, while that with the frequency ν_3 increases asymptotically up to the constant value. These dependencies are illustrated in Figure 7.9.

The experiments with different condenser pressures (0.8, 0.4, and 0.25 MPa) show the increased pressure to stabilize condenser operations. The decreased

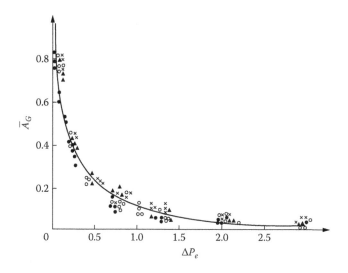

FIGURE 7.8
Variation of the condensate flow rate relative amplitude fluctuations depending on the condenser outlet pressure drop. $P_e = 0.4$ MPa; ●: $G = 3.10^{-3}$ kg/s; ○: $G = 4.10^{-3}$ kg/s; ▲: $G = 5.10^{-3}$ kg/s; X = 6.10^{-3} kg/s.

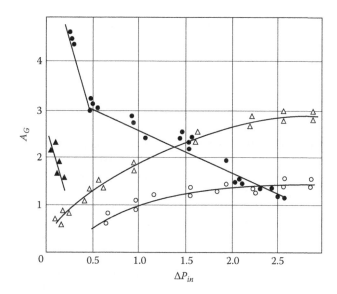

FIGURE 7.9
Variation of the condensate flow amplitude depending on the condenser inlet pressure drop. $P = 0.4$ MPa; ●: $A_G = 3.10^{-3}$ kg/s with v_2; ○ = 3.10^{-3} kg/s with v_3; ▲ = 6.10^{-3} kg/s with v_2; $\triangle = 6.10^{-3}$ kg/s with v_3.

condensation rate (surfactants adding, cooling fluid flow rate decreasing) results in a reduced fluctuation amplitude.

It was shown experimentally that the vertical condenser (steam downflow) operates more stably than the horizontal condenser.

It follows from Figure 7.9 that there are at least two mechanisms of self-oscillation development in the condenser because fluctuations with frequencies ν_2 and ν_3 respond to pressure drop variations at the inlet in a different manner. No such difference was noticed by Wedekind and Bhatt [140] for a horizontal condenser. Only fluctuations of single frequency in the 0.2–0.4 Hz range were observed.

Let us consider the simplest mathematical model with the inner steam condensation [140] for the vertical channel and downflow of the working medium. The dynamic isolation supposes constancy of the pressure at the condenser ends and of the inlet enthalpy.

For a more exact establishment of the mechanism of self-oscillation initiation in the condenser, let us make some assumptions that make it possible to disregard secondary effects:

1. The pressure drop in the condenser steam part is mostly concentrated at the inlet throttle.
2. The liquid film temperature is equal to that of saturation.
3. The total steam void fraction in the condenser changes mainly due to variation of the two-phase section length. The effect of condensate film thickness is of secondary importance.
4. The specific heat flux from steam to film in the disturbed state is constant.

With such assumptions, an oscillatory system composed of such components as the inlet and outlet hydraulic resistances, compressible steam volume, and the liquid section mass may be distinguished in the condenser.

Indeed, with a random increase of the condensate flow rate (G_{con}), the volume of the steam space in the tube (V_G) increases and the steam pressure in the tube (P) reduces below the nominal one. Due to the reduction of P, a force striving appears to decrease the flow rate G_{con} and return the liquid section length H to the initial state. However, due to the liquid section inertia, H passes the neutral position during its reduction and starts compressing the steam cavity volume V_G. The pressure P increases and slows down the further decrease of V_G. When the kinetic energy of the liquid column becomes the potential compression energy, the interphase between the steam and liquid sections starts moving downward and, when passing the neutral state, will cause a positive increment of V_G. Later, the motion repeats periodically.

Thus, the initial change in the flow rate G_{con} repeated itself after a time, which is called the oscillatory period. It should be noted that the influence of the inlet resistance on the oscillatory system behavior is not the determining factor. In principle, ΔP_{in} may be supposed to be large enough so that there would be no effect on steam flow rate fluctuations. A spring with a suspended weight can be named as a mechanical analog of the considered oscillatory system. The role of the spring is played by the steam cavity.

An important member of the oscillatory system is the valve that supplies energy to the oscillatory system and swings it. The function of such a valve is performed by the steam section length (H_G). Indeed, when the steam cavity is compressed (P_G is above the nominal value), H_G, the condensation surface, and the amount of steam condensed per time unit decrease due to the liquid section inertia. When the steam cavity expands due to inertia (P_G is below the nominal value), H_G extends and hence the amount of steam condensed per time unit increases.

Evidently, the increased amount of steam to be condensed on the V_G volume surface when the pressure is decreasing will cause a larger pressure drop, while the decreased amount of steam condensed with the pressure rise will be responsible for a greater pressure rise. Actually, the preceding can be reduced to the Rayleigh criterion [108], the essence of which is as follows: A supply of mass or heat flux to the substance flow excites fluctuations if it is in phase with pressure fluctuations. The steam volume performs positive work during the fluctuation period. The behavior of the considered steam cavity with fluctuation pressure variation resembles behavior of a bubble in the acoustic pressure field (see Section 6.1, Chapter 6): The condensation surface decreases with the increasing pressure, thus causing a larger pressure rise. When the pressure is decreasing, the condensation surface increases and contributes to the further pressure drop. The main difference between the considered phenomena lies in the oscillatory systems. In Section 6.1, the oscillatory system is represented by an acoustically isolated hydraulic channel, while in the case in question the system is made up by a combination of the compressible steam cavity and liquid mass.

A natural way of stabilizing the process in the condenser applying the *swing* mechanism is the liquid section throttling. The larger the pressure drop across the liquid section is, the more difficult it is for the described mechanism to initiate self-oscillations. The increased rate of heat removal from the condenser should destabilize the process, since the same fluctuations of the heat removal may be obtained at lower H_G fluctuations.

The reported mechanism is qualitatively consistent with the results presented in Figures 7.6 and 7.8.

Now, let us present the simplified mathematical description of the mechanism of self-oscillation excitation [140]. Taking the assumptions made into account, the physical picture may be described by the following system of equations:

$$A\frac{d}{dt}\left\{\left[\rho_L(1-\bar{\alpha})+\rho_G\bar{\alpha}\right]H_G\right\}=G_G-\bar{G}_{con}; \tag{7.5}$$

$$A \frac{d}{dt}\left\{\left[i_L \rho_L (1-\overline{\alpha}) + i_s \rho \ \overline{\alpha}\right] H_G\right\} = q \Pi H_G + i_s G_G - i_L \overline{G}_{con};$$ (7.6)

$$\rho_L A \frac{dH_L}{dt} = \overline{G}_{con} - G_{con};$$ (7.7)

$$P_1 - P = K_1 G_G^2;$$ (7.8)

$$P_2 - P_3 = K_2 G_{con}^2;$$ (7.9)

$$P - P_2 = \frac{H_L}{A} \frac{dG_{con}}{dt},$$ (7.10)

where (7.5) is the equation of the two-phase section mass balance; (7.6) is the equation of the two-phase section energy balance; (7.7) is the equation of the condenser liquid section mass balance; (7.8) and (7.9) are the relations describing the quadratic drag at the condenser inlet and outlet, respectively (K_1 and K_2 are proportionality coefficients); and (7.10) is the equation of the momentum for the liquid section (it is assumed that the path hydraulic losses relate to outlet local resistance). $\Pi q / i_{LG}$ is the steam mass amount removed from the steam flow per unit of the condenser length and assumed constant over time and length; $\overline{G}_{con}$ is the condensate flow rate at the two-phase region outlet, measured with respect to the moving interphase; ($H_G(t)$) and G_G are the steam section length and inlet steam flow rate; P is the pressure in the condenser steam section; P_2 is the pressure upstream outlet hydraulic resistance; and H_L is the channel liquid length.

The calculation scheme for the condenser is shown in Figure 7.10. Considering that

$$\delta \rho_G = \frac{d\rho_G}{dP} \cdot \delta P; \quad \delta H_2 = -\delta H_G,$$

upon linearization, the system of (7.5)–(7.10) becomes

$$A \left[\rho_L (1-\overline{\alpha}) + \rho_G \overline{\alpha}\right] \frac{d\delta H_G}{dt} + A H_G \overline{\alpha} \frac{d\rho_G}{dP} \frac{d\delta P}{dt} = \delta G_G - \delta G_{con};$$ (7.11)

$$A \left[i_L \rho_L (1-\overline{\alpha}) + i_s \rho_G \ \overline{\alpha}\right] \frac{d\delta H_G}{dt} + A H_G i_s \overline{\alpha} \frac{d\rho_G}{dP} \frac{d\delta P}{dt}$$

$$= -q \Pi \delta H_G + i_s \delta G_G - i_L \delta \overline{G}_{con};$$ (7.12)

$$A \rho_L \frac{d\delta H_L}{dt} = \delta \overline{G}_G - \delta G_{con};$$ (7.13)

$$-\delta P = \frac{2\Delta P_{in}}{G} \delta G_G \quad (P_1 = const)$$ (7.14)

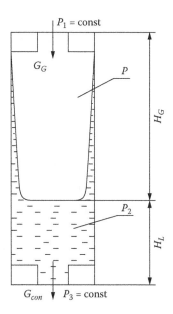

FIGURE 7.10
Condenser calculation scheme.

$$-\delta P_2 = \frac{2\Delta P_e}{G} \delta G_{con} \quad (P_3 = const) \tag{7.15}$$

$$\delta P - \delta P_2 = \frac{H_L}{A} \frac{d\delta G_{con}}{dt}, \tag{7.16}$$

where ΔP_{in} and ΔP_e are the steady-state pressure drops at the inlet (P_1–P) and outlet (P_2–P_3) local resistances, respectively.

The linear system of equations (7.11)–(7.16) has particular solution

$$\delta y = \overline{\delta y} \cdot e^{st}, \tag{7.17}$$

where δy is the disturbance of unknown variables in the system (7.11)–(7.16). The substitution of (7.17) into (7.11)–(7.16) yields a cubic characteristic equation. The negativity condition for the real part s (the condition of the considered system of linear equations' stability) is

$$\left(\frac{\Delta P_e}{\Delta P_{in}} A\rho_L i_{LG} + \overline{\alpha} \frac{d\rho_G}{dP} \frac{Gq\Pi H_L}{2\Delta P_{in}} + (\rho_L - \rho_m)i_s H_L \right)$$
$$\cdot \left[\frac{2\Delta P_{in}}{G} i_{LG} \overline{\alpha}\rho_G + \frac{\Delta P_e}{\Delta P_{in}} q\Pi A \overline{\alpha} \frac{d\rho_G}{dP} + A(\rho_L - \rho_m)i_s \frac{2\Delta P_e}{G} \right] > H_L q\Pi\lambda\overline{\alpha}\rho_L i_{LG}. \tag{7.18}$$

The increased ΔP_e extends the process stability boundary in the condenser, while the increasing ΔP_{in} narrows it (the value of the first term in the second brackets is insignificant), it being qualitatively consistent with the experimental results (see Figures 7.6 and 7.8).

The calculated straight line in Figure 7.6 was obtained using the considered mathematical model.

7.4 Water Hammers in Horizontal and Almost Horizontal Steam and Subcooled Water Tubes

The occurrence of water hammers in almost horizontal tubes with steam and subcooled water started gaining attention after the destructive water hammer in the steam generator feed line at the Indian Point 2 NPP [143].

When the steam and cold water flows are counterdirected in the inclined channel, two different kinds of steady-state flow violations are observed [142,144]: the water hammer at the condensational steam void collapse and flooding.

Figure 7.11 illustrates experimental stability of the counterdirected stratified steam and water flows in an almost horizontal channel. The inclination angle is 4.5°. The channel length between the inlet and outlet water sections is 1.27 m. The channel section is rectangular (0.076 m × 0.038 m). Three regions were found to exist: (1) with a water hammer, (2) with flooding,

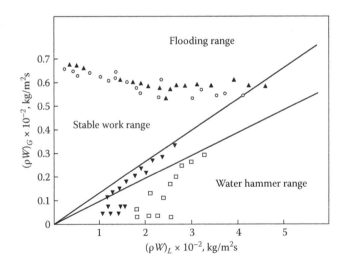

FIGURE 7.11
Stability map of the counterdirected stratified steam and water flows in an almost horizontal channel. ○: T_L is about 90°C; □: T_i is about 70°C; ▼: T_L is about 50°C; ▲: T_L is about 30°C.

and (3) with a stable flow. The straight lines mark the ideal boundaries of complete condensation, derived from the energy balance equation. Above that line, the flow in the entire channel is stratified, while below, steam is completely condensed in the channel. The figure shows that real points of complete condensation differ somewhat from the ideal one.

If the water flow rate is too small for the condensation of all the supplied steam, a water hammer is not possible in the stratified flow. The condensation water hammer is considered in Lee and Bankoff [144] as a thermal-hydraulic instability of the counterdirected stratified flow and it is caused by collapse or almost collapse of local steam voids. The collapse is characterized by formation of junctions with the counterdirected flow in the inclined channel.

7.5 Instability of Bubbling Condensers

The NPP severe accident localization systems widely employ bubbling condensers consisting of a bubbling pool and tubes supplying the steam gas mixture under the liquid surface. Pulsating regimes that occur in such devices may be accompanied by large alternating-sign pressure fluctuations in the steam supplying lines and water of the pool. Two typical regimes of developed fluctuations are distinguished in references 145–147: (1) a regime with the periodic liquid supply to the steam supplying channel, the so-called chugging, and (2) a regime when the interphase boundary does not penetrate into the steam supplying channel but fluctuates at the channel outlet in the bubbling pool water.

Chugging is characterized by frequency of 1–5 Hz and large amplitudes of fluctuations. In some cases, water hammers occur [145]. Therefore, verification of reliability and strength of each new design envisage a great deal of experimentation on models and full-scale bubblers [148]. The analysis of the mechanism of fluctuations' initiation and construction of the mathematical model of the process may contribute to increased reliability and improved operability of such systems.

Chugging may be presented as follows. If the supplying tube pressure exceeds the liquid column weight determined by the tube depth, the steam–air mixture exits the tubes into the liquid and forms a cavity on the surface of which steam is condensing intensely.

With significant subcooling of water in the pool, the steam input from the tube into the cavity will be smaller than the amount of steam used for condensation. The preceding difference causes acceleration of liquid layers in contact with the steam–air cavity when they move to the tube outlet. The void collapses and the liquid enters the tube at high velocity. Due to friction, the velocity of the liquid upflow in the tube reduces, the liquid stops moving and then starts moving in the opposite direction under pressure of the steam–air mixture, leaves the tube, and forms a steam–air cavity in the pool. Then the cycle repeats. Probably, the reduced air content in the steam–air mixture causes increased fluctuation amplitudes. A larger subcooling of the pool water will ambiguously affect the oscillations' amplitude. With a preset steam flow rate, the increased subcooling will at first lead to higher amplitudes, and then to their reduction to some extent. The latter is associated with the decreased steam void in the pool and, hence, a lower possibility of acceleration of the liquid entering the tube in the event of steam void collapse. The reduced steam flow rate influences the fluctuation amplitude in a similar fashion [147].

Aya, Narial, and Kobayashi [147] attempted to describe the preceding phenomenon mathematically. Most difficult was to describe the shape, size of the steam void, and heat transfer between the void and the surrounding liquid. The approximate mathematical model constructed by the authors yields numerical results, which can be compared only qualitatively with the experimental data. Numerous authors indicate that relatively high frequency (100–200 Hz) fluctuations occur in the bubbler along with the low-frequency fluctuations. The fluctuations are of the acoustic nature and may exist independently, in the presence of a steam–gas void past the tube outlet, and together with chugging. The maximum high-frequency fluctuations in the steam–gas flow in the tube appear at passing the condensation boundary through the area of the outlet section of the tube supplying the steam–gas mixture under the water surface. It may be expected that the reasons of excitation of high-frequency fluctuations in the considered case are close to those discussed when considering thermoacoustic fluctuations appearing at subcritical pressures in heated channels.

One more destabilizing factor indicated in Kuznetsov and Bukrinsky [145,146] is the bubbler hydraulic characteristic curve ambiguity. The steam–gas mixture leaving the tube partially condenses and partially moves as bubbles around the tube upward to the liquid surface. The pressure drop across the steam–gas flow path consists of hydraulic losses in the tube and weight of the column of the liquid surrounding the tube flooded portion. The latter component tends to decrease with the increasing mixture flow rate due to a large amount of the lifting bubbles. It leads to formation of the dropping section in the bubbler hydraulic characteristic curve.

Fairly close to the considered phenomenon is the steam jet condensation in the cocurrent flow of cold liquid [149]. In the latter case, high-frequency fluctuations with the acoustic frequency of several kilohertz occur. Such a high frequency is due to the fact that the dominating medium in the tube is liquid. Also, this work presents a mathematical model of the considered phenomenon. The mechanism is based on the model discussed in the section devoted to thermoacoustic fluctuations at subcritical pressures. An account is made only of the feedback due to pressure disturbance. The task considered is associated with the behavior of the jet formed by steam injection into the cold liquid flow. A similar task may be encountered in the analysis of operation of steam–water injectors.

8

Some Cases of Flow Instability in Pipelines

8.1 Self-Oscillations in Inlet Line-Pump System

When pressure decreases at the centrifugal pump inlet, cavitation occurs in the pump flow area (i.e., violation of the fluid flow continuity and formation of cavities filled with steam and gas released from the fluid). In pumps, cavitation initiates when the inlet pressure (P_{in}) substantially exceeds the evaporation pressure at a given fluid temperature. Hence, the minimum pressure regions are located inside the pump flow area. With blade streamlining, the lower pressure regions are formed. Qualitatively, the arrangement of lower pressure regions for pumps with the axial (screw) wheel is presented in Figure 8.1; for pumps with the centrifugal wheel, it is illustrated in Figure 8.2 [150]. These regions are located on the nonoperating side of the blade front side.

When the minimum pressure drops down to the saturated vapor pressure, cavitation initiates on the blade wall. The higher the velocity of the flow around the blade is, the larger is the rarefaction on the blade. With this in view, the points on the blade leading edge, most remote from the axis of rotation, may happen to be the centers of cavitation origin. With some further moderate reduction of pressure P_{in}, the developing cavitation does not change the pump head, but with $P_{in} < P_{cr}$, the head will start dropping sharply in the centrifugal pump. However, in the screw-centrifugal pump it will reduce insignificantly. The screw-centrifugal pump is a combination of axial and centrifugal wheels. For the latter pump, failure in operation (an abrupt head drop) starts when the inlet pressure P_{Br} is substantially lower than P_{cr}. In centrifugal pumps, $P_{Br} = P_{cr}$.

The cavitation characteristics corresponding to the abrupt head drop in both the centrifugal and screw-centrifugal pumps (the dependence of the head on P_{in} at the constant flow rate and rotation) are given in Figures 8.3 and 8.4, respectively. As one can see, the screw is responsible for the delayed failure in operation of the screw-centrifugal pumps as compared to the centrifugal one.

Under certain conditions, cavitation causes self-excitation of pressure and flow rate fluctuations (the so-called cavitation fluctuations) in centrifugal and screw-centrifugal pumps. These fluctuations have been investigated insufficiently in the *pipeline-centrifugal pump* system. Fluctuations have been

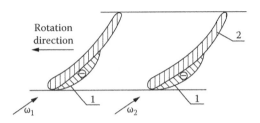

FIGURE 8.1
Region of minimal pressures for the axial screw wheel. w_1: inlet relative velocity; 1: lower pressure region; 2: blade.

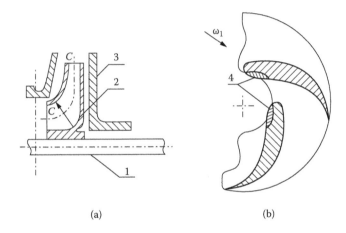

FIGURE 8.2
The minimal pressure region for the centrifugal wheel. 1: shaft; 2: impeller; 3: pump body; 4: lower pressure region; (a) pump longitudinal section; (b) C – C section.

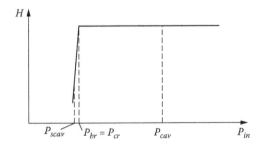

FIGURE 8.3
Cavitation characteristics of the screw-centrifugal pump. P_{cav}: inlet pressure at the beginning of cavitation; P_{cr}: inlet pressure at which cavitation decreases the pump head; P_{Br}: inlet pressure at which a sharp head drop occurs.

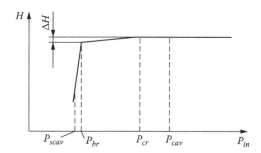

FIGURE 8.4
Cavitation characteristics of the centrifugal pump (designations are the same as in Figure 8.3).

investigated most completely in the *pipeline–screw–centrifugal pump* system [151]. At present, high-speed screw-centrifugal pumps with high anticavitating properties have found quite a wide application in some technical areas. Even in conditions close to the optimal ones, these pumps operate with latent cavitation, which has no marked effect on the pump's head (see Figure 8.4).

In general, all the known investigations of cavitation self-oscillations deal with the high-speed screw-centrifugal pumps. Some results are presented later.

The reasons for cavitation self-oscillations in the low-speed centrifugal pumps and in high-speed screw-centrifugal pumps are the same. Thus, it may be hoped that the stability boundary characteristics and self-oscillation amplitudes of flow parameters reported for the screw-centrifugal pump will be valid for the centrifugal pump.

In principle, cavitating fluctuations may develop in any *pump-upstream pipeline* pair in circuits of nuclear power facilities when the inlet pressure decreases below the pressure at which a developed cavitation appears. Most dangerous for such facilities are cavitation self-oscillations (CSOs) of the main circulating pumps. Water hammers originating at developed CSOs may lead to loop depressurizing in the region where the upstream pipeline is connected to the pump.

The following cavitation stages in the screw in the presence of the decreasing inlet pressure and constant fluid flow rate and pump speed were distinguished in Stripling and Acosta [152]:

1. Initiation of cavitation when the latter originates at the periphery of the blade leading edges (Figure 8.2).
2. Cavitation wandering over the entire leading edge: it propagates over the blade and is observed in most cases only on separate blades.
3. Cavitation fluctuations' regime: the cavity is found to increase and decrease periodically.
4. Cavitation stall regime: the head sharply drops and the cavitation propagates over the entire surface of each blade.

Numerous observations showed that cavities on the nonoperating side of the blade close before they reach the screw outlet. The bubble-free fluid flow moves between the cavity and the blade operating side.

Figure 8.5 [153] presents oscillograms of the basic stages of CSO development in the screw-centrifugal pump with an averaged outlet pressure (P_e) of 13 MPa. It follows from the figure that no CSO occurs when cavitation margins are either large or small; near the stability boundary, CSO has a

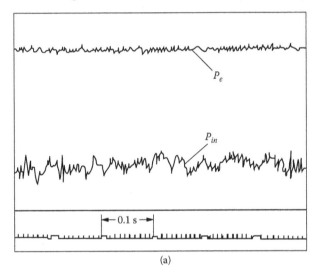

(a)

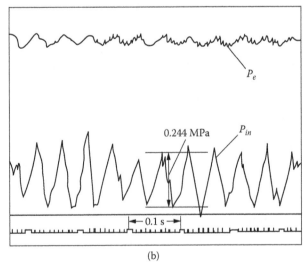

(b)

FIGURE 8.5
(a) Oscillograms of P_{in} and P_e at 21,000 rpm of the pump. P_e = 13.0 MPa; P_{in} = 0.4 MPa; υ = 20 Hz.
(b) Oscillograms of P_{in} and P_e at 21,000 rpm of the pump. P_e = 13.0 MPa; P_{in} = 0.3 ± 0.12 MPa; υ = 20 Hz.

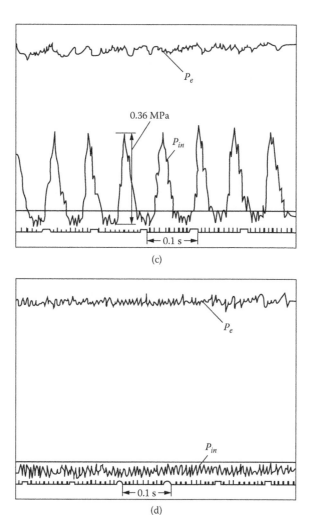

FIGURE 8.5 (*Continued*)
(c) Oscillograms of P_{in} and P_e at 21,000 rpm of the pump. P_e = 13.0 MPa; P_{in} = 0.199 ± 0.018 MPa; υ = 13.3 Hz. (d) Oscillograms of P_{in} and P_e at 21,000 rpm of the pump. P_e = 13.0 MPa; P_{in}=0.104 MPa.

sinusoidal form and in the region of such fluctuations' development, their form changes approaching discontinuity.

CSOs possess a number of peculiarities—for example,

A change of any pump parameter toward the increase of cavitation intensity results in a reduction of oscillation frequency.

The oscillation frequency linearly depends on the inlet pressure.

The oscillation frequency decreases along with the increasing length of the feed pipeline.

The dependence of the oscillation frequency on different lengths of the feed pipeline is shown in Figure 8.6 [151]. With all other quantities unchanged, the decreased fluid flow rate substantially reduces the oscillation frequency due to the increasing cavity area.

The dependence of the pressure fluctuation amplitude ($A_{P_{in}}$) at the pump inlet with CSO on the average inlet pressure (P_{in}) and different speeds of the screw-centrifugal pump shaft is illustrated in Figure 8.7 [151]. With the decreasing rotary speed, the instability region becomes narrower and the maximum pressure fluctuation amplitudes decrease.

Since the discharge line has a larger hydraulic friction than that in the feed pipeline, fluid flow rate fluctuations in the discharge line may be neglected in the first approximation in comparison with those in the feed pipeline:

$$\delta Q_2 \ll \delta Q_1, \tag{8.1}$$

where δQ_2 is the disturbance of the fluid volumetric flow rate in the discharge line and δQ_1 is the disturbance of the fluid volumetric flow rate in the feed pipeline.

Indeed, in the presence of CSO, the pump inlet pressure fluctuation amplitude is much higher than that at the pump outlet (Figure 8.5).

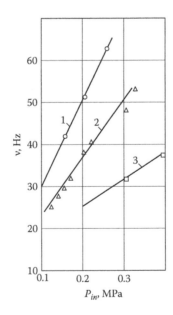

FIGURE 8.6
Dependence of self-oscillations' frequency on the inlet pressure with $Q = Q_0$ and different lengths of the feed pipeline. 1: 0.08 m; 2: 1.2 m; 3: 3.15 m.

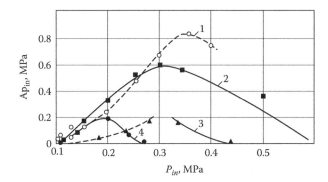

FIGURE 8.7
Dependence of cavitation oscillations' amplitude on the pump inlet pressure at different pump shaft rpm. 1: $n = 1.1\ n_0$; 2: $n = n_0$; 3: $n = 0.82\ n_0$; 4: $n = 0.7\ n_0$.

The rate of the cavity disturbance change may be approximately written as

$$\frac{d\delta V}{dt} = \delta Q_2 - \delta Q_1$$

or, considering (8.1), as

$$\frac{d\delta V}{dt} = -\delta Q_1. \tag{8.2}$$

The earlier noted features of partial cavitation on the screw blades are acceptably described by the model of jet cavitating flow around the pump blades [152]. For the screw-centrifugal pump, positive angles of incidence (the angle between the flow direction in relative motion when hitting the blade and the blade leading edge direction) should be chosen. The dependence of the fluid flow rate (Q_1) on the angle of incidence (γ) is expressed by

$$\gamma = \beta - arctg\,\frac{C_{1m}}{u}$$

$$C_{1m} = \frac{Q_1}{\frac{\pi}{4}\left(D_{ou}^2 - d_{in}\right)},$$

where D_{ou} and d_{in} are the maximum and the minimum screw diameters, respectively; u is the leading edges' mean diameter peripheral velocity; and β is the screw leading edges' angle of inclination.

With the increasing pump inlet flow rate, the angle of incidence decreases, as well as the volume of cavities located on the nonworking sides of the blade. This dependence is very important for the CSO initiation and development. In addition, the cavity area also decreases with the increasing pump inlet pressure due to the steam cavity compressibility and the decreasing cavitation intensity. In this case, the cavity volume perturbation may be represented in the form [151] that follows:

$$\delta V = -\left|\frac{\partial V}{\partial P}\right| \cdot \delta P_{in} - \left|\frac{\partial V}{\partial Q_1}\right| \cdot \delta Q_1. \tag{8.3}$$

If the absence of the feed pipeline is supposed (i.e., the pump is directly connected to the feed tank with the constant pressure), then (8.3) may be written as

$$\delta V = -\left|\frac{\partial V}{\partial Q_1}\right| \cdot \delta Q_1. \tag{8.4}$$

The system of (8.2) and (8.4) is statically unstable. For instance, in the case of an occasional increase of the cavity area, the fluid flow rate will decrease to follow (8.4). Later, the negative flow rate disturbance will cause the time derivative increase with versus the volume disturbance (see 8.2). Thus, the initial positive cavity volume disturbance yields a positive time derivative versus the volume disturbance (i.e., the process aperiodically drifts from its initial state).

In real systems there always exists a feed pipeline of a definite length. The linearized equation of liquid movement in the pipeline is

$$\delta P_e - \delta P_{in} = K_1 \frac{d\delta Q_1}{dt} + K_2 \delta Q_1, \tag{8.5}$$

where the first term of the right-hand part takes the disturbance due to fluid flow inertia into account, while friction losses are taken into account by the second term,

$$K_1 = \rho_L H/A, \quad K_2 = 2(P_e - P_{in})/Q_1,$$

where A is the channel flow area, and P_e is the feed tank pressure that will henceforth be assumed constant ($\delta P_e = 0$).

The connection between the cavity pressure disturbance and the cavity volume perturbation (8.3) and the connection between the latter and the flow rate disturbance (8.2) and the dependence of the volumetric flow rate on the pressure disturbance in the cavity area (8.5) form an oscillatory system:

$$(8.3) \quad (8.2) \quad (8.5)$$

$$\delta Pin \rightarrow \delta V \rightarrow \delta Q1 \rightarrow \delta Pin.$$

The omitting of the dependence of the cavity volume on the fluid flow rate (the second term in the left-hand part of Equation 8.2) and the neglect of friction in the feed pipeline yield a neutral oscillatory system described by three simple equations:

$$\delta V = -\left|\frac{\partial V}{\partial P}\right| \cdot \delta P_{in},$$

$$\frac{d\delta V}{dt} = -\delta Q_1,$$

$$-\delta P_{in} = K_1 \frac{d\delta Q_1}{dt}, \tag{8.6}$$

or by one equation of the form

$$M\frac{d^2\delta W_1}{dt^2} + \frac{A^2}{\left|\frac{\partial V}{\partial P}\right|}\delta W_1 = 0, \tag{8.7}$$

with the oscillation frequency of

$$\omega = \sqrt{\rho_L H \left|\frac{\partial V}{dP}\right|}.$$

It is taken into account that $\delta Q_1 = A\delta W_1$. (W_1 is the fluid velocity in the feed pipeline.) $M = \rho_L HA$ is the mass of the fluid participating in oscillations.

Thus, we have an oscillatory system presented by a fluid flow in a tube with the length H, "resting" on an elastic steam–gas cavity with the volume V. A weight with the mass M, supported by a spring with linear rigidity of $A^2 / \left|\frac{dV}{dP}\right|$ may be suggested as a mechanical analog of the oscillatory system described by (8.7).

The block diagram of the oscillatory system described by (8.6) is shown in Figure 8.8. The transfer functions corresponding to (8.6) are

$$\delta\tilde{V} = \left|\frac{\partial V}{\partial P}\right|\delta\tilde{P}_{in},$$

$$s\delta\tilde{V} = -\delta\bar{Q}_1,$$

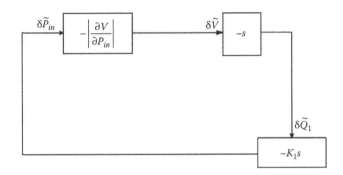

FIGURE 8.8
Block diagram of the oscillatory system according to (8.6).

and

$$\delta \tilde{P}_{in} = -K_1 s \delta \tilde{Q}_1.$$

In this case, the sign "~" denotes the Laplace transform at zero initial conditions.

With the inclusion of friction into the preceding system (the second term of the right-hand parts of Equation 8.5), a system with time-decaying oscillations is obtained; when the dependence of the cavity volume on flow rate (the last term of Equation 8.3) is introduced, a system with the oscillations' amplitudes increasing with time may be obtained.

In the general case, (8.7) will look like

$$\frac{d^2 \delta Q_1}{dt^2} + \left(\frac{K_2}{K_1} - \frac{|\partial V/\partial Q_1|}{K_1 |\partial V/\partial P|} \right) \cdot \frac{d \delta Q_1}{dt} + \frac{1}{|\partial V/\partial P| K_1} \cdot \delta Q_1 = 0 \qquad (8.8)$$

and the stability condition will be written as

$$\frac{2(P_e - P_{in})}{Q_1} > \frac{|\partial V/\partial Q_1|}{|\partial V/\partial P|}. \qquad (8.9)$$

The increased hydraulic loss in the feed pipeline stabilizes the process, while the increased length of the feed pipeline destabilizes the process.

The block diagram of the oscillatory system described by (8.2), (8.3), and (8.5) is depicted in Figure 8.9. The inner loop is in fact a pump with a cavity and without the feed pipeline.

Let us first consider a purely oscillatory system ($\partial V/\partial Q_1 = 0$; $K_2 = 0$). With the random rise of fluid velocity before the pump, the volume of the steam–gas cavity decreases and the kinetic energy of the disturbed fluid flow in

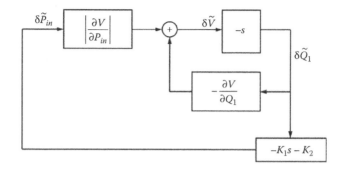

FIGURE 8.9
Block diagram of the oscillatory system according to (8.2), (8.3), and (8.5).

the tube transforms into the potential energy of cavity compression. The cavity pressure reaches the maximum and the flow velocity disturbance becomes zero. Then, the back transition occurs: The cavity expands, its potential energy transforms into the kinetic energy, and the negative velocity disturbance increases. When the potential energy of cavity compression has transformed into the kinetic energy of the flow disturbed motion (the flow velocity is under the maximum negative disturbance), the flow inertia exhibits itself by extending the cavity. The kinetic energy of the flow disturbed motion transforms into the potential energy of the extended cavity, etc.

The dependence of the cavity volume on the inlet flow rate of the working medium swings the oscillatory system, since the volume compression (due to the increased flow rate) causes a greater flow rate; each volume expansion (resulting from a reduced flow rate) leads to a still larger flow rate reduction. Evidently, the friction pressure drop across the feed pipeline leads to the flow kinetic energy loss and eventually to system stabilization.

The previously described mechanism of CSO initiation in screw-centrifugal pumps reflects the basic aspects only. A comprehensive analysis of this phenomenon is presented in Pelipenko, Zadontsev, and Natanzon [151]. Some results of experimental investigations of CSO from these authors are considered next.

The screw design parameters substantially influence the CSO frequency. An increased screw adjustment angle causes a reduction in the CSO frequency. In the considered case, the increased β promotes the cavity volume increase. With the increasing screw outer diameter and the decreasing number of screw starts, the oscillation frequency was found to decrease.

The results of experiments involving the screw-centrifugal pumps with two-start screws of constant pitch (h) are as follows [151]:

Pump no. 1: $D_{ou} = 0.12$ m; $d_{in} = 0.063$ m; $h = 0.054$ m; $\tau_0 = 0.43$; $\gamma_0 = 4°31'$

Pump no. 2: $D_{ou} = 0.56$ m; $d_{in} = 0.028$ m; $h = 0.0252$ m; $\tau_0 = 0.54$; $\gamma_0 = 3°56'$

Figures 8.10 and 8.11 [153] present the experimental stability boundaries for a screw-centrifugal pump in the P_{in} – (n/n_0) plane of parameters, respectively, for two values of τ(0.54 and 0.27), where τ is the so-called flow coefficient equal to the ratio between the current flow rate and at which the angle of incidence at hitting the screw equals zero. For the axial screw primary pumps with constant pitch, the flow coefficient is determined in the form of Chebaievsky and Petrov [154] as follows:

$$\tau = \frac{4Q}{\pi h \left(D_{ou}^2 - d_{in}^2 \right) \cdot n}.$$

It follows from these authors [154] for the considered pumps that with $\tau < 0.5$ before the screw, reverse currents initiate stepwise at τ of about 0.5 and immediately occupy over half of the feed pipeline section area. In the case of strong reverse currents, the cavity forms not only in the interscrew passage, but also before the screw. In Figures 8.10 and 8.11, clear circles denote the stable and dark circles denote the self-oscillating regime of pump operation.

From the presented experimental data it follows that the self-oscillations' region expands with decrease in the flow rate. At $\tau = 0.27$, fluctuations are observed up to the cavitation stall [151].

The stability region for pump no. 1 in the $(Q/Q_0) - P_{in}$ coordinates is shown in Figure 8.12. The stability region narrows as the flow rate increases. In this case, no cavitation occurs on the centrifugal wheel of the pump.

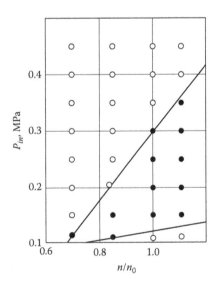

FIGURE 8.10
Experimental stability region boundaries of the screw-centrifugal pump–pipeline system in the $(n/n_0 - P_{in})$ parameters' plane for pump no. 1 with $\tau = 0.54$. ○: stable regime; ●: unstable regime.

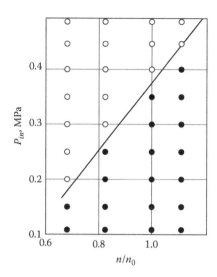

FIGURE 8.11
Experimental stability region boundary of the system in the $(n/n_0) - P_{in}$ parameters' plane for pump no. 2 with $\tau = 0.27$. ○: stable regime; ●: unstable regime.

In the case when cavitation occurs on the screw and centrifugal wheel, the stability region looks quite different (Figure 8.13). In pump no. 2, the centrifugal wheel operates at the nominal flow rate with partial cavitation. The instability region first decreases with the increasing flow rate, and then it starts to expand due to the appearance of cavitation cavities on the centrifugal wheel. The increased pressure before the centrifugal wheel due to the variable pitch screw (cavitation on the centrifugal wheel is eliminated) radically changes the instability region location in the $(Q/Q_0) - P_{in}$ coordinates (Figure 8.14).

The blade angle has a significant effect on stable operation of the *pump–feed pipeline system*. For instance, when β is reduced from 8°10′ down to 6°52′, the self-oscillations' region decreases substantially (Figure 8.12) both with respect to inlet pressure and inlet flow rate. In the case of centrifugal wheel operation in the partial cavitation regime, decreasing β may destabilize the system. A decrease of the screw outer diameter from 0.12 to 0.11 m stabilizes pump operation essentially. This result, like the previous one, is valid for the case when the centrifugal wheel operates without cavitation.

The experiments with pump no. 2 without the primary screw showed [151] that the self-oscillations' region in the system with a centrifugal pump shifts toward larger values of pump inlet pressure, and that the maximum inlet pressure peak-to-peak fluctuations increase almost over two times even with no screw; the maximums are shifted toward larger average inlet pressures. The installation of a primary screw before the centrifugal pump

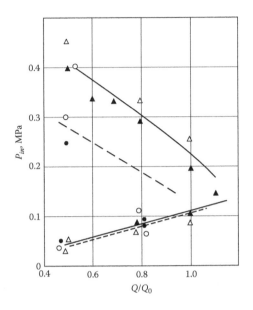

FIGURE 8.12
Experimental stability regions boundary of the system in the $(Q/Q_0) - P_{in}$ parameters plane for pump no. 1 with the blade angles $\beta = 8°10'$ (solid lines) and $\beta = 6°25'$ (dashed lines); $\triangle$ and $\bigcirc$: stable regimes; $\blacktriangle$ and $\bullet$: unstable regimes.

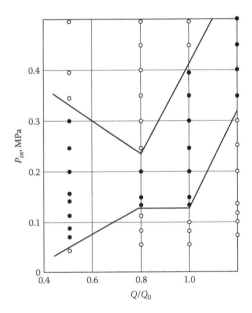

FIGURE 8.13
Experimental stability region boundaries of the system in the $(Q/Q_0) - P_{in}$ parameters plane for pump no. 2 with the blade angle $\beta = 8°09'$. $\bigcirc$: stable regime; $\bullet$: unstable regime.

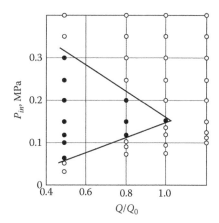

FIGURE 8.14
Experimental stability region boundaries of the system in the $(Q/Q_0) - P_{in}$ parameters plane for pump no. 2 with the variable pitch screw. ○: stable regime; ●: unstable regime.

substantially stabilizes the system in question with respect to the cavitation self-oscillations and shifts the self-oscillations' region toward lower inlet pressures.

8.2 Instability of Condensate Line-Deaerator System

Presently, both nuclear and thermal generation employ deaerators of two types, schematics of which are shown in Figures 8.15 and 8.16.

Figure 8.15 illustrates schematics of a direct-contact jet-type deaerator designed at CKTI [155]. The condensate is fed via nozzles (1 and 14) to a mixer (2) and, through the holes in the neck, is drained to a perforated tray (4). The condensate jets are crossed by the steam flow and are heated up by the condensing steam. From the water overflow, the plate (5) condensate is drained to the bubbler via a segment-shaped hole (13). The bubbler where final deaeration takes place is made of a perforated plate (6), steam relief pipes (12), and drain pipes (7). The heating steam is supplied via a tube (10). At the maximum load, tubes (12) are also used to let the steam pass through. A sump (11) is connected to the perforated sheet, thus forming the hydraulic lock. The deaerated water is drained into a deaeration tank (8) via its neck (9). The gases with noncondensed steam are removed through a nozzle (15).

The basic diagram of the Ural VTI-designed deaerator is shown in Figure 8.16 [156].

The main difference of the design is in the use of the steam control valve (SCV). This substantially reduces hydraulic friction between the lower and

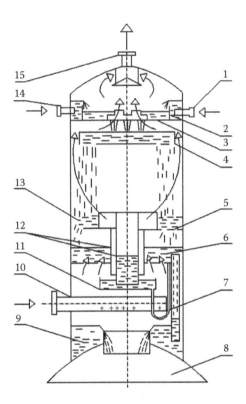

FIGURE 8.15
Basic diagram of the CKTI deaerating column. See description in text.

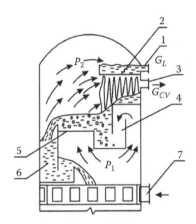

FIGURE 8.16
Basic diagram of the URAL VTI deaerating column. 1: Water supply; 2: water distributor; 3: flash steam outlet; 4: relief valve; 5: bubbler; 6: water drain; 7: steam supply.

upper steam spaces under the increasing steam load in the deaerator column, as compared to the other design.

There are certain conditions where deaerators operate unstably. It is manifested in periodic water hammers in deaerators causing strong vibration there and of the associated pipelines. In time, unstable operation of deaerators may lead to misarrangement and impaired integrity of column components [157,158].

The mechanism of initiation of deaerator unstable operation may be represented as follows [157]. An accidental increase in condensate flow rate (G_{con}) may cause fluid temperature decrease in the jet compartment and increased jet velocity (Figure 8.15). The latter will cause steam condensation in the space above the water overflow plate and hence a steam pressure decrease. The moving pressure drop in the condensate line increases, thereby increasing condensate flow rate. The decreased pressure in the column upper portion induces steam flow rise from below via the bubbler. If the same pressure drop P_2 (see Figure 8.15) will cause a larger disturbance of steam flow rate via the bubbler and steam relief tubes than disturbance of the flow rate of steam used for jet condensation (with the increased condensate flow rate), then deaeration will be stable. If, on the contrary, the disturbance of the flow rate of steam used for condensation will exceed that of steam input to the column upper portion via the bubbler and steam relief tubes, the pressure will keep decreasing. It follows from this that the increased friction in the condensate feed pipeline stabilizes deaeration, while the increased friction in the bubbler and steam tubes of the heating steam feed line destabilizes it.

Thus, the initial increase in the condensate flow rate under certain conditions causes further increase in the condensate flow rate and the process proceeds in an avalanche-like manner, flooding the water overflow and later the perforated plates. Because of the P_2 pressure drop, the pressure drop across the bubbler grows and stops operation of the downspout; the pressure drop across the bubbler exceeds the hydrostatic pressure of cup liquid in the downspout.

The two-phase layer height on the perforate plate keeps increasing and at some point its pressure becomes bigger than the pressure drop across the bubbler. The cold water drop occurs (the flow rate was above the nominal one). The pressure under the perforated plate drops. The steam makeup increases due to self-evaporation of the fluid in the deaerator tank and the increased heating steam makeup (the pressure drop across the steam line has increased), and so on.

The described process of static instability development later leads to variable force impacts on the water overflow and perforated plates in the first place. Aperiodic deviation of P_2 from its nominal value develops less intensively at $P_1 = $ const. and more intensively at $P_1 \neq$ const. The steam makeup will be stabilizing the pressure in the column lower portion (the more there is, the smaller is the friction in the steam line) and, due to water evaporation, in the dearation tank.

During startups, the water in the tank is subcooled down T_s and the column vibrations are stronger than during the nominal operation.

A random condensate flow rate decrease causes an increase in P_2 (steam condensation decreases), and with low friction across the condensate line, the process will at first develop aperiodically toward lower condensate flow rate and higher pressure in the column upper portion. Later, the process will be governed by nonlinear relations between the heat and mass transfer in the deaerator and its hydraulic parameters. We shall concentrate on the initial phase of instability initiation only.

The mechanism of static instability initiation in bubbler columns of the Ural VTI-designed deaerator (Figure 8.16) is determined by the same causes as in the case with the previously considered design. The Ural VTI-designed columns operate more stably, and the reason is as follows.

First, hydraulic resistance in the steam bubbler of the Ural VTI design is less than that in the CKTI case [155,156], which is due to the use of the SCV in the former design.

Second, the SCV accumulates a great volume of water, which functions as a thermal damper (see Figure 8.16). The jet water temperature drops with the increasing condensed water flow rate. This temperature reduction is smoothed in the SCV and reaches the perforated plate with a delay. Therefore, at every time moment with the aperiodic rise of the condensed water flow rate, the steam input to the column upper portion via the perforated plate will be greater, the larger the SCV capacity is, since steam condensation in the bubbler will be lower. Thus, the steam flow rate for condensation in jets will increase greatly without the SCV and at a higher water flow rate, and the steam input via the bubbler will reduce even more. This will result in a more intense pressure drop in the column upper portion. If the SCV is available, the steam input via the bubbler will be larger as compared to the previous case and the pressure drop will be less intense.

A simplified mathematical model of the considered static instability for small relative parameter deviations may be constructed (neglecting inertia of heat and mass transfer processes) as is shown later.

The equation for the condensate line from the constant pressure feed point up to the deaerator column inlet is written as

$$P_1 - P_2 = \Delta P_f(G_{con}) + \frac{H}{A}\frac{dG_{con}}{dt}$$

or, for the disturbed state,

$$\delta P_1 - \delta P_2 = \frac{d\Delta P_f(G_{con})}{dG_{con}}\delta G_{con} + \frac{H}{A}\frac{d\delta G_{con}}{dt}, \tag{8.10}$$

where $\Delta P_f(G_{con})$ are the pressure losses due to the line and local hydraulic resistances, H is the line length, A is the condensate line flow passage, and G_{con} is the condensate mass flow rate.

The first and second terms in the right-hand part of (8.10) describe the disturbance of hydraulic losses and of fluid flow inertia, respectively.

In the context of the considered instability development, the pressure change under the bubbler is of secondary importance. Further, P_1 will be regarded as constant.

The rate of steam mass change in the column upper portion is determined by the difference in the flow rate of steam that passed through the bubbler and SCV (G_{Bc}), on the one hand, and the flow rate of steam condensing in jets (C_{Gc}) and the vent steam flow rate (G_{ev}), on the other hand. This relation is written as the equation of steam mass balance in the column upper portion:

$$\frac{d\left[V\rho_G(P_2)\right]}{dt} = G_{Bc} - G_{Gc} - G_{ev}$$

or, for the disturbances near the steady-state conditions ($\delta G_{ev} \approx 0$),

$$\frac{d\rho_G}{dP} V \cdot \frac{d\delta P_2}{dt} = \delta G_{Bc} - \delta G_{Gc}, \qquad (8.11)$$

where V is the steam volume above the bubbler.

The steam flow rate via the bubbler and the SCV to the column upper steam space depends on the water temperature in the bubbler and in the SCV, water flow rate from the jet compartment, and also the P_2 pressure. Approximately, let us assume that the water temperature is defined by the water flow rate only. It is also assumed that the time constant of the processes in the bubbler and the SCV is small and the processes are described by the quasisteady relation

$$G_{Bc} = G_{Bc}(G_{con}, P_2).$$

In a similar way, the relation for the flow rate of steam used for jet heating is

$$G_{Gc} = G_{Gc}(G_{con}).$$

For disturbances, it will be

$$\delta G_{Bc} = \frac{\partial G_{Bc}}{\partial G_{con}} \delta G_{con} + \frac{\partial G_{Bc}}{\partial P_2} \delta P_2 \qquad (8.12)$$

$$\delta G_{Gc} = \frac{dG_{Gc}}{dG_{con}} \delta G_{con} \qquad (8.13)$$

In (8.10)–(8.13), the coefficients applied before the disturbances are written for the steady-state conditions.

The second term in the right-hand part of (8.12) accounts for the change of G_{Bc} due to pressure drop variation across the bubbler with constant P_1. From (8.10)–(8.13), the equation for the condensate flow rate disturbance with $\delta P_{L.del} = 0$ is written as

$$a_1 \frac{d^2 \delta G_{con}}{dt^2} + a_2 \frac{d\delta G_{con}}{dt} + a_3 \delta G_{con} = 0, \qquad (8.14)$$

where

$$a_1 = V\frac{H}{A}\frac{d\rho_G}{dP_2}; \quad a_2 = \frac{d\rho_G}{dP_2}V\frac{d\Delta P_f}{dG_{con}} - \frac{H}{A}\frac{dG_{Bc}}{dP_2};$$

$$a_3 = \frac{d\Delta P_f}{dG_{con}}\left|\frac{\partial G_{Bc}}{\partial P_2}\right| - \frac{\partial G_{Gc}}{\partial G_{con}} - \frac{\partial G_{Bc}}{\partial G_{con}}.$$

A necessary and sufficient condition for stability of the solution of the linear differential equation with the second-order constant coefficients is that the coefficients shall be positive. In this case, the stability condition (static stability) is

$$a_3 > 0. \qquad (8.15)$$

In the steady-state conditions, $\delta G_{Bc} \approx \delta G_{Gc}$. By equalizing the right-hand parts of (8.12) and (8.13), we obtain the dependence of pressure variation in the deaerator head upper portion on the water flow rate:

$$\frac{dP_2}{dG_{con}} = \frac{1}{\partial G_{Bc}/\partial P_2}\left(-\frac{\partial G_{Bc}}{\partial G_{con}} + \frac{\partial G_{Gc}}{\partial G_{con}}\right).$$

It follows from this that

$$a_3 = \left|\frac{\partial G_{Bc}}{\partial P_2}\right| \cdot \left(\frac{d\Delta P_f}{dG_{con}} + \frac{dP_2}{dG_{con}}\right),$$

and condition (8.15) becomes

$$\frac{d\Delta P_f}{dG_{con}} + \frac{dP_2}{dG_{con}} > 0. \qquad (8.16)$$

The relation of the (8.16) kind from Nesterov [157] should be regarded as a postulated condition.

From the last inequality it follows that within the framework of the simplified model for the column stable operation, an increase/decrease of condensate water flow rate causes a greater increase/decrease of pressure loss across the fluid line than the pressure drop/rise in the column because of the subsequent increasing steam condensation and G_{Bc} variation.

Stability condition (8.15) may be obtained by considering an analog of the hydraulic characteristic used in the analysis of the steam generator static instability. The hydraulic characteristic of the condensate line in Figure 8.17 is shown as a parabolic curve

$$P_{L.del} - P_{2f} = \Delta P_f(G_{con}).$$

The condensate line connects the point of condensate distribution under constant pressure with the upper steam space of the deaerator column with the pressure P_2. Let us call the pressure drop $(P_{L.del} - P_2) = \Delta P_p$ the available or driving pressure drop.

In the case of a steam-generating tube, the available pressure drop across the headers is regarded as independent from the flow rate in a single steam-generating tube (see Chapter 1). In the case in question, the $(P_{L.del} - P_2)$ difference depends on the condensate flow rate value, since a change of G_{con} causes a change in G_{Bc} and G_{Gc} and finally leads to a change in pressure P_2.

In the situation presented in Figure 8.17, two steady-state regimes with flow rates G_{1con} and G_{2con} are possible. $\Delta P_f = \Delta P_p$ in points "a" and "b," and $P_{2f} = P_2$.

The flow rate G_{1con} corresponds to the unstable regime, while G_{2con} corresponds to the stable one.

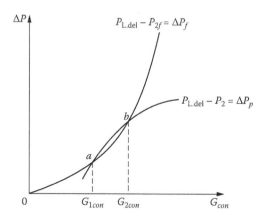

FIGURE 8.17
Thermal-hydraulic characteristic curves of a bubbling column.

Indeed, with a random increase of G_{1con} up to G'_{1con}, the pressure drop due to hydraulic losses across the condensate supply line will increase less than the rise in the available pressure drop. This will lead to further increase in the condensate flow rate and to the aperiodic drift of the regime from the "a" point. With the flow rate increase from G_{2con} up to G'_{2con}, the pressure drop across the condensate line due to hydraulic losses becomes greater than the available (driving) pressure drop, and the condensate flow rate will be forced to acquire the previous value. Similar reasoning can be made with the initially decreased G_{1con} and G_{2con}.

At points "a" and "b," respectively, we have

$$\frac{d\Delta P_f}{dG_{con}} < \frac{d\Delta P_P}{dG_{con}}, \quad \frac{d\Delta P_f}{dG_{con}} > \frac{d\Delta P_P}{dG_{con}},$$

or, considering that $\Delta P_f = P_{L.del} - P_{2f}$, $\Delta P_P = P_{L.del} - P_2$, $P_{L.del} = const.$, the equations may be rewritten as

$$\frac{d\Delta P_{2f}}{dG_{con}} > \frac{dP_2}{dG_{con}}, \quad \frac{dP_{2f}}{dG_{con}} < \frac{dP_2}{dG_{con}}.$$

The two deaerator tank columns may be considered as two systems connected in parallel. The point of these systems' connection (see Figure 8.18) is the condensate line branch point A and the steam space under the bubbler. If a plant employs two or more deaerators, they may be considered as the systems connected in parallel and incorporating the previously mentioned subsystems. In this case, the connection points of deaerators as parallel systems are the branch points of the condensate lines B and steam lines C.

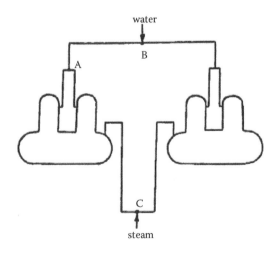

FIGURE 8.18
Schematic of connection of two deaerators.

Similarly to the multichannel steam generators, the following types of static instabilities are possible in the case with identical parallel deaerators:

1. The intercolumn instability: there is no deviation of flow parameters from the steady-state values above point A, while in the columns of one deaerator and in respective condensate feed pipelines deviations of flow parameters are in antiphase; dynamically, deaerators are not interconnected; the pressure at point A remains constant as the column instability develops.

2. The interdeaerator instability: deviation of flow parameters from the steady-state values in the columns of one deaerator are in phase, while in columns of two parallel deaerators they are in antiphase; there are no disturbances of flow parameters above points B and C.

The pressure values at B and C points remain constant as the instability in deaerators develops. The intercolumn and interdeaerator instabilities may coexist.

The foregoing is valid for small disturbances of flow parameters when the processes involved are described by linear equations.

When parameters in deaerators and supply lines sufficiently deviate from their steady-state values, the suggested classification of instabilities will not hold true. The disturbances of parameters will penetrate the system-wide lines.

The unit operation under the sliding pressure may cause initiation of undesirable deaerator fluctuations under certain conditions. Figure 8.19 illustrates fluctuations of the weight level and of the total resistance across the bubbler at the variation of the flow rate of steam coming into the deaerator after the load on the unit has been shed [159]. Periodic level surges may be explained by the drain device periodic failure.

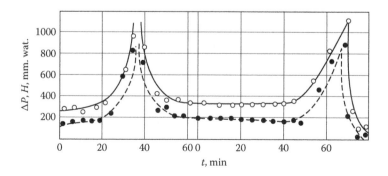

FIGURE 8.19

Time history of the weight level and total resistance in the bubbler. Dark circles: water weight level; clear circles: pressure drop.

With a smooth reduction of the load on the unit, the bleed pressure drops, as does the pressure in the deaerator tank. The water in the latter starts boiling and the $P_1 - P_2$ pressure drop increases (see Figure 8.16) because of the decreased steam density. The faster the power (deaerator pressure) decrease is, the faster is the pressure drop increase. As soon as $P_1 - P_2$ becomes larger than the fluid column weight in the drain device, the drain terminates and the fluid column height above the perforated plate increases. Then the fluid break through the perforated plate and a decrease of the $P_1 - P_2$ pressure drop occur. The drain device resumes its work and the column begins operating normally. Due to the further unit power reduction, deaerator pressure drop, and steam flow rate rise, the process repeats.

8.3 Vibration of Pipelines with Two-Phase Adiabatic Flows

8.3.1 Examples of Vibration of Industrial Pipelines with Two-Phase Flows

Vibration of two-phase incipient boiling flows at nuclear power plants (NPPs) and thermal power plants (TPPs) has turned into a serious problem that in some cases complicates operation of the equipment. The problems mainly applies to auxiliary pipes such as blowdown pipes (for continuous and periodical blows), starting pipes used for the condensed steam removal from steam pipelines during the power plant startup, and drain pipes, including bleed pipes and condensate transfer lines of steam separator-superheaters (SSHs) and high-pressure heaters (HPHs). In the following, some cases of vibration in such pipes are illustrated.

The starting lines used for collecting the steam–water mixture from the live steam lines to the condenser have a diameter of 200–500 mm and an operating pressure of 2.0–2.5 MPa. Significant vibration is a serious problem in pipes' operation that quite often leads to their failure. As an example, the experience of operating the starting pipes of the 800 MW unit of the Slavyanskaya Thermal Power Plant [160] may be offered. The pipeline layout is shown in Figure 8.20.

Vibrations in all sections of the pipeline were characterized by relatively low frequency (3–5 Hz) and large amplitudes (up to 250 mm). The authors of reference 160 share an opinion that the appearance of large amplitudes was induced by the insufficient fastening of pipes, which had been mounted on spring hangers with flights of up to 9–10 m and a limited number of stationary supports. For instance, there was not a single fixed support along a 60 m section from the built-in separators up to the external expander, while the 70 m long section from the external expander up to the condensate drain header was equipped with spring hangers only. The additionally installed braces (point A) did not improve the situation. The continuous vibration

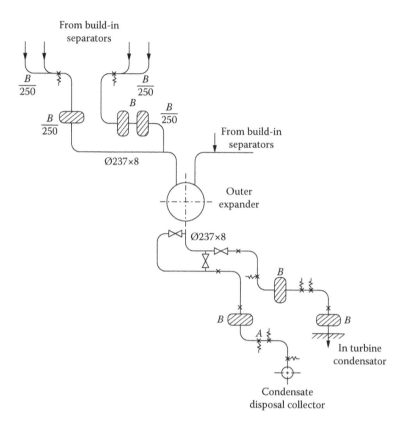

FIGURE 8.20
Basic diagram of the startup pipeline of the 800 MW unit at the Slavyanskaya thermal power plant. X: fixed support; ⚡: movable support (spring suspension); ▨: vibration damper.

led to the appearance of cracks in the section past the drain expander in a T-piece (point T) connecting the 426 and 273 mm diameter pipes. To eliminate dangerous vibrations of the drain pipes, an antivibration friction support designed for damping oscillatory energy in pipelines by friction forces in the support was suggested [161]. The test results showed that the installation of supports in B points substantially reduced the vibration level.

According to the ORGRES Co., the main causes of vibrations of the pipelines of such type have not been established yet, though the tube bend reactive forces originating at large velocities of the two-phase flows (45–130 m/s) are presumed to be the probable source. The failures of starting pipes were reported at other power stations, NPPs included.

In this case, according to the experience, vibration may cause pipe failure due to superposition of other factors: increased static pressure and erosion of pipe inner surfaces. So, at a 300 MW unit with starting of one of the boilers, the starting pipe designed for discharge of the steam–water mixture from the live steam pipe to turbine condenser broke [162]. As a result of inspection, the direct cause was found to be sharp rise of vibrations at fivefold operating pressure rise.

Vibrations associated with the steam–water mixture flow occur in steam lines and pipes of pressure-regulating and cooling installations and fast-operating pressure-regulating and cooling installations with the absence of adequate steam condensate drain [163,164]. Most frequently, water hammers are observed at the startup of the equipment, when large amounts of condensate accumulate in cold equipment and pipes.

It should be noted that condensate formation can also occur in a fully heated isolated steam pipeline. For instance, when using the saturated steam heated up to 180°C at the ambient temperature of 30°C, the amount of condensate that forms in the adequately isolated steam pipeline is 0.5–1 kg/(h·m²) [164]. This rate of condensate formation is sufficient for the liquid to accumulate periodically in the pipeline at the vertical upflow section lower point and thus create conditions for recurrent water hammers. The amount of condensate that forms in a nonisolated saturated steam pipeline (e.g., a steam extraction line or a process drain line) is shown in Figure 8.21 [165].

Condensate accumulation in a pipeline in the case of a sudden opening of the steam valve causes closing of the pipe section by condensate, which is equal to the instantaneous valve closing. Though the magnitude of water hammers occurring in steam is generally considered negligible due to steam compressibility, estimates indicate that at steam velocity of 80 m/s and pipe parameters of P = 14 MPa, T = 550°C, the force of the water hammer is 15% of the nominal pressure [163]. Therefore, in the case of inadequate drainage of the steam pipelines and discharge pipes, and with the presence of upflow sections in the pipeline, all conditions exist for the recurrent pipeline flooding accompanied by pressure fluctuations and acceleration of the liquid

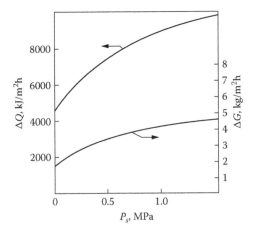

FIGURE 8.21
Saturated steam heat loss and condensate formation in the noninsulated steam line at 20°C ambient air.

plug up to the velocities close to those of steam. Quite often, such a "liquid hammer" is the very cause of pipe rupture, flange opening, and damage of pipeline fixtures [164].

Most frequently, vibrations of pipelines are observed in the systems of either gravity or cascade (in terms of pressure) condensate drainage. Flow fluctuations may appear in such pipelines with the slug motion of the steam–water mixture with the characteristic fluctuation of the flow parameters (pressure, flow rate) near their nominal values. Fluctuations during the gravity condensate drainage from one tank to another with periodic reversal of the circulation, which had been observed in General Electric designs, may be suggested as examples [166]. In this case, the boiling up of condensate in drain pipes (Figure 8.22) promoted formation of steam "cavities," which, in presence of the steam equalizing line, create conditions for circulation reversal in the drain pipe.

Recently, the problem of ensuring vibration-free operation of process condensate removal lines from SSHs and HPHs to the deaerator or drainage expander has become more urgent at NPPs. The pipes are 100–400 mm diameter and conduct the flashing condensate at the initial pressure of 2.0–6.5 MPa. According to experimental data [160,167], the pipelines are exposed to strong vibrations reaching at some plants the peak-to-peak values of 250–500 mm. A specific feature of the considered pipes is the presence of sections with pre-dominantly vertical-upward flow and the resulting gravity pressure drop, causing condensate flashing even in the case of negligible friction pressure losses and flow acceleration. Investigations carried out by a number of authors [167–169] have yielded vibration characteristics of such pipelines.

An investigation of pipe strength characteristics in the case of superposition of random flow-generated disturbances has been described [168]. Measurements made on the U-shaped section of the condensate pipeline similar to that in Figure 8.23(a) allowed construction of the relative dispersion spectra of vibra-tory displacements, from which the fluctuation fundamental harmonic with a frequency of about 5 Hz and the largest amplitude of vibration displacements has been determined. Other harmonics were observed at frequencies close to

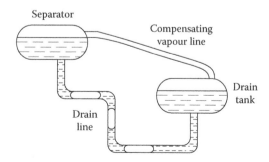

FIGURE 8.22
Cascade drain of flashing condensate.

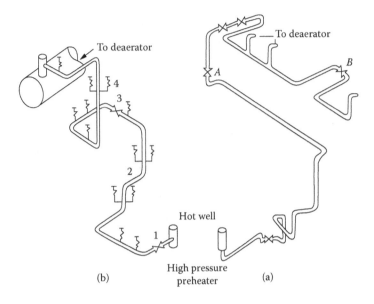

FIGURE 8.23
Condensate drainage pipeline diagrams. 1, 2, 3: Pressure gauge locations; 4: spring suspension (hanger); A and B: vibration sensor locations.

multiples of the fundamental harmonic (12.4, 16.4, 23.7 Hz). Investigations of dynamic strain in the pipe metal showed the maximum value of 12.7 MPa at 12.4 Hz. Thus, the disturbing effect of the flow is of a continuous nature and represents a range of random disturbances (i.e., is of the impact type).

The condensate flow inner characteristics have been investigated [169]. Steam qualities, flow rate, and pressure at fixed points have been measured for the pipe shown in Figure 8.23(b). For 257 mm diameter, 61 m long pipes operating under 0.88 MPa, the inlet pressure variable component was approximated by a 5.24 Hz harmonic. The amplitudes of forced fluctuations were measured by rapid filming and then these were compared with predictions made applying a technique suggested by the author. The experiment showed the quality to vary from zero at the inlet up to 0.67 at the outlet of the pipes, and pressure fluctuations at points 2 and 3 (in Figure 8.23a) were respectively 0.27 of the inlet pressure fluctuations. The model developed in Leschinsky [169], taking the boundary condition at the pipe inlet into account showed, in particular, that quality fluctuations decrease along the pipeline length due to damping of pressure fluctuations.

The essence of investigations on the drainage pipeline shown in Figure 8.23(a) was in recording pipeline fluctuation amplitudes at different operating modes using a low-frequency electromagnetic transducer [167]. It was revealed that flow-generated disturbances in the pipeline led to pipeline "shaking" with subsequent fluctuations damping at frequencies of the fundamental harmonics. An analysis of oscillograms of the pipeline characteristic fluctuations at nominal operation conditions showed that the disturbances occur at an infralow frequency of the order of 0.1 Hz and manifest themselves in the pulse action. The damping happened at the natural frequency of the pipeline section in question and was close to 4 Hz.

The investigation has shown that the level height in the hotwell was of practically no effect on the pipeline vibration characteristics. Opening of the level controller substantially increased the pipeline vibration before and past the controller. When the level controller was open from 20% to 80%, the peak-to-peak amplitude of fluctuations of the *HW–deaerator* condensate pipeline increased five times upstream of the controller and two times downstream from the controller. The peak-to-peak amplitude was found to be 30 and 18 mm, respectively. This fact allowed the authors to make a conclusion about the dynamic origin of the flow-generated disturbances influencing the pipeline, which increase in proportion to the square of the flow rate, which in turn depends on the rate of controller opening.

8.3.2 Vibration of Gas–Liquid Pipelines

The physical phenomena in adiabatic flows are to a certain extent simpler as compared to the diabatic two-phase flows. Nevertheless, the mechanisms of pulsating phenomena initiation in adiabatic two-phase flows are not completely clear. The greater part of work [160,167,172] has been dedicated to the reduction of vibrations in pipelines carrying two-phase flows causing such vibrations, and only a few researchers [167,170,171] have addressed the initial causes of vibrations. The knowledge of these causes will make it possible to develop effective measures against condensate pipeline vibrations.

8.3.2.1 Pulsations of the Adiabatic Two-Phase Flow Pressure in Pipelines

Gerliga, Korolev, and Gerliga [170] and Prudovsky and Radionov [171] have proved the dynamic nature of the two-phase flow disturbances influencing the pipeline. The oscillograms obtained for steam–water [170] and air–water [171] flows are presented in Figures 8.24 and 8.25. The first figure shows the dependencies of section variation of the liquid-occupied pipelines and pressure fluctuations upstream of the orifice plate. The pipeline diameter was 34 mm and that of the orifice plate was 12.5 mm. The condensate flow rate at the pipeline inlet was constant in time, and flashing was due to the pressure drop. The pipeline was located horizontally and the liquid plug boundaries were "fuzzy." At $\rho W = 200–900$ kg/(m²s), the water hammers reached 0.5 MPa and were observed when liquid plugs approached the orifice plate (Figure 8.24). In this case, the pulse load acting upon the plate was about 400 N. No water hammers were observed in the absence of slug regime. In Prudovsky and Radionov [171], the forces acting on the 90° pipeline bend in the case of slug regime were investigated. With 70 mm diameter pipelines, the force reached 500 N and acted along the bend radius (Figure 8.25). It may

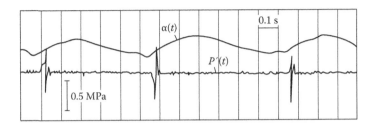

FIGURE 8.24
Oscillogram of the flashing condensate flow and pressure fluctuations on the orifice plate [170].

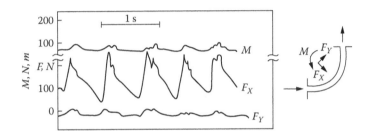

FIGURE 8.25
Force of the two-phase flow acting against the pipeline bend [12].

be expected that with the increasing pipeline diameter (D), these forces will be increasing proportionally to D^2.

The results of an experiment on determining the relative pressure fluctuations upstream a local resistance provided by the inlet pressure for the 40 mm diameter, 105 m long pipeline depending on the diameter of the outlet orifice plate are shown in Figure 8.26. The maximum of fluctuations occurs in the region with the slug-plug regime of the two-phase flow. With the decreasing orifice plate diameter, the blockage of the flow area by the plate increases, whereas the reduction of the liquid plugs' average velocity causes a decrease in the rate of fluctuations. Thus, for each pipeline there exists a definite hydraulic size of the local resistance at which the pressure fluctuations' amplitude reaches its maximum. It is particularly characteristic of control valves where the flow area varies with time, passing through the region of "critical" sizes. Valve operation under varying conditions leads to a situation when the slug-plug regime of the adiabatic condensate flow moves along the pipeline acting in a variable manner upon the pipeline sections of variable geometry (e.g., branches, T-pieces, diameter changes, bends).

The effect of relative water subcooling on the pressure fluctuations' amplitude upstream of the plate ($D_1 = 8$ mm) is illustrated in Figure 8.27. Relative subcooling (the ratio between the saturation pressure at the inlet water temperature and inlet pressure) is plotted on the X-axis and the average pressure fluctuation amplitudes upstream of the plate on the Y-axis.

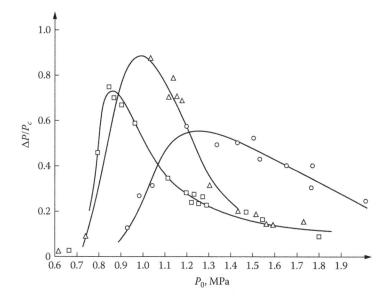

FIGURE 8.26
The effect of flow throttling at the pipeline outlet on the relative pressure fluctuations ampli-tude upstream the orifice plate ($T_s = 198°C$). Orifice plate diameter (Dt): O: 32 mm; ∆: 12.5 mm; □: 8 mm; $\Delta P/P_C$: ratio between the peak-to-peak pressure fluctuations and static pressure mea-sured before the orifice plate.

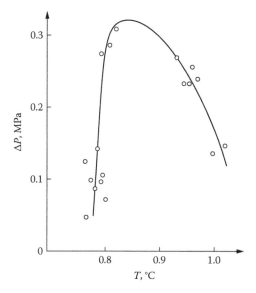

FIGURE 8.27
The effect of the degree of liquid subcooling on the local water hammer amplitude in the flashing flow. $D_1 = 8$ mm.

With $P_{so}/P_0 = 0.7$, which corresponds to $(T_s - T_{in}) < 19°C$, the slug-plug regime is absent and pressure fluctuations are insignificant. Evidently, reliable fluctuation-free operation of drain pipelines requires condensate depression to be determined in each particular case. The results of a detailed experimental investigation of the liquid plug passing through the local resistance (straight bend) are given in Korolev [167]. The oscillograms for both air–water and steam–water flows (Figure 8.28) showed the presence of four characteristic stages of the *gas slug–liquid plug* body passage through the local resistance.

The first phase consisted of the gas slug expansion at flowing through the local resistance and the associated pressure drop before the local resistance. The nature of the pressure drop is similar to that of the compressed gas outflow from the confined volume (i.e., obeys a law that is close to the adiabatic

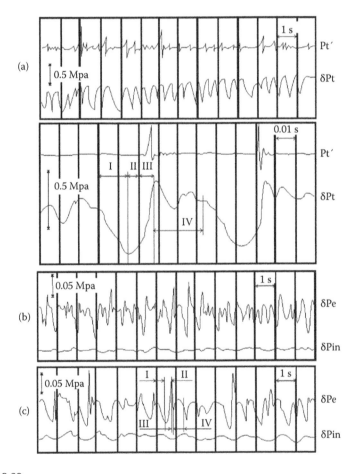

FIGURE 8.28
Typical oscillograms of pressure fluctuations in the two-phase flow pipeline. (a) Steam–water, $\beta = 0.33$, $w_m = 2.4$ m/s, $P_0 = 1.3$ MPa; (b) water–air, $\beta = 0.48$, $w_m = 3.0$ m/s, $P_0 = 0.2$ MPa; (c) water–air, $\beta = 0.74$; $w_m = 5.0$ m/s, $P_0 = 0.2$ MPa.

expansion law). In the considered case, as in that with the compressed gas outflow through the local necking, a reactive force applied to the local resistance develops and acts along the pipeline axis in the direction opposite to the flow. The rate of pressure drop was 0.4–0.5 MPa/s at the pipeline outlet static pressure of 0.15–0.2 MPa (at the air–water flow test conditions). The pressure drop is determined by the flow inertia properties and local resistance geometry and is commensurate with the difference in pressure drops across the local resistance during the liquid flow.

The second phase represented an insignificant pressure rise upstream of the local resistance due to closing of the flow area by the skew front of the liquid plug. With a steep front of the gas–liquid interphase, this phase is absent, which is the case with vertical pipes. For horizontal tubes with increasing gas content, duration of this phase increases.

The third phase is represented by the local water hammer that occurs in the liquid plug when it suddenly blocks the flow passage in the local resistance. The magnitude of water hammer depends on the form of the gas–liquid interphase and design features of the local resistance. Propagation of a pressure surge characterizing the water hammer hitting backward along the pipeline was not observed (i.e., the water hammer was localized within a single liquid plug).

The fourth phase was characterized by damping high-frequency pressure fluctuations upstream of the local resistance, followed by an insignificant pressure rise due to the pressure gradient restoration in the pipeline with the liquid plug flowing through the outlet. Duration of this phase was determined by the time it took the liquid to pass through the local resistance.

The preceding cycle was repeated with the next slug-plug flow. The pressure drop that occurred during the first phase propagated in the counterflow direction toward the pipeline inlet, causing pressure fluctuations there and, hence, changes in liquid and gas flow rates. The rate of the negative pressure disturbance propagation coincided with that of sound velocity in the two-phase mixture [167]. The previously mentioned feedback may facilitate the development of the slug flow to a certain extent.

The basic mechanism of steam slug formation in the adiabatic flashing flows is governed by the bubbles' growth and coalescence and may approximately be represented as follows. The bubbles moving within the liquid flow grow mostly at the expense of liquid evaporation. The difference in bubble and liquid velocities depending on the bubble volume (V_G) has its maximum. For small bubbles, the bubble velocity (W_G) is equal to that of the liquid (W_L). Further, the difference in ($W_G - W_L$) velocities increases along with the growing (V_G), reaches its maximum, and then starts decreasing due to the increasing resistance in the rear part of the bubble [110] and tube wall effect. The bubbles with a reduced velocity grow due to the mass exchange with liquid, to the absorption of the higher velocity bubbles catching up with the former ones, and also to steam expansion in the pressure field varying along the pipeline. The process continues until formation of the steam slug; synchronization of the process is ensured by pressure and flow rate fluctuations.

The existence of similar causes of vibrations' initiation in oil and gas pipelines was shown in references 173–175. The investigations have shown that pressure fluctuations on the pipeline walls are determined by flow gas content and Froude number comparing inertia and gravitational forces in the channel. The maximum values of such pressure fluctuations are in the gas content region $\beta = 0.6$–0.9, which corresponds to the region of slug regime and transition from the slug to either annular or dispersed regime [175]. The use of data on pressure fluctuations at the wall of a horizontal pipeline in a gas–liquid flow [188] and of a map of two-phase flow regimes in the horizontal pipeline [177] makes it possible to compose a map of dynamic loads exerted on the pipeline wall by the two-phase flow under different regimes (Figure 8.29b). One can see that maximum pressure fluctuations occur in the region of slug regimes and the region of transition from the slug to the annular regime, which is in good agreement with references 171 and 175–177.

Further, if we assume that the flashing condensate in the drainage pipeline is characterized by a quality rise as in Figure 8.29(a), which corresponds to Miropolsky, Shneerova, and Mekler [178] for incipient boiling flows, then, having assumed flow parameters to correspond to those of the HW-I drainage stage of the K-500-65/3000 turbine (G = 20 kg/s, P = 2.0 MPa), the A line will denote the *flashing length* [179] for the given pipeline (Figure 8.29b). Since this line crosses the lines of equal fluctuation amplitudes of pressure P, these points may be related to the pipeline length. In this case, the B line (Figure 8.29a) that indicates the pressure fluctuations at different coordinates of a horizontal pipeline forms. Thus, when draining the flashing condensate in long pipelines at inlet condensate velocities ≤ 6 m/s, there always occurs a transition through the slug regime region characterized by strong pressure fluctuations at the inner wall and fluctuations across local resistances exceeding the latter ones by two orders [170]. A similar phenomenon will also take place in vertical pipelines, where the region of slug regime existence is wider [167].

In the considered case, the main cause of flow parameters' fluctuation in the two-phase flow and pipeline vibration is the dynamic interaction of liquid plugs and steam slugs with the pipeline components (bends and local resistances). Naturally, the methods of flow stabilization are concentrated on the elimination of the slug-plug flow [167,180]. A device capable of transferring the slug regime into either the annular or annular-dispersed regime for reducing fluctuations' amplitude has been suggested [167]. The device represents a linear axial screw-type swirler designed for the shockless flow inlet (Figure 8.30). The experiment was conducted with a swirler (H/D = 10, the outer diameter equal to the tube inner diameter, the exit angle of 45°), installed 2 m downstream of the orifice plate. According to Hubbard and Ducler [181], the flow regime identification by a conductivity probe has shown that, downstream of the swirler, the flow regime is annular. The installation of a swirler has caused a two- to four-times reduction of the

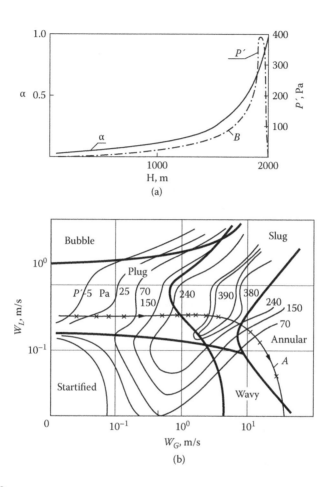

FIGURE 8.29
(a) Quality variation along the pipeline length. (b) Change of flow patterns and pressure fluctuations on the pipeline wall at quality variation along the pipeline length. W_L: liquid velocity related to the pipe open section; W_G: steam related to the pipe open section velocity.

average level of fluctuations, and the maximum fluctuation amplitudes were reduced eight to ten times. In this case, the additional pressure drop induced by the screw did not exceed 0.01 MPa. The disadvantage of the method is in the finite coverage range that equals $H/D = 50$–100 of the pipeline [167].

An interesting mechanism of flashing condensate initiation in drain pipelines and a device for fluctuations' elimination have been suggested [182]. It is supposed that with the plug regime downstream from the orifice plate because of the pressure drop below saturation pressure, a sharp coolant flashing occurs, and a pressure surge occurs that later is destroyed when passing by the slug through the plate. The process repeats itself with a certain periodicity, causing channel flow fluctuations, deterioration of the HW level controller stable operation, and vibration of pipelines. In this case, the main

FIGURE 8.30
External view of the screw swirler with shockless input [167].

contribution among the disturbances on the part of the flow is made by the reactive force varying in magnitude and direction and originating on the orifice plate in the case of critical blowdown [180]. It is stated by the authors that simultaneous initiation of the periodic cavitating collapse of steam bubbles results in considerable wear of pipelines in the surge area.

To eliminate the previously mentioned disadvantages and increase pipeline reliability, it was suggested to install cylindrical channels with sharp edges, relative length of $H/D = 10$–12, and such a diameter downstream from the HW level controller that will ensure the preset flow rate in the critical regime of the flow of flashing liquid. Unfortunately, the information in references 180 and 182 on the mechanism of two-phase flow fluctuations was not supported by experimental verification and comprises a number of incorrect statements—in particular on the irreversibility of a homogeneous and finely dispersed two-phase flow along the entire length of the pipeline downstream from the cylindrical channel. Also, a question remains open about the increasing potential of erosion failures due to high-velocity two-phase dispersed flow as a result of the support orifice plate elimination [162]. Because of this, the device proposed by Fisenko, Alferov, and Makukhin [180] was not recommended by VTI [183].

8.3.2.2 Transport Delay-Based Instability Mechanism

Under certain conditions in a pipeline with adiabatic flashing water, self-oscillations may be initiated by the mechanism based on transport delay [185].

Most simply, the mechanism may be illustrated by a dynamically isolated pipeline (the pressure upstream of (P_{in}) and downstream from (P_e) the pipeline and water enthalpy at the inlet are constant in time. The pressure drop is concentrated across the inlet and outlet local resistances.

The water at saturation temperature is fed to the inlet local resistance, flashes partly, and then bubbles grow due to mass transfer in the nonequilibrium flow (the pipeline pressure is below that at the inlet and the water in the flow is superheated). Let us first assume that the variation of pressure drop across the outlet local resistance is defined by steam flow variation.

When the tube pressure rises randomly (P(t)), the number of bubbles formed per unit of time at the inlet local resistance reduces. During the time θ required for the bubbles to reach the outlet, the pressure drop across the outlet local resistance reduces. At a constant P_e, it corresponds to the decreased P(t + θ). Increasing tube pressure at time moment (t + θ) will increase the flow of the bubbles formed at the inlet, which, in time θ, will lead to tube pressure rise since the quality by the moment (t + 2 θ) upstream of the outlet local resistance increases, and so on. Thus, the initial variation of tube flow parameters in 2 θ repeats (i.e., the considered tube adiabatic flashing flow is an oscillatory system with a period of 2 θ).

The increased ΔP_{in} stabilizes while the increased ΔP_e destabilizes the oscillatory system. Indeed, the larger the ΔP_{in} is, the less sensitive to the pipeline pressure variation is the variation of the number of formed bubbles. The larger ΔP_e is, the larger is the module disturbance due to pipeline pressure when passing of the same module steam flow disturbance through the local outlet resistance. In this case, module increase in tube pressure disturbance in time θ causes module increasing of the steam flow disturbance, and so on.

The preceding relationship of flow parameter disturbances may be approximately described by the following system:

$$\delta\alpha(t) = \delta n(t - \theta) \cdot V_{Ge}, \tag{8.17}$$

$$\delta n(t) = \frac{\partial n}{\partial P} \delta P(t), \tag{8.18}$$

$$\delta P(t) = \frac{\partial \Delta P_e}{\partial \alpha} \delta\alpha(t), \tag{8.19}$$

where the following assumptions were made: The quality disturbance upstream of the outlet local resistance ($\delta\alpha(t)$) is proportional to the disturbance of the number of bubbles formed at the tube inlet $\delta n(t - \theta)$; the breakup and coagulation of bubbles are insignificant.

The block diagram corresponding to the system of equations (8.17) and (8.19) is

$$\delta P \rightarrow \delta n \rightarrow \delta\alpha > \delta P.$$

The oscillatory system will be stable with a gain factor of the broken chain below unity

$$V_{Ge} \cdot \frac{\partial n}{\partial P} \cdot \frac{\partial \Delta P_e}{\partial \alpha} < 1. \tag{8.20}$$

The tube pressure variation causes liquid flow fluctuations at the inlet and of the two-phase mixture at the outlet. If such relations are taken into account, the condition of (8.20) will be changed. Considering that the closed loop of the internal relations with disturbance α is the major loop, while the loop with the flow disturbance is the auxiliary one, it may be supposed that the transfer function between the liquid inlet and two-phase mixture outlet disturbances is close to unity. This makes it possible to establish that the system under investigation will be stable if the following condition is satisfied:

$$\frac{\partial \Delta P_e}{\partial a} V_{Ge} \frac{\partial n}{\partial P} \left(1 + \frac{\Delta P_e}{\Delta P_{in}}\right)^{-1} < 1. \tag{8.21}$$

In Morozov and Grabovich [185], the relation (8.21) is reduced to

$$\frac{\Delta P_e}{\Delta P_{in}} < \frac{1 + (\rho_L/\rho_G) x^{3/2}}{(\rho_L/\rho_G) x^{3/2} - 1}, \tag{8.22}$$

which has been proved experimentally using the physical model (Figure 8.31). The main conclusion from (8.22) and Figure 8.31 is as follows: The increased ΔP_e and x destabilize while the increased ΔP_{in} stabilizes the process in the considered pipeline.

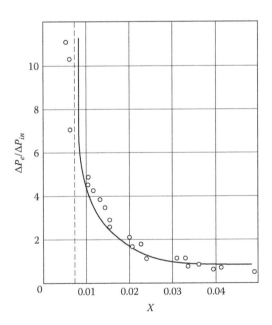

FIGURE 8.31
Stability region boundary. x: mass steam quality.

8.3.2.3 Hysteresis of the Hydraulic Discriminant in Pipelines

With definite geometries and flow parameters, the two-phase adiabatic liquid–gas flows have ambiguous characteristics. Figure 8.32 shows the hydraulic characteristic curve of a 3.1 m long glass tube with an inner diameter of 3.18 mm and horizontal liquid–gas flow in it [184]. The experimental results have been obtained for three liquid velocity values reduced to the entire tube section. The reduced gas velocity values are plotted along the X-axis. The entire $\Delta P(W_G)$ curve may be divided into three sections. At the first section (small W_G values), the tube pressure drop is substantially larger than that across the third section (large W_G values). The bubble flow was observed at the first section, while the ordered air flow in the form of large bubbles was registered at the third section. Obviously, the flow pattern at the third section is close to the slug one. The curve connecting the first and third sections has a negative slope.

When the tube is completely or partially dynamically isolated from other components of the test facility, the static instability in the small originates in it (if the initial regime point is located on the dropping section). Further, the limiting cycle of self-oscillations becomes established [184]. Similar oscillations may be expected in power plant pipelines of significantly larger size.

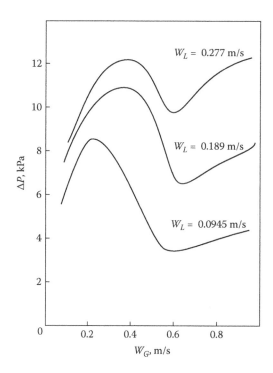

FIGURE 8.32
Hydraulic characteristic curves of the pressure loss at a change of flow pattern. W_L, W_G: liquid and gas velocities with reference to the total tube cross-section area, respectively.

8.3.2.4 Slug and Plug Regimes of the Two-Phase Flow in Pipelines

The two-phase slug and plug regimes are characterized by alternate passing of water and steam through the channel fixed section. Taking into account this phenomenon, Baldina et al. [116] established the relation between the intensity of variation of coolant by weight and development of fluctuations in tubes. Here, as a flow stability determining parameter, the quantity, called dynamic coefficient of tank, is assumed equal to the ratio between variations of coolant by weight and tube inlet flow per unit of time.

More clearly, the properties of slug regime can be seen when considering a support-free pipeline with the two-phase flow. Korolev [167] presents the results of qualitative investigations indicating that, with the gas slug length equal to or larger than the distance between the pipe line supports, the observed pipeline vibrations resemble the response to pulse disturbance. At a smaller size of slugs, as compared to supports, the resulting vibrations may be characterized as whipping. The vibrations of such a pipeline increase substantially in the coincidence of the liquid phase repetition rate and natural frequency of cross vibrations of the two-phase flow pipeline.

8.3.2.5 Oscillatory Processes in the Two-Phase Flow in the Externally Controlled Pipelines

The oscillatory process in a two-phase flow may be substantially complicated if the pipeline used for controlling the drain of the flashing flow is provided with a control member with play. The development of fluctuations in a system with a single-phase flow may be illustrated by using a tank with no level self-equalizing [186,187]. The real level controller has a dead zone caused by clearances between contacts, friction, plays, etc. The liquid level variation within the dead zone ($2\varepsilon H$) will not be accompanied by changing over the contacts of the actuator. That is why the control member starts moving to initiate closing—not at time τ_0, but rather in $\Delta\tau_0$, during which the level varies within the dead zone εH (Figure 8.33). The equality between the flow and inflow will be later than with an ideal controller (by $\Delta\tau$), with resultant level deviation from the average values being larger. The delay in reverse of the control member motor will become longer ($\Delta\tau_0 < \Delta\tau_2 < \Delta\tau_3 < ...$) causing level fluctuation to increase. The process becomes divergent and unstable, and the deviations of the tank level from the average value H_0 will inevitably reach impermissible values.

An analysis of oscillatory processes in the two-phase flow transport system equipped with a controller should take into account both the processes characterizing the two-phase flow intrinsic properties and characteristics of the controller included in the system.

Undoubtedly, other mechanisms of vibration generation that have not been covered by the present work exist in pipelines. However, in our opinion, the most important and frequently encountered cases have been presented.

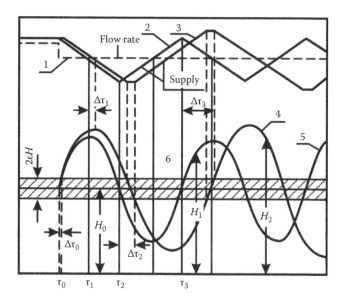

FIGURE 8.33
Variation of the tank liquid level with the presence of a controller insensitive zone. 1: Time variation of the tank flow rate; 2: time variation of the tank water intake with the ideal controller, 3, 4: time variation of the tank water intake and level with the presence of a controller insensitive zone; 5: time variation of the tank level (H_2) with the ideal controller; 6: the controller insensitive zone $2 \in H_0$; τ_0: the time moment at which a disturbance of the tank water flow rate occurs; H_0: the controller level set point; $\Delta\tau_0$: the time delay of the beginning of control member closing; τ_2: the time moment at which the ideal controller is switched to increase the tank water inflow; τ_3: the time moment at which the ideal controller is switched to decrease the tank water inflow; $\Delta\tau_2$, $\Delta\tau_3$: respective delays in operation of the real controller as compared to the ideal controller.

8.4 Two-Phase Flow Instabilities and Bubbling

During accident situations, low-velocity and low-pressure natural circulation and steam bubbling through vessels filled with free-surface water may occur in the main and auxiliary coolant circuits.

Such flow systems can be hydrodynamically unstable in a certain parametric range. "Geysering" is one of the main types of instability that is caused by volume boiling of coolant in adiabatic risers. This type of instability is characterized by a periodic ejection of the two-phase mixture from a riser into the upper collector with the free-surface water, followed by the riser steam space filling with water.

The results of geyser instability studies have been quite widely reported in published sources. In Jiang, Wu, and Zhang [215], experimental studies of a flow in a long adiabatic riser of a natural circulation loop are described. A bundle of electrically heated rods was installed at the bottom part of the riser. Geyser-like periodic flow oscillations were observed at low inlet subcooling. The oscillations were caused by rapid volume boiling of superheated water in the adiabatic riser top section.

A similar instability was observed in experiments [216] conducted at an experimental facility that modeled the Water-Water Power Reactor (VVER) external cooling system during in-vessel molten corium retention in conditions of a severe accident involving core melting. The experimental findings show that at low coolant subcooling at the Reactor Pressure Vessel (RPV) lower head inlet, coolant flow remains stable until some boundary value of power removed from the vessel is reached. As soon as this boundary value is exceeded, periodic parameter fluctuations occur. As is shown in Jiang et al. [215], this power boundary value should be such that the single-phase coolant temperature at the heated section outlet should exceed the saturation temperature at the pressure existing at the outlet from the adiabatic riser into the expansion vessel.

Insufficient experimental data are available on an extreme case of such circuit operation when the heated section of the loop is cooled without circulation flow. In this situation, steam generates upstream of the loop riser and bubbles through the water into the expansion vessel. A similar mode can be realized in an NPP safety system comprising a bubbler. Numerical analyses [217] demonstrate that, under some boundary conditions, geyser-generated steam bubbling may also occur and resemble flow pulsation regimes observed in natural circulation loops.

To explore the flow instability mechanism in the case of steam bubbling through a vertical water-filled pipe to a free-surface tank above, experimental results are presented in Verbitskiy, Efimov, and Migrov [218,219] and a generalized dependence for defining the flow stability boundary is proposed.

The experimental investigations of steam bubbling described in Verbitskiy et al. [218,219] have been carried out on a thermophysical facility, a diagram of which is presented in Figure 8.34. The experimental setup includes the following main units: electrical steam generator (1), bubble column (2), expansion vessel (3), and condenser (4). In the experiments, steam pressure upstream of control valves (5) was two times higher than that downstream of the valves. This condition ensured critical flow of steam through a valve (5) and maintained steam flow rate constant at the bubble column inlet during pressure fluctuations in the lower section of the column caused by pulsations during the bubbling process. Pressure in the expansion vessel was held constant by controlling steam flow rate at the expansion vessel outlet.

The bubble column (2) is made from pipes with the outer diameter of 89 mm and a 4.5 mm thick wall. Experiments were performed with two columns of different height. Bubble column 1 was made from two tandem pipes and had the total height of ~5.63 m from a perforated sheet (6) to an expansion vessel bottom (3). Bubble column 2 was made from one pipe with a total height of 3.18 m. The expansion vessel was 2.9 m high and 0.44 m³ in volume.

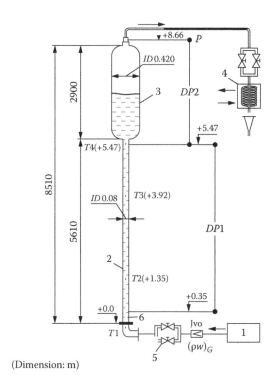

(Dimension: m)

FIGURE 8.34
Experimental setup.

The experimental setup was equipped with 40 sensors for automatic parameters recording at a sampling frequency of 5–50 Hz (see Figure 8.35). The following parameters were measured:

- Temperature (T1), pressure (P1), and steam volume flow (G_V) at the steam generator outlet
- Coolant temperature (T2, T3) and pressure drop (DP1) upstream of the steam-distribution perforated sheet
- Coolant temperature along the column height (T4–T7) and pressure drops at selected measurement points (DP2–DP4)
- Coolant temperature along the expansion vessel height (T8–T13), relative pressure in the vessel steam space (P2), and pressure drop along the entire vessel length (DP5) for measuring the mass level in the expansion vessel

The results of experimental investigations [218, 219] are briefly described next.

In this experimental series, two parameters were varied—that is, pressure (P2) in the expansion vessel and steam mass velocity $(\rho W)_G$ at the working section entrance. In each mode, values of these parameters were kept constant.

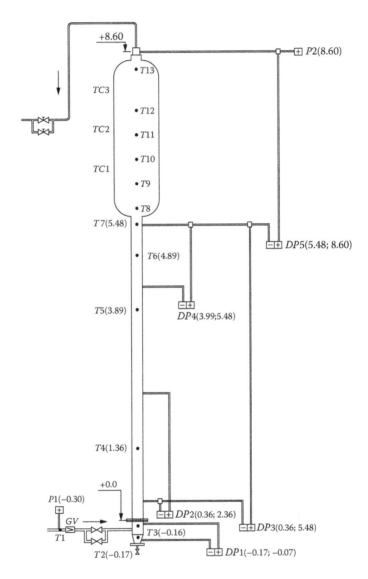

FIGURE 8.35
Measurement system.

Combinations of (P2, $(\rho W)_G$) values were chosen so that steam bubbling could be performed in stable and pulsating regimes for determining the experimental boundary curve of stability.

Several dozens of experimental regimes were tested on the experimental setup for each bubble column within the following ranges of parameter variation: P2 ∈ (100/450) kPa for pressure and $(\rho W)G$ ∈ (1.0/3.25) kg/m^{2}c for steam mass velocity.

Figures 8.36 and 8.37 offer examples of parameters' behavior in pulsating and stable bubbling regimes. Figure 8.36 demonstrates that the pulsating (or geysering) mode is characterized by periodic large-amplitude inharmonious low-frequency oscillations of coolant parameters.

The stable bubbling conditions (Figure 8.37) are illustrated by low-amplitude, high-frequency chaotic oscillations, which usually occur in discontinuous, two-phase flows.

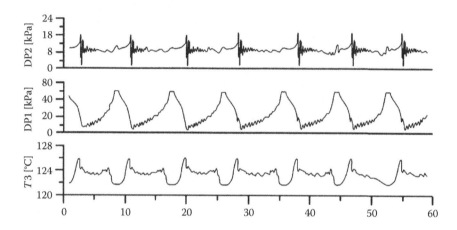

FIGURE 8.36
Parameters' behavior in the geysering mode. P = 190 kPa, steam mass velocity $(\rho W)_G = 2.1$ kg/(m2s), column N1.

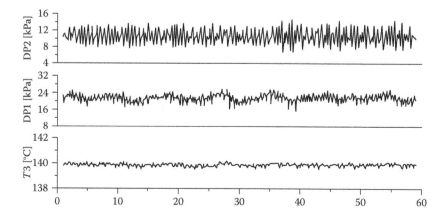

FIGURE 8.37
Parameters' behavior under stable bubbling conditions. P = 330 kPa, steam mass velocity $(\rho W)_G = 7.7$ kg/(m²s), column 1.

Findings [218,219] demonstrate that under the reported experimental conditions, the geysering instability was observed within the expansion vessel pressures ranging from 100 to 450 kPa for column 1, and up to 250 kPa for column 2. At higher pressures in the expansion vessel, no pulsating regimes were observed within the whole range of inlet steam mass velocities. The difference between expansion vessel pressure ranges, within which the pulsating regime of bubbling can occur in columns 1 and 2, is due to the difference in columns' height and, consequently, difference in hydrostatic pressure at the column inlet. The higher the column is, the more heat accumulates in the column bottom part and the more vigorous is the self-boiling process in the rising liquid.

Under any instability-inducing pressure from the range defined before, the geysering mode with the longest period of duration was observed at the bubble column inlet starting at the lowest steam flow rate realized in the experimental setup. With the increasing steam flow rate, the oscillation period kept decreasing until a certain limit was reached, at which the geyser-like regime was replaced with the stable one. An increased steam flow rate induced slug bubbling that caused transition to the stabilized regime and made the liquid fraction for increasing steam quality at the expense of self-evaporation insignificant.

According to Verbitskiy et al. [218,219], *the mechanism of oscillations during the geyser-like bubbling* can be presented as described next. Flow instability in the bubble column is characterized by the following repeated phases of the process (Figure 8.38):

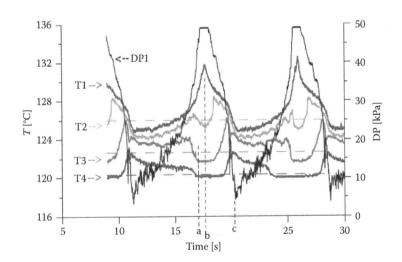

FIGURE 8.38
Bubbling phases in the geysering mode. P = 190 kPa, steam mass velocity $(\rho W)G$ = 2.1 kg/(m^{2}s), column 2.

- The incubation period of the single-phase coolant heating along the column height up to the temperature of saturation due to steam condensation

- Steam pressure increase in the column lower part because of steam accumulation and the beginning of the single-phase coolant move toward the column outlet

- Rapid self-boiling of the coolant upon entering a lower hydrostatic pressure region, followed by the steam–water mixture ejection into the expansion vessel

- Column flooding with water from the expansion vessel

- Coolant heating and cycle repetition

It has been reported [219] that the mechanism of parameters' oscillation during the geyser-like bubbling is significantly different from that of other known instabilities in natural circulation loops. The main difference is that volume boiling begins in the bubble column lower part. The liquid superheated relative to the saturation temperature corresponding to the pressure in the expansion vessel is forced out into the bubble column top, and it leads to rapid self-boiling and ejection of the two-phase mixture into the expansion vessel.

This process is illustrated in Figure 8.38 [218] by the curves reflecting the coolant temperature change along the working section height (T3, T4, T5, T7) and the pressure drop curve across the column (DP1). The horizontal dashed lined represents water saturation temperature for T4, T5, and T7 measurement points. The saturation temperature is calculated from pressures at the temperature measurement points (i.e., taking into account the hydrostatic pressure head in the column fully filled with saturated liquid).

At the initial stage, after the last entry of water from the expansion vessel, the bubble column is filled with the single-phase coolant (see DP1 in Figure 8.38) with a temperature below the local saturation temperature (see curves T4, T5, and T7 in Figure 8.38). The DP1 sensor indicates the presence of steam upstream of the steam distribution sheet. The volume flow of steam at the column inlet remains constant due to high resistance of the steam distribution sheet. Therefore, at this stage of the process, absolute pressure builds up in the space under the sheet, while the pressure drop does not change. The steam partially passes through holes in the sheet and condenses at the bottom of the column. As soon as the water above the steam distribution sheet heats up close to the saturation temperature, condensation terminates and water is forced from the column lower part upward, where it becomes superheated and boils up. The subsequent spikes on temperature curves T4, T5, and T7 (Figure 8.38) show that the hot steam region front moves up to the bubble column top with considerable acceleration. The pressure drop sensor DP3 indicates the column steaming, and the cycle of column flooding with water from the expansion vessel repeats.

According to Verbitskiy et al. [218,219], the decisive role in the appearance of geysering instability in bubbling systems is played by the thermodynamic nonequilibrium of a steam–water mixture caused by pressure changes. To verify this suggestion, special-purpose experiments [218,219] were conducted on the previously described setup with air supplied to the bubble column. The air flow rate and pressure were within the same ranges as those used in steam bubbling experiments. Only stable two-phase flows were observed under the mentioned conditions.

In Verbitskiy et al. [218,219], the stability boundary during bubbling was determined by plotting experiment data on the $(\rho W)_G$, P coordinate plane (Figure 8.39) where $(\rho W)_G$ is the steam mass velocity at the test section inlet and P is the absolute pressure at the exit from the bubble column into the expansion vessel. Each experimental regime can be represented by one point. The results of experimental data processing in the $(\rho W)_G$, P coordinates are shown in Figure 8.39.

Figure 8.39 shows two regions of stable and unstable steam bubbling distinctly separated by a sufficiently clear boundary line. In a shorter column, the instability region is smaller in size and is shifted toward the region of lower steam pressures and flow rates at the column inlet.

A numerical correlation for the stability boundary is obtained, assuming that the critical value of steam mass velocity at the bubbling system inlet, which corresponds to the stability boundary value, is proportional to steam generation rate during volume self-boiling of superheated liquid in the bubble column as the hydrostatic pressure decreases.

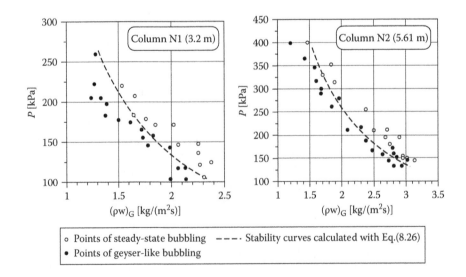

FIGURE 8.39
Steam bubbling stability curves. P = pressure in the expansion vessel steam volume (see Figure 8.34).

Assuming the steam displaces the saturated liquid column up for a dz distance, the amount of the liquid column superheating above the saturation temperature can be evaluated [218] as follows:

$$dQ = A \cdot H\rho_L' di_L',$$ (8.23)

where A is the liquid column cross section, H is the liquid column height from the steam distribution sheet to the expansion vessel bottom, and i_{LG}' is the liquid enthalpy change when the hydrostatic pressure changes by the $\rho_L g dz$ value.

This amount of heat is consumed for the evaporation of the dm_G mass of water:

$$dm_G = \frac{dQ}{i_{LG}} = \frac{AH\rho_L' di_L'}{i_{LG}}.$$ (8.24)

A change of liquid enthalpy di_L' with a decrease in hydrostatic pressure can be defined as

$$di_L' = \frac{\partial i_L'}{\partial P}\frac{\partial P}{\partial z}dz = \frac{\partial i_L'}{\partial P}\rho_L' g dz.$$ (8.25)

Equations (8.24) and (8.25) yield a correlation (within the accuracy of a dimensional constant) for calculating the critical mass velocity of steam, $(\rho W)^{bou}_G$, at the stability boundary:

$$(\rho w)^{bou}_G = \frac{1}{A}\frac{dm}{d\tau} = CH(\rho_L')^2 \frac{g}{i_{LG}}\frac{\partial i_L'}{\partial P},$$ (8.26)

where C = 0.18 m/s is a dimensional constant determined from generalization of experimental data presented in Figure 8.39.

In (8.26), the thermodynamic properties of water along the saturation line are taken at the bubble column outlet pressure.

Relationship (8.26) can be used to generalize the data from the experiments aimed at determining the two-phase flow stability boundary during bubbling in bubble columns of different height. Figure 8.39 illustrates the use of (8.26) for calculating the stability boundary that separates the regions of stable and unstable regimes with sufficient accuracy.

References

1. Kramerov A. Ya., Shevelev Ya. V. 1984. *Engineering calculations of nuclear reactors.* Moscow: Energoatomizdat, 736 pp.
2. Morozov, I. I., Gerliga, V. A. 1969. *Boiling apparatuses' stability.* Moscow: Atomizdat, 280 pp.
3. Goryachenko, V. D. 1977. *Methods of investigation of nuclear reactor stability.* Moscow: Atomizdat, 296 pp.
4. Artamonov, K. I. 1982. *Thermohydroacoustic instability.* Moscow: Mashinostroenie, 261 pp.
5. Mitenkov, F. M., Motorov, B. I., Motorova, E. A. 1976. Stability of natural heat and mass transfer. Moscow: Atomizdat, 96 pp.
6. Mitenkov, F. M., Motorov, B. I. 1981. *Mechanisms of unstable processes in thermal and nuclear power engineering.* Moscow: Energoatomizdat, 88 pp.
7. Mitenkov, F. M., Kutyin, L. N., Motorov, B. I., Samoilov, O. B. 1983. *Stability of apparatuses with low exit steam quality.* Moscow: Energoatomizdat, 96 pp.
8. Boure, J. A. 1978. Oscillatory two-phase flows. In *Two-phase flows and heat transfer with applications to nuclear reactor design problems.* New York: Hemisphere Publishing. Co., pp. 211–239.
9. Bailey, N. 1977. Introduction to hydrodynamic instability. In *Two-phase flow and heat transfer.* Oxford, England: Oxford University Press.
10. Khabensky, V. B., Baldina, O. M. 1969. Investigation of equations of steam-generating channel dynamics. *CKTI Proceedings* 98:44–59.
11. Khabensky, V. B. 1969. Investigation of flow stability in heated parallel tube system. Master degree dissertation, CKTI, Leningrad Division, 112 pp.
12. Lokshin, V. A., Peterson, D. F., Shvarts, A. L., eds. 1978. *Boiler unit hydraulic design, standard method.* Moscow: Energia, 256 pp.
13. Saha, P., Ishii, M., Zuber, N. 1976. An experimental investigation of the thermally induced two-phase flow oscillations in two-phase systems. *Journal of Heat Transfer, Transactions of ASME* 98 (4):616–622.
14. Fucuda, K., Kabori, T. 1979. Classification of two-phase flow instability by density wave oscillation model. *Journal of Nuclear Science Technology* 16 (2):95–108.
15. Yadigaroglu, G. 1981. Two-phase flow instabilities and propagation phenomena. In *Thermohydraulics of two-phase systems for industrial design and nuclear engineering,* ed. J. M. Delhaye, M. Giot, M. L. Riethmuller. New York: Hemisphere Publishing Co., McGraw–Hill, pp. 353–404.
16. Fucuda, K., Kabori T. 1978. Two-phase flow instability in parallel channels. *6th International Heat Transfer Conference,* Toronto, 1:369–374.
17. Hayama, S. 1967. A study on the hydrodynamic instability in boiling channels. *Bulletin of Japan Society of Mechanical Engineers* 10 (38):320–327.
18. Fukuda, K. M., Hasegawa, S. 1979. Analysis of two-phase flow instability in parallel multichannels. *Nuclear Science Technology* 16 (3):190–197.
19. Zavalsky, V. P., Kobzar, L. L., Leppik, P. A., Maimistov, V. V. 1983. Investigation of coolant circulation stability in the nuclear reactor AST-500 model. *Atomnaya Energia* 55 (4):202–208.

20. Urusov, G. L., Treshchev, G. G., Sukhov, V. A. 1985. Investigation of boiling water flow thermal-hydraulic stability boundaries in natural-circulation loops. *Teploenergetika* 4:66–68.

21. Kotani, K., Nakao, T., Sumida, I., Yokomizo, O., Matsumoto, T. 1980. Experimental study of effect of geometry on density-head-driven instability in boiling channels. *Journal of Nuclear Science Technology* 17 (10):791–793.

22. Takitani, K., Fakemura, T. 1978. Density wave instability in once-through boiling flow system (I). *Journal of Nuclear Science Technology* 15 (5):355–364.

23. Takitani, K., Fakemura, T. 1978. Density wave instability in once-through boiling flow system (II). *Journal of Nuclear Science Technology* 15 (6):389–399.

24. Takitani, K., Fakemura, T. 1979. Density wave instability in once-through boiling flow system (III). *Journal of Nuclear Science Technology* 16 (1):16–29.

25. Khabensky, V. B., Baldina, O. M. 1969. Analysis of flow rate fluctuations in the parallel steam-generating tube system. *Inzhenerno-fizicheskiy Zhurnal* 27 (5):819–828.

26. Khabensky, V. B., Baldina, O. M., Kalinin, R. I. 1973. Fluctuations mechanism and effect of design and operating parameters on the flow stability boundary. In *Achievements in the research on heat transfer and hydrodynamics of two-phase flows in power equipment components.* Leningrad: Nauka, pp. 48–66.

27. Khabensky, V. B., Peterson, D. F. 1970. Determination of flow stability in a system of parallel horizontal steam generating tubes. *Teploenergetika* 7:78–80.

28. Khabensky, V. B., Baldina, O. M., Kalinin, R. I. 1971. Flow stability in vertical parallel heated tubes. *Energomashinostroenie* 3:17–19.

29. Akyuzlu, K., Veziroglu, T., Kakac, S., Dogan, T. 1980. Finite difference analysis of two-phase flow pressure-drop and density wave oscillations. *Waerme- und Stoffuebertragung* 14:253–267.

30. Dogan, T., Kakac, S., Veziroglu, T. 1982. Lumped parameter analysis of two-phase flow instabilities. *International Heat Transfer Conference*, Munich, Germany, 5:213–218.

31. Gerliga, V. A., Mokhrachev, L. P., Shelkhovskoy, R. D. 1976. Experimental investigation of thermal-hydraulic stability of steam generator loop. In *Voprosy atomnoy nauki i tekhniki, Series: Fizika i tekhnika yadernykh reaktorov. Dynamika yadernykh energeticheskih ustanovok,* Moscow, 1 (9):40–46.

32. Koshelev, I. I., Surnov, A. V., Nikitina, L. V. 1969. Investigation of the fluctuations onset using the vertical screen model. *Energomashinostroenie* 10:13–16.

33. Gerliga, V. A., Dulevsky, R. A., Mokhrachev, I. P. 1976. Investigation of stability of multichannel steam-generating systems with identical or almost identical channels. *Voprosy atomnoy nauki i tekhniki, Series: Fizika i tekhnika yadernykh reaktorov. Dynamika yadernykh energeticheskih ustanovok,* Moscow, 2 (10):29–38.

34. Namestnikov, G. I. 1985. On the possibility of nonlinear interaction between intercoil and loop oscillations in the parallel channel system. *Voprosy atomnoy nauki i tekhniki, Series: Fizika i tekhnika yadernykh reaktorov. Dynamika yadernykh energeticheskih ustanovok,* Moscow, 1:19–25.

35. Proshutinsky, A. P., Lobachev, A. G., Trofimova, N. N. 1983. On the thermal-hydraulic stability of a channel with the nonheated bypass. *Teploenergetika* 6:64–66.

36. Gerliga, V. A., Dulevsky, R. A. 1966. On the thermal-hydraulic stability of multichannel steam-generating systems. *Izv. Akad. Nauk SSSR, Energetika i transport,* 5:81–89.

37. Proshutinsky, A. P., Lobachev, A. G. 1981. Investigation of the effect of the number of parallel channels on thermal-hydraulic interchannel stability. *Teploenergetika* 11:58–61.

38. Aritomi, M., Aoki, S., Narabayashi, T. 1981. Instabilities in parallel channels of forced-convection boiling upflow system (IV). Instabilities in multi-channel system and with boiling in downcomer. *Journal of Nuclear Science Technology* 18 (5):329–340.

39. Plyutinsky, V. I., Fishgoit, L. L. 1968. On the derivation of equations of steam quality dynamics in steam generating channels with sub-cooled water boiling. *Atomnaya Energia* 25 (6):474–479.

40. Molochnikov, Yu. S., Batashova, G. N. 1973. Steam void fraction with tube subcooled water boiling. In *Achievements in heat transfer and hydraulics of two-phase flows in power equipment components.* Leningrad: Nauka, pp.79–96.

41. Filin, R. D. 1988. Comparative study of mathematical models of non-equilibrium boiling for prediction of unsteady processes. *Voprosy atomnoy nauki i tekhniki, Series: Fizika i tekhnika yadernykh reaktorov. Dynamika yadernykh energeticheskih ustanovok,* Moscow, 1:32–36.

42. Baldina, O. M., Zinkevich, V. G., Kalinin, R. L, Khabensky, V. G. 1970. Flow oscillations in steam boiler tubes. *CKTI Proceedings* 101:165–175.

43. Unal, H. 1981. Density-wave oscillations in sodium-heated once-through steam generator tubes. *Journal of Heat Transfer, Transactions of ASME* 103:485–491.

44. Unal, H. 1985. Two simple correlations flow the immersion density wave oscillations in long sodium-heated steam generator tubes. *International Journal of Heat Mass Transfer* 28 (7):1385–1392.

45. Goldman, K. 1954. Heat transfer to supercritical water and other fluids with temperature dependent properties. *Chemical Engineering Progress Symposium Series* 50 (11):106–110.

46. Krasyakova, L. Yu, Glusker, B. N. 1965. Stability investigation in parallel coils with upflow-downflow fluid movement at subcritical and supercritical pressures. *CKTI Proceedings,* Leningrad, 59:198–217.

47. Krasyakova, L.Yu., Glusker, B. N. 1963. On the flow stability in the U-shaped channels of once-through boilers. *Teploenergetika* 11:41–46.

48. Khabensky, V. B. 1971. Computer study of flow fluctuations in the parallel heated tube system at supercritical pressure. *Application of Mathematical Methods and Computation Means in Heavy Machine Building Industry,* NIIINFORMTYAZHMASH, Moscow, 15-71-3:17–28.

49. Dany, D. E., Ladtke, P. R., Jones, M. K. 1979. Experimental investigation of thermally induced oscillations in helium flow at supercritical pressure. *Transactions of ASME, Heat Transfer* 101:2–4.

50. Labuntsov, D. A., Mirzoyan, P. A. 1983. Analysis of stability boundary of helium flow in heated channels at supercritical parameters. *Teploenergetika* 3:2–4.

51. Labuntsov, D. A., Mirzoyan, P. A. 1985. Investigation of stability of helium flow in heated channels at supercritical parameters with due account for pressure losses along the channel length. Teploenergetika 11:57–59.

52. Labuntsov, D. A., Mirzoyan, P. A. 1986. Stability of helium flow at supercritical pressure with nonuniform heat load distribution along the channel length. *Teploenergetika* 4:53–56.

53. Treshchev, G. G., Sukhov, V. A., Shevchenko, G. A. 1974. Flow self-oscillations in heated channel at supercritical parameters. *Proceedings of the IV All-Union Conference on Two-Phase Heat Transfer and Hydraulics,* Leningrad. Heat and Mass Transfer in Phase Changes. Minsk, ITMO, Part 2, pp. 166–175.

54. Dykhuizen, R., Roy, R., Calra, S. 1986. A linear time-domain two-fluid model analysis of dynamic instability in boiling flow systems. *Journal of Heat Transfer, Transactions of ASME* 108 (1).

55. Golovan, O. V., Nikonov, A. A. 1965. Experimental investigation of pulsating flows initiation in coil steam generators. *CKTI Proceedings* 59:155–161.

56. Malkin, I. G. 1966. *Theory of stability of motion.* Moscow: Nauka, 532 pp.

57. Neimark, Yu. A. 1949. *Stability of linearized systems.* Leningrad: LKVVIA, 327 pp.

58. Vdovin, S. I. 1986.On the possibility of practical instability of BWR. *Voprosi atomnoi nauki i tekhniki. Series: Fizika i tekhnika yadernykh reaktorov. Dynamika yadernykh energeticheskih ustanovok,* Moscow, 1:59–63.

59. Hamming, R. W. 1962. *Numerical methods for scientists and engineers.* New York: McGraw–Hill Book Co.

60. Roache, P. 1976. *Computational fluid dynamics.* Albuquerque, NM: Hermosa Publishers.

61. Malkin, S. D., Filin, R. D. 1984. Particularities of mathematical models of space effects in nuclear reactors. *Voprosy atomnoi nauki i tekhniki. Series: Fizika i tekhnika yadernykh reaktorov. Dynamika yadernykh energeticheskih ustanovok,* Moscow, 2(39):29–33.

62. Patankar, S. 1980. *Numerical heat transfer and fluid flow.* New York: Hemisphere Publishing Co.

63. Kutateladze, S. S., Styrikovich, M. A. 1976. *Hydrodynamics of steam–liquid systems.* Moscow: Energia, 296 pp.

64. Shannon, R. E. 1975. *Systems simulation: The art and science.* Englewood Cliffs, NJ: Prentice Hall Inc.

65. Yarkin, A. N., Kornienko, Yu. N., Kulikov, B. I., Shvidchenko, G. I. 1983. Determination of boundaries and period of self-oscillations in the system of parallel steam generating channels. Preprint FEI/1394, Obninsk, 15 pp.

66. Shvidchenko, G. I., Kulikov, B. I., Sudnitsin, O. A. et al. 1983. Determination of boundaries of interchannel fluctuation regions in a system of parallel steam generating channels. Preprint FEI/1494, Obninsk, 22 pp.

67. Komyshny, V. N. Kornienko, Yu. N., Kulikov, B. I. et al. 1983. Particularities of boundary behavior in the interchannel fluctuation regions. *Atomnaya Energia* 54 (3):173–175.

68. Yarkin, A. N., Kulikov, B. I., Sudnitsin, O. A., Shvidchenko, G. I. 1984. On the nature of steady-state conditions in parallel steam-generating channel systems. Preprint FEI/1591, Obninsk, 32 pp.

69. Yarkin, A. N., Kulikov, B. I. Shvidchenko, G. I. 1985. A conservative model of interchannel fluctuations. Preprint FEI/1716, Obninsk, 30 pp.

70. Yarkin, A. N., Kulikov, B. I., Shvidchenko, G. I. 1986. Boundaries of instability regions and fluctuation period in a system of parallel steam generating channels. *Atomnaya Energia* 60 (1):19–23.

71. Petrov, P. A. 1960. *Once-through boiler hydrodynamics.* Moscow, Leningrad: GEI, 168 pp.

72. Wallis, G. B. 1982. Review—Theoretical models of gas-liquids flows. *Journal of Fluids Engineering* 104:279–283.

73. Nakoryakov, V. E. 1981. Two-phase flow hydrodynamics. In *Two-phase hydrodynamics and heat transfer, Proceedings of the 2nd All-Union School on thermalphysics.* Novosibirsk: ITF AN SSSR, Siberian Branch, pp. 5-30.

74. Mishima, K., Ishii, M. 1984. Flow regime transition criteria for upward two-phase flow in vertical tubes. *International Journal of Heat Mass Transfer* 27 (5):723–737.

75. Petukhov, B. S., Genin, L. G., Kovalev, S. A. 1986. *Heat transfer in nuclear power plants.* Moscow: Energoatomizdat, 472 pp.

76. Delhaye, J. M. 1981. Two-phase flows regime. In *Thermohydraulics of two-phase systems for industrial design and nuclear engineering,* ed. J. M. Delhaye, M. Giot, M. L. Rithmuller. New York: Hemisphere Publishing Co., McGraw–Hill Book Company.

77. Delhaye, J. M. 1981. Local instantaneous equations of conservation. In *Thermohydraulics of two-phase systems for industrial design and nuclear engineering,* ed. J. M. Delhaye, M. Giot, M. L. Rithmuller. New York: Hemisphere Publishing Co., McGraw–Hill Book Company.

78. Delhaye, J. M. 1981. Combined-average equations. In *Thermohydraulics of two-phase systems for industrial design and nuclear engineering,* ed. J. M. Delhaye, M. Giot, M. L. Rithmuller. New York: Hemisphere Publishing Co., McGraw–Hill Book Company.

79. Delhaye, J. M. 1981. Two-phase flow models. In *Thermohydraulics of two-phase systems for industrial design and nuclear engineering,* ed. J. M. Delhaye, M. Giot, M. L. Rithmuller. New York: Hemisphere Publishing Co., McGraw–Hill Book Company.

80. Blekhman, I. I., Myshkis, A. D., Panovko, Ya. G. 1983. *Mechanics and applied mathematics. Logics and specifics of mathematics application.* Moscow: Nauka, 328 pp.

81. Khasina, E. N. 1980. On the connection of asymptotic stability of differential and discrete systems. *Automation and Telemechanics* 12:165–167.

82. Syu, Yu. 1981. Programs and predictions of two-phase flow heat transfer. In *Thermohydraulics of two-phase systems for industrial design and nuclear engineering,* ed. J. M. Delhaye, M. Giot, M. L. Rithmuller. New York: Hemisphere Publishing Co., McGraw–Hill Book Company.

83. Kuznetsov, Yu. N. 1989. *Nuclear safety heat transfer.* Moscow: Energoatomizdat, 296 pp.

84. Gerliga, V. A., Dulevsky, R. A., Shelkhovskoy, R. D., Khudyakov, V. F. 1972. Experimental investigation of coil-type channels. *Proceedings of Chelyabinsk Polytechnical Institute* 115:112–116.

85. Ivyansky, S. I. 1965. Conditions of flow stability in tubular elements at supercritical pressure with multivalued hydraulic characteristics. *CKTI Proceedings,* Leningrad, 59:218–222.

86. Belyakov, I. L., Kvetnyi, M. A., Loginov, D. A. 1985. Analysis of thermal hydraulic stability in steam generating elements with convective heating. *CKTI Proceedings,* Leningrad, 217:10–19.

87. Belyakov, I. I., Breus, V. I., Loginov, D. A. 1988. Investigation of thermal hydraulic stability in steam generating elements with convective heating. *Atomnaya Energia* 65 (1):12–17.

88. Waszink, R. P., Efferding, L. E. 1974. Hydrodynamic stability and thermal performance test of a 1-MW sodium-heated once-through steam generator model. *Journal of Engineering for Power* 96 (3).

89. Belyakov, I. I., Kvetny, M. A., Loginov, D. A., Mochan, S. I. 1984. On static instability of straight flow steam generators with convective heating. *Atomnaya Energia* 56 (5):317–319.

90. Belyakov, I. I., Kvetny, M. A., Loginov, D. A. 1985. Effect of ballast zone on hydraulic stability of the straight flow steam generator. *Atomnaya Energia* 58 (3):155–159.

91. Proshutinsky, A. P., Lobachev, A. G. 1980. Static instability of coolant downflow in parallel-channel system. *Teploenergetika* 10:18–21.

92. Proshutinsky, A. P., Timofeeva, N. N. 1982. Calculation of circulation reversal in coolant downflow channel. *Teploenergetika* 12:18–21.

93. Malkin, S. D., Khabensky, V. B., Migrov, Yu. A., Efimov, V. K. 1984. Single-phase static instability in the parallel-channel system at low rates of loop circulation. *Teploenergetika* 11:20–24.

94. Peterson, D. F., Krasyakova, L. Yu., Glusker, B. N. 1968. Some aspects of reliability of differently arranged heating surfaces in once-through boilers under supercritical pressure. *CKTI Proceedings,* Leningrad, 90:70–93.

95. Semenovker, I. E., Golberg, Yu. A. 1968. Hydraulic maldistribution in multipass waterwall circles of supercritical pressure. *CKTI Proceedings,* Leningrad, 90:94–104.

96. Bussard, R. W., de Lauer, R. D. 1958. *Nuclear rocket propulsion.* New York: McGraw–Hill, p. 370.

97. Reshotko, E. 1967. An analysis of the "laminar instability" problem in gas cooled nuclear reactor passages. *AIAA Journal* 5 (9):1606–1615.

98. Antonyuk, N. I., Korolev, A. V. 1980. On thermal hydraulic instability of heated gas flows. In Dynamics of machines and working processes, *Chelyabinsk,* 237:34–40.

99. Bendat, J. S., Piersol, A. G. 1971. *Random data: Analysis and measurement procedures.* New York: Wiley-Interscience.

100. Gerliga, V. A., Pogosov, A. Yu., Rogovsky, V. T. 1987.Author Certificate No. 1307375, USSR, MKI 01 23/18. An instrument for measuring matrix elements of two signals power spectral density. Issued 30.04.87, Bull. 16.

101. Gerliga, V. A., Pogosov, A. Yu., Khabensky, V. B. 1989. Author Certificate No.1513305, USSR, MKI G228 35/18. A diagnostic system for diagnosing the thermal hydraulic stability margin of a steam generating channel. Issued 07.10.89. Bull. 37.

102. Kafengaus, N. L. 1974. Overview of experimental investigation of thermal acoustic oscillations with heat transfer to turbulent channel liquid flow. *Proceedings of G. Krzhizhanovsky Power Engineering Institute,* Moscow, 19:106–130.

103. Dorofeev B. M. 1985. Sound phenomena in boiling. Rostov University, Rostov, 88 pp.

104. Gerliga, V. A., Prokhorov, Yu. F. 1972. Some specifics of heat transfer with water during surface boiling. *Heat and Mass Transfer,* Minsk, 2 (Part 1):159–163.

105. Hines, W. S., Wolf, H. 1962. Pressure fluctuations with heat transfer to liquid hydrocarbon at supercritical pressures and temperatures. ARS (American Rocket Society) 32 (3).

106. Gerliga, V. A., Prokhorov, Yu. F. 1974. Mechanism of initiation of thermal acoustic fluctuations at subcritical pressures. *Izv. Akad. Nauk SSSR, Energetika i Transport* 6:125–134.

107. Assman, V. A., Dorofeev, B. M. 1979. Experimental investigations of thermal acoustic self-oscillations in boiling. *Investigations of Boiling Physics,* Stavropol, 5:52-61.
108. Raushenbah, O. V. 1961. *Vibration burning.* Moscow: GIFL, 500 pp.
109. Hayama, S. 1970. The modes of oscillation in multi-channel systems. *Bulletin of Japan Society of Mechanical Engineers* 13 (63):132–141.
110. Levich, V. G. 1959. *Physicochemical hydrodynamics.* Moscow: Fizmatgiz, 780 pp.
111. Nigmatulin, R. I. 1978. *Fundamentals of mechanics of heterogeneous media.* Moscow: Nauka, 336 pp.
112. Kroshilin, A. E., Kroshilin, V. E., Nigmatulin, B. I. 1984. Effect of relative motion and volume concentration of bubbles on the interface heat and mass transfer in steam–liquid flows. *Teplofizika Vysokih Temperatur* 26 (2):335–362.
113. Galitseisky, B. M., Ryzhov, Yu. A., Yakush, E. V. 1977. *Thermal and hydrodynamic processes in oscillating flows.* Moscow: Mashinostroyeniye, 256 pp.
114. Belyaev, N. M., Belik, N. G. Polshin, A. V. 1985. *Thermal acoustic oscillations of gas–liquid flows in power plant complex pipeworks.* Donetsk: Vyshcha Shkola, 160 pp.
115. Prokhorov, Yu. F., Gerliga, V. A. Prokhorov, M. F. 1976. Effect of dissolved gas on thermal acoustic pressure fluctuations in heated channels. Optimization of parameters of machines and production processes. *Proceedings of Chelyabinsk Polytechnical Institute* 80:3–76.
116. Baldina, O. M., Kalinin, R. I., Saburova, R. I., Baitina, Ts. M. 1968. Flow fluctuations in steam generator horizontal components. *Teploenergetika* 8:35–39.
117. Kafengaus, N. L. 1983. Heat transfer to turbulent liquid flow in tubes at supercritical pressures. *Inzhenerno-Fizichesky Zhurnal* 44 (1):14–19.
118. Gerliga, V. A., Vetrov, V. I. 1978. Experimental investigations of thermal acoustic fluctuations in heated channels at supercritical pressures. *Izv. Vuzov. Aviacionnaya Tekhnika* 1:31–36.
119. Goldman, K. 1961. Heat transfer to supercritical water at 5000 psi flowing at high mass flow rates through round tubes. In *International developments in heat transfer.* Washington, DC: ASME, Pt. 3, pp. 562–568.
120. Aladiev, I. T., Vasyanov, V. D., Kafengaus, N. L., et al. 1976. Experimental investigation of the mechanism of pseudoboiling of heptane. *Inzhenerno-Fizichesky Zhurnal* 31 (3):389–395.
121. Beschastnov, S. P., Petrov, V. P. 1973. Heat transfer by free convection from horizontal cylinders to CO_2 at near-critical conditions. *Teplofizika Vysokih Temperatur* 11 (3):588–593.
122. Vetrov, V. I., Gerliga, V. A., Razumovsky, V. G. 1977. Experimental investigations of thermal acoustic fluctuations in heated channels at water supercritical parameters. *Voprosy Atommoy Nauki i Tekhniki. Series: Fizika i Tekhnika Yadernyh Reaktorov. Dynamika Yadernykh Energeticheskih Ustanovok* 2:51–57.
123. Gerliga, V. A., Vetrov, V. I. 1975. Mechanism of thermal acoustic fluctuations' development at supercritical pressures. *Proceedings of Uralsk Polytechnical Institute,* Chelyabinsk, 162:4–15.
124. Vukalovich, M. P., Rivkin, S. L., Alexandrov, A. A. 1960. *Tables of thermal physical properties of water and water steam.* Moscow: State Standard Publishers, 408 pp.
125. Vetrov, V.I. 1984. Mathematical model for investigation of thermal acoustic instability in heated channels at coolant supercritical pressures. *Voprosy Atomnoy Nauki i Tekhiniki, Series: Fizika i Tekhinka Yadernykh Reaktorov* 2 (39):34–42.

126. Bogovin, A. A. 1984. Specifics of heat transfer at supercritical fluid pressures. Abstract PhD dissertation, Odessa, OPI, 32 pp.

127. Kalinin, E. K., Dreitser, G. A., Yarkho, S. A., et al. 1972. *Heat transfer enhancement in channels.* Moscow: Mashinostroyeniye, 200 pp.

128. Bogovin, A. A., Gerliga, V. A., Prokhorov, Yu. F. 1983. Effect of dissolved gas on heat transport by free convection at pentane supercritical pressures. *Inzhenerno-Fizichesky Zhurnal* 44 (4):533–536.

129. Kalbaliev, F. I. 1978. On the wall temperature conditions at variable substance properties. *Thesis of papers of 6th All-Union Conference on Heat Transfer and Hydraulic Friction in Two-Phase Flows in Energy Machine and Power Equipment Components,* Leningrad, pp. 159–161.

130. Jackson, T. W., Pordy, K. R. 1965. Resonant pulsating flow and convective transfer. *Transactions of ASME, Heat Transfer* 87 (4):93–100.

131. Kulandin, A. A., Timashov, S. V., Ivanov V. P., et al. 1987. *Fundamentals of theory and design of space nuclear power plants.* Leningrad: Energoatomizdat, 327 pp.

132. Lokshin, V. A. 1974. Formation of free level in straight-flow tubular heat exchangers. *Teploenergetika* 11:16–19.

133. Lokshin, V. A. 1974. Hydraulic instability of cooling surfaces of steam heat exchangers. *Teploenergetika* 12:27–30.

134. Artemov, P. N. Desyatun, V. F., Kreidin, B. L., et al. 1977. Parallel operation of separator-superheaters at the Novo-Voronezh NPP unit 4. *Teploenergetika* 12:18–23.

135. Desyatun, V. F. Kreidin, I. L., Kreidin, B. L., et al. 1985. Test results for the Chernobyl NPP separator-superheaters. *Energomashinostroyeniye* 9:30–34.

136. Desyatun, V. F., Artemov, P. N., Kreidin, I. A., et al. 1981. Investigation of parallel operation of vertical straight-flow heat exchangers with condensation. In *Increasing efficiency of heat transfer in power generating equipment,* 132–141. Leningrad: Nauka.

137. Sorokin, U. L. NPP turbine separator-superheaters. Thermal and hydraulic calculations. RTM-108, Leningrad, *CKTI Proceedings,* 1984, p. 217.

138. Gerliga, V. A., Shelkhovskoy, R. D. 1975. On the stable operation of condensers. *Dynamics of Machines and Operating Processes, Proceedings of Chelyabinsk Polytecnical Institute,* Chelyabinsk, 162:33–40.

139. Wedekind, G., Bhatt, B. 1977. An experimental and theoretical investigation into thermally covered transient flow surges in two-phase condensing flow. *Transactions of ASME, Journal of Heat Transfer* 99 (4):561–567.

140. Wedekind, G., Bhatt, B. 1960. Liquid flow self-oscillations in two-phase condensing flows. *Transactions of ASME, Journal of Heat Transfer* 4:113–121.

141. Accidents at facilities supervised by Kotlonadzor and preventive measures. Information letter. Moscow, Nedra, 1965, 173 pp.

142. Lee, S., Bankoff, S. 1983. Stability of steam-water counter-current flow in an inclined channel: Flooding. *Transactions of ASME, Journal of Heat Transfer* 105 (4).

143. Bjorge, R., Griffith, P. 1984. Initiation of water hammer in horizontal and nearly horizontal pipes containing steam and subcooled water. *Transactions of ASME, Journal of Heat Transfer* 106 (4).

144. Lee, S., Bankoff, S. 1984. Stability of a steam-water countercurrent flow in an inclined channel: Part 2—Condensing—induced water hammer. *Transactions of ASME, Journal of Heat Transfer* 4:201–203.

145. Kuznetsov, M. V., Bukrinsky, A. M. 1984. On the mechanism of instability of bubbling steam condensation in accident localization systems at NPP. *Teploenergetika* 12:56–58.

146. Kuznetsov, M. V., Bukrinsky, A. M. 1988. Frequency of pressure fluctuations during bubbling condenser instable operation. *Teploenergetika* 10:61–65.

147. Aya, I., Narial, H., Kobayashi, M. 1980. Pressure and fluid oscillations in vent system due to steam condensation, (I). *Journal of Nuclear Science Technology* 17 (6):499–515.

148. Rubashkin, V. N. Mathematical modeling of chugging processes. *Energetika i elektrifikacia, Series: Atomnaya energetika za rubezhom,* Moscow, 1982, Issue 1, pp. 34–36.

149. Ustinov, A. K., Solodov, A. P. 1988. Investigations of the condensation cone fluctuations. *Teplotekhnika* 7:62–65.

150. Ovsyannikov, B. V., Borovsky, B. L 1971. *Theory and prediction of power systems for liquid propellant rocket engines.* Moscow: Machinostroyeniye, 539 pp.

151. Pelipenko, V. V., Zadontsev, V. A., Natanzon, M. S. 1977. *Cavitating self-oscillations and dynamics of hydrosystems.* Moscow: Machinostroyeniye, 352 pp.

152. Stripling, L. B., Acosta, A. J. 1962. Cavitation in blade pumps, part 1. Cavitation in blade pumps, part 2 (Stripling, L. B.). *Transactions of ASME, Series. D, Journal of Basic Engineering* 3:29–55.

153. Natanzon, M. S., Baltsev, N. I., Bazhenov, V. V., et al. 1973. Experimental investigation of cavitating oscillations of screw-centrifugal pump. *Izv. Acad. Nauk SSSR, Series: Energetika i Transport* 2:151–157.

154. Chebaievsky, V. F., Petrov, V. I. 1973. *Cavitating characteristics of high-speed screw-centrifugal pumps.* Moscow: Mashinostroyeniye, 152 pp.

155. Margulova, T. Kh. 1984. *Nuclear power stations.* Moscow: Vysshaya Shkola, 304 pp.

156. Kurnyk, L. N. 1980. Dissolved oxygen removal in bubble column deaerators. *Elektricheskiye Stantsii* 2:27–31.

157. Nesterov, Yu. V. 1986. Analysis of the "condensate line–deaerator" system stability. *Teplotekhnika* 3:67–70.

158. Kondratiev, A. D. 1984. Causes of deaerator hydrodynamic stability during power unit load reduction. *Elektricheskiye Stantsii* 9:32–36.

159. Kondratiev, A. D., Kostylev, V. F. 1966. Causes of water hammers in power plant deaerators and ways of eliminating them. *Electricheskiye Stantsii* 6:33–36.

160. Prevention of vibrations in dump pipes of power plants start-up systems (overview), Moscow, *Informenergo,* 1972, 22 pp.

161. Antivibration friction support, Newsletter. PO Soyuztekhenergo, Moscow, VDNKH SSSR, "Electrification of the USSR" Pavillion, 1981, p. 2.

162. *Guides and regulations on power system operation* (thermal section). Moscow: Energoizdat, 1981, 320 pp.

163. Robozhev, A. V. 1984. *Reducing cooling units of thermal and nuclear power plants.* Moscow: Energoatomizdat, 224 pp.

164. Zavadovsky, B. A. 1934. *Process steam lines.* Moscow-Leningrad: ONTI, 231 pp.

165. Balke, G. 1931. *Thermal facilities rationalization.* Moscow-Leningrad: OGIZ, 432 pp.

166. Troyanovsky, B. N. 1978. *NPP turbines.* Moscow: Energia, 232 pp.

167. Korolev, A. V. 1988. Vibration of pipelines with two-phase flows. Abstract PhD dissertation, Odessa, Odessa Polytechnical Institute, 22 pp.

168. Verezemsky, V. G., Gorbachev, S. I. Nikitin, B. E. 1983. Experience of studying vibration of process pipes of a powerful thermal power plant. In *Hydrodynamics and vibration in NPP components*, 215–224. Obninsk: FEI.

169. Leschinsky, G. A. 1983. Unsteady steam-water flows in pipes and vibration thereof. Abstract PhD dissertation, Kharkov, Kharkov Aviation Institute, 23 pp.

170. Gerliga, V. A., Korolev, A. V., Gerliga, A. V. 1985. Pressure fluctuations in the incipient boiling condensate drain pipes. *Energetika i Elektrifikacia,* Kiev 2:17–19.

171. Prudovsky, A. N., Radionov, V. B. 1985. Experimental investigations of two-phase flow power influence on pipe bends. *Teploenergetika* 5:58–61.

172. Serov, E. P. 1965. Toward flow pulsatory fluctuation in steam generating components. *CKTI Proceedings* 62:7–14.

173. Lutoshkin, G. S. 1977. *Collection and preparation of oil, gas and water.* Moscow: Nedra, 192 pp.

174. Silash, A. P. 1980. *Production and transport of crude oil and gas, part 1.* Moscow: Nedra, 375 pp.

175. Mamaev, V. A. 1969. Pipeline transport of gas–liquid mixtures. Subject overview. Moscow, VNIOENG, 102 pp.

176. Fokin, B. S., Akselrod, A. F. 1980. Prediction of pressure fluctuation intensities and gas content in adiabatic two-phase flows. *Inzhenerno-Fizichesky Zhurnal* 39 (5):806–810.

177. Mandhame, I., Gregory, G., Aziz, K. 1974. Flow pattern map of gas–liquid flow in horizontal pipes. *International Journal of Multiphase Flow* 1:537–553.

178. Miropolsky, Z. L., Shneerova, R. I., Mekler, V. Sh. 1984. Experimental investigation of hydraulic friction and quality with fluid flashing in long pipelines. *Teploenergetika* 2:54–62.

179. Petukhov, B. S., Shikov, V. K., eds. 1987. *Heat exchanger handbook,* vol. 1 (English trans.) Moscow: Energoatomizdat, 560 pp.

180. Fisenko, V. V., Alferov, A. V., Makukhin, A. A. 1985. Method of elimination of drain pipeline vibration and failure. *NPP Operation and Repair,* Moscow, Minenergo SSSR, TSNTI 2:15–19.

181. Hubbard, M. G., Ducler, A. E. 1966. The characterization of flow regimes for horizontal two-phase flow. In *Proceedings of the 1966 heat transfer and fluid mechanics institute,* ed. M. S. Saad and I. A. Miller, 100–121, Stanford University Press, Stanford, CA.

182. Fisenko, V. V. 1987. *Coolant compressibility and NPP circulation loops' operation efficiency.* Moscow: Energoatomizdat, 200 pp.

183. Generalization and analysis of the data on vibration and erosion of two-phase coolant pipes at NPP. VTI report no. 13186, FTO-941, Moscow, VTI, 1986, 98 pp.

184. Ozava, M., Akagawa, K., Sakaguchi, T., Takagi, S. 1984. Fluctuations of two-phase air-water flow instability. *Transactions of Japan Society of Mechanical Engineers,* Series B49 448:2715–2724.

185. Morozov, I. I., Grabovich, A. P. 1975. Investigation of adiabatic two-phase flow stability. In *Dynamics of machines and processes.* Chelyabinsk, pp. 16–26.

186. Perelman, R. G., Pryakhin, V. V. 1986. *Erosion of steam turbine components.* Moscow: Energoatomizdat, 184 pp.

187. Syromyatnikov, V. F. 1983. *Fundamentals of automatics and complex automation of marine steam power systems.* Moscow: Transport, 312 pp.

188. Klapchuk, O. V., Elin, N. I. 1978. Statistical characteristics of gas–liquid flow pressure fluctuations in horizontal tubes. Transport of gas and associated equipment. Moscow, VNIIGaz, pp. 82–87.

189. Dulevsky, R. A. 1988. Determination of steam generator stability taking account for thermal inertia of heating cooling. In *Papers to the All-Union Conference Steam Generator Heat Transfer.* Novosibirsk, pp. 203–211, 1988.

190. Aya, I., Narial, H. 1985. Boundaries between regimes of pressure oscillation induced by steam condensation in pressure suppression containment. *Proceedings of Third International Topical Meeting on Reactor Thermal Hydraulics,* Newport, RI, Oct. 15–18, 1:E1–E8.

191. Malkin, S. D., Khabensky, V. B., Migrov, Yu. A., Efimov, V. K. 1987. Investigation of heat and mass transfer in the channel with upper plenum. *Inzhenerno-Fizicheskiy Zhurual* 44 (5):714–720.

192. Malkin, S. D., Khabensky, V. B., Migrov, Yu. A., Efimov, V. K. 1986. Investigation of single-phase flow stability in slot heated channels at low velocities and coolant downflow. In *NPP heat transfer and hydrodynamics,* 93–106. Leningrad: Nauka.

193. Jannello, V., Todreas, N. 1989. Mixed convection in parallel channels with application to the liquid-metal reactor concept. *Nuclear Science and Engineering* 101:315–329.

194. Khabensky, V. B., Migrov, Yu. A., Efimov, V. K., Volkova, S. N. 1990. Specifics of unsteady processes in parallel heated channels at low velocities and single-phase coolant flow reversal. *TVT* 28 (3):518–529.

195. Khabensky, V. B., Migrov, Yu. A., Efimov, V. K. 1991. Coolant flow patterns and critical phenomena at low velocities of the loop rates. *Teploenergetika* 12:41–46.

196. Saha, P., Zuber, N. 1974. Point of net vapor generation and vapor void fraction in subcooled heat transfer. *Proceedings of 5th International Conference,* Paris, IV:175–179.

197. Malkin, S. D., Khabensky, V. B., Migrov, Yu. A., Efimov, V. K., Volkova, S. N. 1987. Effect of static instability on single phase flow dynamics in parallel heated channels at low coolant velocities. *Voprosy atomnoi nauki i tekhniki. Series: Dinamika yademykh ustanovok.* 2:27–33.

198. Sakai, K., Yano, M., Tezuka, H. 1982. Reactor core thermohydraulic transients with thermohydrodynamic coupling. *Nuclear Engineering and Design* 73 (3):373–404.

199. Singer, R., Betteu, P., Dean, E. 1980. Studies of thermal-hydraulic phenomena in EBR-II. *Nuclear Engineering and Design* 62:219–232.

200. Zvirin, Y., Rabinovich, Y. 1982. On the behavior of natural circulation loops with parallel channels. *International Heat Transfer Conference,* Munich, Germany, 2:299–304.

201. Singer, R., Mohr, D., Gillette, J. 1982. Transition from forced to natural convection flow in an LMFBR under adverse thermal conditions. *International Heat Transfer Conference,* Munich, Germany, 5:537–542.

202. Chato, J. C. 1963. Natural convection flows in parallel-system channels. *Transactions of ASME, Heat Transfer* 4:61–68.

203. Malkin, S. D., Khabensky, V. B., et al. 1987. Thermal-hydraulics in research pool-type reactor WER-2 in heat-up from cold state with non-loop circulation. *Teploenergetika* 12:39–42.

204. Gershuni, G. Z., Zhukhovitsky, E. M. 1972. *Convective stability of compressible fluid.* Moscow: Nauka, 392 pp.

205. Petukhov, B. S., Polyakov, A. F. 1982. *Heat transfer in mixed turbulent convection.* Moscow: Nauka, 192 pp.
206. Malkin, S. D., Khabensky, V. B., Migrov, Yu. A., and Efimov, V. K. 1983. Study of coolant mixing in the mixing chamber, *TVT* 21 (2):349–357.
207. Miropol'skii, Z. L. 1965. Flow rate oscillations in evaporative channels in the presence of elastic fluid upstream of the flow train. *Trudy TsKTI* 59:169.
208. Morozov, I. I., Malkov, V. A. 1963. The effect of elastic containers on the stability of the working process in steam-generator pipes. *Inzhenerno-fizicheskiy Zhurnal* 6 (1):54.
209. Ozava, M., Nikanishi, S., Jshigai, S., et al. 1979. Flow instabilities in boiling channels. *Bulletin of Japan Society of Mechanical Engineers* 22 (170):1113.
210. Godik, I. V., Eskin, N. V., Grigor'ev, A. S., Levin, G. S. 1985. The experimental study of the hydrodynamic stability of working fluid in a once-through steam generator with an attached tank. *Teploenergetika* 11:20.
211. Mentes, A., Kakac, S., Veziroglu, T., Zhang, H. 1989. The effect of inlet subcooling on two-phase flow oscillations in a vertical boiling channel. *Wärme Stoffübertrag* 24:24.
212. Yuncu, H., Yildirim, O., Kakac, S. 1991. Two-phase flow instabilities in a horizontal single boiling channel. *Applied Scientific Research* 48 (1):83.
213. Gurgenci, H., Veziroglu, T., Kakac, S. 1983. Simplified nonlinear descriptions of two-phase instabilities in a vertical boiling channels. *International Journal of Heat Mass Transfer* 26 (5):671.
214. Rabinovich, M. I., Trubetskov, D. I. 1984. *Introduction to the theory of oscillations and waves.* Moscow: Nauka.
215. Jiang, S. V., Wu, X. X., Zhang, Y. J. 2000. Experimental study of two-phase flow oscillation in natural circulation. *Nuclear Engineering and Design* 135:177–189.
216. Volkova, S. N., Granovski, V. S., Efimov, V. K., Kuvshinova, O. V. 1999. Investigations into thermal-hydraulic effects of external VVER reactor vessel cooling for in-vessel retention under severe core melt accidents. All-Minatom Conference on Hydrodynamics and NPP Safety. Abstracts of papers, Obninsk (in Russian).
217. Migrov, Yu. A., Yudov, Yu., Danilov, I., et al. 2001. KORSAR: A now generation computer code for numerically modeling dynamic behavior of nuclear power installations. ICONE-9, Nice, France, April 8–12, Report 545.
218. Verbitskiy, Yu. G., Efimov, V. K., Migrov, Yu. A. 2003. An experimental investigation of steam bubbling instability in a long vertical pipe at low pressures. 11th International Conference on Nuclear Engineering, Tokyo, April 20–23, ICONE 11-36151.
219. Verbitskiy, Yu. G., Efimov, V. K., Migrov, Yu. A. 2005. An experimental investigation of steam bubbling instability in a long vertical pipe at low pressures. *Thermal Engineering (Teploenergetika)* 52 (4):1–6.

Index